ARCHAEOLINGUA

Edited by
ERZSÉBET JEREM and WOLFGANG MEID

Volume 28

Explorations in Salt Archaeology in the Carpathian Zone

Anthony Harding and Valerii Kavruk

with contributions by

Krzysztof Bukowski, Andrea Chiricescu, Richard Brunning, Adela Kovács, Dan Buzea, Marion Uckelmann, Dan Ştefan, Eszter Bánffy, Tomasz Ważny, Ionel Popa, Anca Semeniuc, Gelu Ţiculeanu, Rowena Gale, Nyssa Mildwaters and Steven J. Allen

BUDAPEST 2013

This volume is published with financial support from the Leverhulme Trust within the research programme *Tracing Networks: Craft Traditions in the Ancient Mediterranean and Beyond* and the National Cultural Fund of Hungary

Front Cover

The salt stream at Săsarm, Bistrița-Năsăud county, Romania,
seen from the north and showing the brine well (photo by Anthony Harding)

Back Cover

Wood installations uncovered in excavation at Băile Figa,
Bistrița-Năsăud county, Romania (photo by Anthony Harding)

ISBN 978-963-9911-44-4
HU-ISSN 1215-9239

2013

ARCHAEOLINGUA ALAPÍTVÁNY
H-1250 Budapest, Úri u. 49

Desktop editing and layout by Rita Kovács

Printed by Prime Rate Kft.

Contents

List of Figures

List of Plates

List of Tables

Acknowledgements

The work in the Carpathian zone was made possible by means of financial support from various organisations, and through the help of many individuals. Fieldwork in Romania has been funded by grants from the Romanian Ministry of Cults and Culture and by the British Academy. Rescue work at the site of Băile Figa was funded also by the municipality of Beclean and the developers of the spa. The work in Slovakia and Ukraine was funded by a grant from the Society of Antiquaries of London. Radiocarbon dates were provided by means of grants from the Natural Environment Research Council of the UK through the Oxford Radiocarbon Accelerator Unit.

The work of Marion Uckelmann to create the GIS and organise the finds catalogue was supported by the Leverhulme Trust as part of the project *Tracing Networks: Craft Traditions in the Ancient Mediterranean and Beyond.*

The authors express their thanks to all these organisations, and to the following individuals who have provided help and support of various kinds: Dr Ioan Chintăuan (Bistriţa), Mr Gheorghe Marinescu (Bistriţa), Mr Nicolae Moldovan (Beclean), Dr Bryn Morris (Exeter), Professor Tony Brown (Southampton), Dr Josip Kobal' (Uzhgorod), Professor Janusz Czebreszuk (Poznań), Mgr Anna Majchrzycka (Poznań), Dr Eva Horvathová (Košice), Dr Klára Marková (Nitra), Ing. Jaroslav Kotul'ák (Košice), Professor Sławomir Kadrow (Rzeszów/Kraków), Dr Jarosław Fraś (Wieliczka). The two principal authors also thank all the contributors to the volume for their collaboration.

The work of Seán Goddard (Department of Archaeology, University of Exeter) in preparing many of the illustrations for publication, and in improving the descriptions of the wooden objects, is gratefully acknowledged.

The publication process was greatly assisted by the kind attention and professionalism of Dr Erzsébet Jerem and her staff at Archaeolingua, while Professor Eszter Bánffy gave much assistance with the search for grant aid and helped in many other ways. The publication is published with the aid of financial support from the *Tracing Networks* project, funded by the Leverhulme Trust.

To all of these institutions and individuals the authors express grateful thanks.

Chapter 1
Introduction

ANTHONY HARDING – VALERII KAVRUK

The study of ancient salt has become a topic of lively debate in recent years. An increasing number of conference proceedings, books and articles have appeared in the literature, especially since 2000, reflecting both the academic interest in the topic but also a considerable amount of fieldwork on sites or areas where salt is present. This has been particularly apparent in, though is not restricted to, Europe, where work in France, England, Germany, Austria, Spain, Italy and Romania has been undertaken by various groups and individuals. The reason for this is not hard to seek: an intake of salt is a necessity of human and animal life, and is also used in traditional societies in a variety of other ways. One may assume that all ancient societies regarded the provision of salt as very important for the maintenance of health, for the preservation of foodstuffs, medicinal use, and a number of industrial processes. We do not discuss these further here, as they have been presented in the literature many times.

Salt is not evenly distributed across the earth, however. Some areas are very rich in salt resources; others have little or none, which means the communities living in salt-poor areas had to take steps to obtain it. The production and movement of salt by trade thus becomes a matter of great importance in our understanding of ancient or traditional economies.

This book describes work carried out in a number of countries in the Carpathian zone of Europe – mostly inside the Carpathian ring, but with excursions outside it to the north. The work has principally been conducted in Transylvania, but has also embraced adjacent parts of Ukraine, Slovakia and Poland. Other teams have worked on the area immediately to the east of the mountains in Moldavia (the Romanian province, to be distinguished from the modern state of Moldova); we refer to this work, and have visited some of the sites involved, but we do not consider it in detail. The sites that we discuss are in many ways remarkable because of the preservation of wood on them and the unusual nature of the installations involved; these have enabled us to obtain a detailed (though still incomplete) picture of the production process, and to speculate on the way in which the salt produced was moved around the eastern half of Europe, particularly in the Bronze Age but potentially also in later periods.

After a brief presentation of salt sources, we present information on how salt is (or has been) used in traditional societies in the area under investigation, and then the site information itself, along with an account of experiments carried out in order to test theories about how the wooden installations worked in practice. This is followed by a discussion of how the different production methods might have contributed to the movement of salt in trade and exchange. These latter remarks are relevant to all salt-producing areas, not just to the Carpathian zone.

Background to the project

It had been known for some time when we set out on this venture that there was evidence for salt production in prehistory in the Carpathian zone, though that evidence was scanty and hard to understand. The discovery in 1938 of wooden objects, including a trough with perforations in the base, in an old mine shaft under the railway lines at Valea Florilor north of Turda, was not appreciated at the time for what it was, and was only published in 1971, when it was suggested that it belonged to the Dacian Iron Age (MAXIM 1971). Subsequently Wollmann connected this find with one made in the early part of the 19th century at a location known as Királyvölgy in the then Hungarian county of Máramaros (now the

Sub-Carpathian Ukraine) (WOLLMANN 1996). A trough and other implements were also found there in an old salt mine shaft (PREISIG 1877). The Valea Florilor trough was subsequently shown to belong in all likelihood to the Bronze Age (WOLLMANN – CIUGUDEAN 2005), but this was not known at the time we first became interested in salt. These finds are discussed in more detail in the relevant chapters (below).

Meanwhile fieldwork in Moldavia had uncovered briquetage of Neolithic date in close connection with salt sources (ANDRONIC 1989; URSULESCU 1977); further fieldwork was already underway in this province when our own attention was drawn to the potential of salt research in Transylvania. A similar situation prevailed in southern Poland, where vessels believed to represent briquetage had long been known from the Wieliczka area (JODŁOWSKI 1968). There was thus an existing story about prehistoric salt production in the Carpathian zone prior to our arrival on the scene, but it was far from clear how the rich salt resource of the area had been exploited, when, and with what consequences.

The genesis of our project lay in a small colloquium held in Sfîntu Gheorghe, Romania, in 1998, when Harding presented some ideas relating to salt as an economic commodity of importance in the Bronze Age, and speculating on the situation in Romania. At that time he was already interested in salt in Bronze Age Europe, an interest which had been made manifest already in a conference paper presented in Berlin in 1997 and published soon afterwards (HARDING 1998); As a consequence, a joint project was initially proposed with Dr Horia Ciugudean of Alba Iulia; it was as part of that exploratory visit that the visit to Sf. Gheorghe occurred. Kavruk and Harding both attended a conference in Baia Mare later the same month and were able to discuss the question of salt more fully. This paper was published later (HARDING 2001), as was a section on salt in the book *European Societies in the Bronze Age* (HARDING 2000, 249 ff.).

Kavruk had already been excavating a *zolnik* (ash-pit) site of the Noua culture at Zoltan in Covasna county, in a region where there are many salt springs (locally at Olteni; see below, chapter 4). A connection with salt seemed plausible; hence our subsequent inclusion of this area in our studies.

In 2001 we spent some days travelling to visit salt sources in Transylvania, Moldavia and Muntenia, in order to obtain a better picture of the extent of the sources and the possibility of conducting fieldwork. We rapidly became convinced that an approach using modern ethnographic data would be the best starting point; traditional ways of using brine or rock salt might provide a window into how ancient people operated. With this in mind, Kavruk and his staff undertook extensive investigations in central and eastern Transylvania, interviewing local people and recording the artifacts and technologies they used (see Chapter 3). We continued our initial exploration of the Transylvanian sites in 2002 and 2003, also visiting the excavations at Lunca-Poiana Slatinei (jud. Neamț) in Moldavia, and discussed how we could set up a project to make progress in developing our understanding of ancient, as well as modern, salt production and exploitation. At this stage, most of our work was dependent on the examination of salt use in peasant villages, but significantly we saw the wooden piles in the salt stream at Sânpaul (below, chapter 4), which apparently lay close to the site of the former salt mine (its precise position no longer evident). We had no means of knowing their age at that time, but their existence seemed significant. We were also aware of the increasing amount of work in Moldavia, both ethnographic, toponymic and archaeological, and in particular the evidence from Lunca, where Neolithic material was present in a small mound beside the salt well, and from Cacica (jud. Suceava), where large amounts of briquetage had been found in excavation and can still be seen eroding from a bank beside the modern mine. These factors combined to suggest to us that more could be obtained from the available evidence in Transylvania.

Following the award of a grant from the British Academy, survey work began in 2005 on sites in Covasna and Harghita counties, using a Bartington gradiometer to examine possible salt production sites, and differential GPS to provide mapping. At the end of that season we were taken to Băile Figa (and later Săsarm) and saw the wooden remains on site, as well as the trough (no. 1) in Bistrița Museum (Complexul Muzeal Bistrița-Năsăud). Knowing of the Valea Florilor find, but not at that time of its Bronze

Age date, we took samples for radiocarbon dating; and when the results came back in autumn 2005, we realised fully for the first time the enormous potential importance of the Figa site. We subsequently obtained more funds from the British Academy, as well as from the Ministry of Cults and Culture of Romania, to undertake a full survey at Băile Figa and a partial survey at Săsarm in 2006; we took more samples at Figa, and of the six results obtained, four fell in the Bronze Age. This remarkable result meant that we resolved to conduct excavations at the site; these started in April 2007 and have continued each year to the present. During subsequent years we were also able to examine other places in the area where modern salt production occurs; this was when the importance of the Caila site was noted.

A brief presentation of some of our work up to 2006 was given in the exhibition catalogue *Sarea, Timpul şi Omul*, describing a presentation of salt-related material from all over Romania that was put on in the Museum of the Eastern Carpathians in Sfîntu Gheorghe in autumn 2006 (Cavruc – Chiricescu 2006). This book represented a major advance in the discussion of ancient, medieval and modern salt production, but it was for us only a start, as no excavatio,n had occurred on salt sites in northern Transylvania at that time. Apart from this, annual summary reports have been presented in the *Cronica Cercetărilor Arheologice din România* as well as a number of preliminary reports on the work in scientific journals and conference papers (most fully in Harding – Kavruk 2010).

At the time of writing, work continues at Băile Figa, but only on a small scale. A complicating factor has been the development of the area as a leisure park and spa. While this has generated funds for rescue excavation, it has also changed priorities for further research at the site, as effort has had to be devoted to salvaging material prior to destruction. Fortunately it has been possible so far to prevent damage to the most sensitive and archaeologically informative areas, but the prospects for the future are not good. Salty mud is constantly being removed from the site for mud baths without any archaeological supervision, including in areas which we have not been able to investigate; outside the enclosed spa area, the site is open to all, though the wooden remains uncovered in excavation are covered over. The site remains threatened, even though it has now been inscribed on the list of protected monuments in the Romanian state. Only a large injection of funds will safeguard it in the medium term; in the long term, matters are even more uncertain.

The concentration on Romania, and specifically Transylvania, was supplemented in 2008 by a visit to Ukraine and Slovakia, with the specific aim of examining the area where a trough, ladder and other material had been extracted in the early 19th century at a mining site at the place then called Királyvölgy. This locality, and others in the same region, were visited, as were known salt extraction sites in Slovakia. While no specific evidence for prehistoric working was discovered, much useful information on the situation in those countries was obtained. Equally important was the rediscovery of the wooden objects from Királyvölgy in the Hungarian Central Mining Museum at Sopron, followed by their dating to a period that ties in closely with what we had found in Transylvania. This work is presented in Chapters 5 and 6.

In 2007 and 2009, Harding paid visits to Poland to investigate the possibility of prehistoric salt exploitation in that country, already suggested for the Neolithic and supplemented by some Bronze Age material in the south of the country. This work is presented in Chapter 6.

Funding and Personnel

The work in Transylvania has been carried out with a series of grants from two sources: the British Academy, and the Ministry of Cults and Culture of Romania, through annual grants to the Museum of the Eastern Carpathians, Sfîntu Gheorghe; and from rescue funds obtained via the Mayor of Beclean, Mr Nicolae Moldovan, from the developer and from local resources.

The 2005 survey was carried out by the authors with Andrea Chiricescu, Liliana Lazea, Darryl Freer and Julia Heeb; logistical support was provided by Marius Domboşi. In 2006 the survey was conducted by Brynmor Morris, and the wood remains at Figa and Săsarm catalogued by the authors and George Marinescu of the Complexul Muzeal Bistriţa-Năsăud, with Andrea Chiricescu and other members of Sfîntu Gheorghe staff. From 2007 on, survey work has been in the hands of Dan Ştefan and Magdalena Dutescu (now Ştefan), of S.C. Digital Domain srl and S.C. Vector Studio srl. On the Romanian side, excavation has been carried out by Kavruk and Marinescu with Andrea Chiricescu, Adela Kovács, Dan Buzea, Björn Briewig and Radu Zagreanu. On the British side, work in 2007–9 was carried out by Harding with Brynmor Morris (2007–8) and Kayleigh Hamilton (2009), and a succession of students from the universities of Exeter, Cluj and Sibiu. In 2009 specialist help was obtained from Dr Ionel Popa (Câmpulung Moldovenesc), Professor Tony Brown (Southampton) and Dr Richard Brunning (Taunton), whose reports appear below (or have already appeared).

Salt sources in the Carpathian-Danube-Pontic region

Salt sources are unevenly spread in the Carpathian-Danube-Pontic region (*Fig. 1.1*): on the one hand, there are many salt-rich localities in the area, on the other, many regions are salt-poor or salt-less. The most reliable and accessible natural salt sources in the region – rock salt deposits found at shallow depths or on the surface, brine springs and lakes, salty soils and halophile vegetation – are available in the Carpathian saline province that includes southern Poland, north and east Slovakia, most of Transylvania, Ukrainian Transcarpathia (Zakarpatiia), and the eastern and southern Sub-Carpathians.

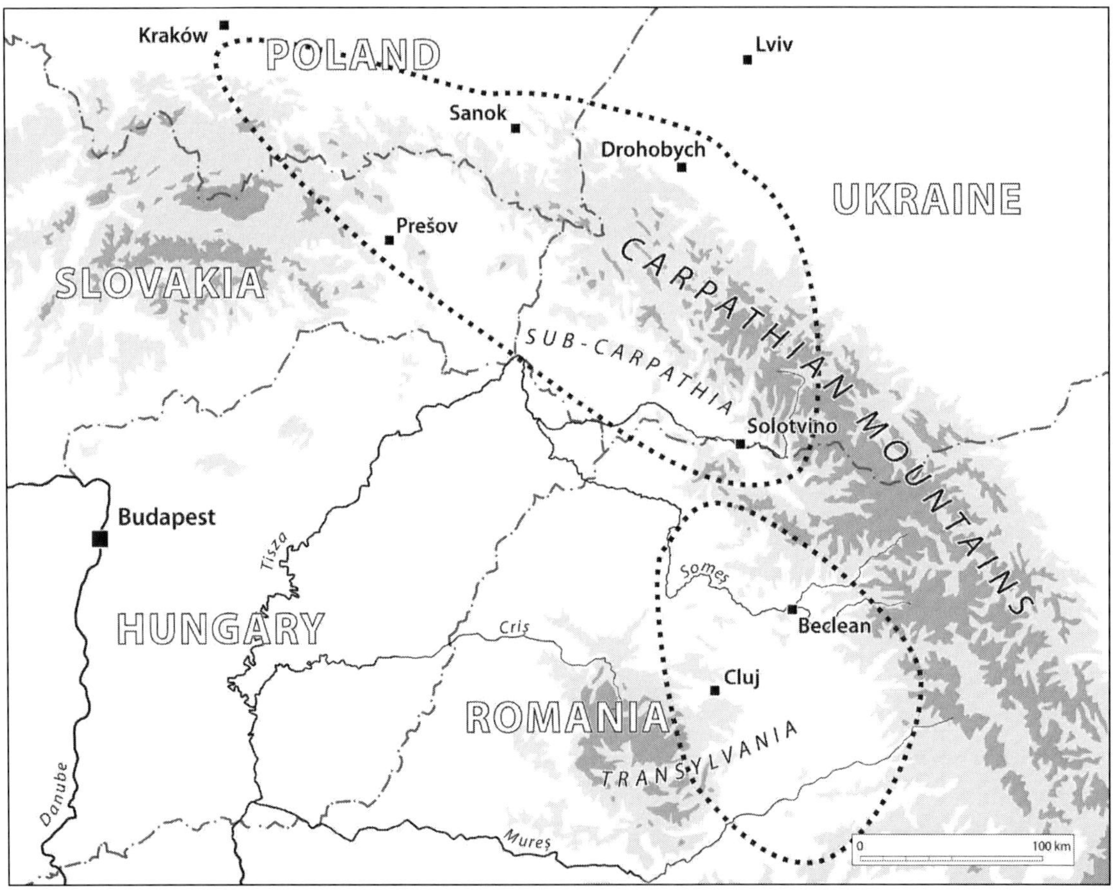

Fig. 1.1. Map showing the areas investigated in this book.
The dotted line shows the approximate extent of the major salt deposits in the region.

A salt-rich strip about 3000 km long and some tens of km wide stretches from Kraków to the valley of the river Jiu, along the eastern slope of the Eastern Carpathians (the Polish, Ukrainian and Moldavian sub-Carpathians) and southern slope of the Southern Carpathians (sub-Carpathian Wallachia and Oltenia). There are many rock salt sources and brine springs north of the Carpathian Basin on the territory of present-day Poland. However, on the territories south, east and west of the Carpathian saline province, the salt sources are less frequent and poorer, if they are present at all.

Poland

A convenient summary has been provided by CZAPOWSKI – BUKOWSKI (2010). Rock salts occur in two evaporite formations: in southern Poland in the Badenian stage of the Middle Miocene, and in central and northern Poland in the Zechstein formation of the upper Permian.

The Zechstein salts cover most of Great Poland, Kujavia and Pomerania. Salt diapirs (domes) are present in many places, but only some of them pierce the Mesozoic cover deposits; it is these that have been exploited in modern and historical times. A band of such diapirs runs from north-west to south-east across the country, from the Sczeczin area to Kłodawa and beyond. In addition, rock salt deposits are present in the north on the Puck Bay and at Łeba (west of Gdańsk); and in the south-west of the country at Sieroszowice-Kazimierzów west of Wrocław.

The Badenian deposits are located in the Carpathian Foredeep, an area which stretches into Ukraine and northern Romania. Geologically these are more complicated than the Zechstein salts, which makes them less attractive for modern exploitation. They are extensively folded, and of very variable depths. From an archaeological point of view, however, the fact that they appear on or near the earth's surface in a number of places makes them important as potential sources areas in ancient times.

Slovakia

There are few rock salt deposits accessible for non-industrial exploitation in Slovakia, but there are natural brine springs in the eastern part of the country. Very concentrated salt water has been industrially extracted at Solivar (Prešov) from the Middle Ages up to the present day. Because of its considerable depth it is unlikely though that this source was accessible for prehistoric exploitation.

Romania

Over 300 large salt deposits are situated in Romania, covering altogether around 30,000 square km. The salt deposits from Transylvania include a surface of about 16.200 square km (cited by MONAH 2007 [2008]: 87; STOICA – GHERASIE 1981, 61 ff., not seen) (*Fig. 2.5*). At the same time, there are some areas in the inter-Carpathian region that are extremely poor in salt sources: the depressions in south-east and east Transylvania (from Ţara Bârsei down to the Giurgeu Depression). A recent unpublished study has reviewed much of the geological literature relating to salt in Transylvania (TOMA 2012).

Bulgaria

In Bulgaria, it has usually been thought that salt was obtained by the evaporation of sea salt in brine lagoons. Salt pans are present around Burgas (notably at the Pomorie Lake, Pomoriysko ezero), and this is where salt was obtained in the Graeco-Roman period.

There are a number of isolated brine springs in the country (GAYDARSKA – CHAPMAN 2007), but only one accessible rock salt deposit, at Mirovo in north-eastern Bulgaria (near Provadia), at a depth of about 10–20 m (NIKOLOV 2008). It is overlain by a huge and very salty waterbed (the concentration of the salt is about 30%), which meant the rock salt could not be exploited in prehistoric times. On the other hand access to salt water at Provadia is very easy (NIKOLOV 2008; 2009).

Dobrudja and Moldova

A few isolated rock salt deposits are known east of the Carpathian saline province, between the Siret and Prut rivers. Two deposits are known between the Prut and the Dniester. One of these, the larger, is found at the mouths of the Danube, around Izmail, and the other one, much smaller, is found in the north of the Republic of Moldova, at Ocnița (ICHIM 2006). Sources of brine are found in the Lower Danube region. Among these the most reliable are the lakes at Brăila and Amara in Wallachia, Techirghiol and Tuzla in Dobrudja, and Cahul, Cetatea Albă-Soani and Martaza in southern Bessarabia (DRĂGĂNESCU 2006). Some of these salt lakes are relatively recent in terms of geological age. They are supplied by brine deposits found at considerable depths that were inaccessible to prehistoric people, but which have come to the surface from time to time. In the Lower Danube region and in the north-west Pontic area one can frequently find springs and wells with brackish water. There also wide areas where salty soils and halophile plants occur.

The Black Sea is an important source of salt, especially its shores and lagoons. The water in the lakes, springs and wells in the Lower Danube region and southern Bessarabia is less salty than in the Carpathian saline province, and the Black Sea is one of the least salty seas in Europe. On the other hand, it has been suggested that the lagoons (*limany*) of the northern Black Sea coast were easily exploitable in ancient times, as they have been in modern (MIRCEA – ALEXIANU 2007).

No salt sources are known in the Banat, Crişana and the Hungarian Plain.

Conclusion

From all this it is apparent that there is plentiful salt in the Carpathian zone, though in varying degrees of accessibility. Today, while industrial concerns usually mine rock salt at considerable depths, peasant communities exploit salt on or near the surface in the form of brine, or – where it is accessible – as rock salt. In this case the technology involved is usually extremely simple. While prehistoric salt production may not have had all the tools that modern peasants use (heavy iron sledge-hammers, for instance), in essence the methods of production are likely to have been rather similar – though we discuss some notable exceptions to that generalization in this book. Roman mining in the Dacian province and elsewhere made use of a range of advanced technologies that we have not discovered in the Carpathian zone in prehistoric contexts (WOLLMANN 1996), nor have we found any other direct evidence for Roman salt mining, but this is surely a factor of survival and research rather than anything else. We do not believe it stems from our own primary affinity as prehistorians, since we have treated all the evidence equally, and would have reported Roman material if it had been present. As it happens, the majority of what we have discovered relates to the Bronze Age, with some parts lying in the Iron Age and medieval periods; in due course we trust that this later material will be the subject of more intensive research than has been possible within the limitations of our project.

Chapter 2
Salt sources and salt springs in the Carpathian zone

KRZYSZTOF BUKOWSKI

The production of salt in the Carpathian regions in prehistoric times until the Middle Ages was based on evaporation of brine originating from salt-water springs, and the mining of rock salt.

Brine in this context refers to water with a mineralization of at least 15 g NaCl/l (MAŚLANKIEWICZ 1985). In addition to NaCl, such brines could also contain potassium chlorides, as well as Mg^{+2}, $SO4^{-2}$, Ca^{+2} Br^{-}, J^{-} and other ions. Salt springs occur frequently in the vicinity of salt rock, potash salt, salt clays, gypsum or anhydrite outcrops. Precipitation (meteoritic) waters penetrating the ground dissolve salt and flow out back to the surface in the form of salt springs. This process is accelerated when the terrain is cut by ravine systems and river valleys.

Not all brines originate from salt deposit leaching. The sedimentary rocks developed in the seas of earlier geological periods could preserve relics of seawater that had originally occurred as pore waters, within the sediments created on the sea bottom. Such brines occur especially in porous sands and sandstones, which often accompany crude oil deposits (MAŚLANKIEWICZ 1985). Salt springs in the Carpathians (Mid-Carpathian salt springs) are probably supplied by such brines, e.g. at Sól near Żywiec, Poland (RAJCHEL *et al.* 2004) (*Fig. 2.1a*).

However, a large majority of brines are associated with the Miocene salt-bearing formation. Those brines flowed out to the surface in the form of salt springs over a wide area of the Carpathian Foredeep (Poland, Ukraine, and Romania) and in the inner-Carpathian basins: eastern Slovakia, Transcarpathia (Ukraine), and Transylvania (*Fig. 2.1b*).

The occurrence of salt springs and later of brine wells in the Małopolska region has been the subject of notes and descriptions for many years (ALTH 1868; BUKOWSKI, Z. 1985; JODŁOWSKI 1971; PUSCH 1833–36; STASZIC 1815 (reprint 1955); SZAJNOCHA 1891; 1893; WINDAKIEWICZ 1926; ZARĘCZNY 1894). We know from historical data that salt springs were found in 196 locations in the Carpathian regions on the territory of Galicia (Małopolska and Western Ukraine; WINDAKIEWICZ 1926). In the western part of Małopolska (up to the San valley), 23 such locations were identified. Others (173) occurred further east along the Carpathians up to the Romanian Bukovina. The phenomenon was explained by the fact that the eastern Miocene salt-bearing formation, supplying these springs, covers a wide strip of territory. In addition, rivers and streams cut the area densely.

In 1773, 92 active salt-works were registered in Poland (the called Ruthenian salt-works), producing over 40 000 tonnes of salt per year (ibid.). They were often privately owned, exploiting brine in what is now western Ukraine. Rock salt was mined underground in several mines, of which the largest were the Royal Saltworks in Wieliczka and Bochnia (*Fig. 2.1c–d*).

Miocene salt-bearing series

The Miocene salt-bearing series are complexes of sediments containing halite (NaCl) and other chemically precipitated minerals (gypsum, anhydrite, and carbonates). Those minerals and rocks are called evaporates or evaporites (Latin *evaporatio* – evaporation) as they developed as a result of the evaporation and chemical precipitation of salt contained in seawater or salt lakes. Evaporites contain various admixtures of pellite and clastic rocks (silts, claystones, marls, shales, mudstones, sandstones and conglomerates). Such rocks occur jointly in various proportions, in the form of rock salt, mixed

a

b

c

d

Fig. 2.1. Occurrence of salt in the Carpathian region.
a: Mid-Carpathian salt spring "Slanica" in Sól near Żywiec (Poland), mineralization 40 g/l (photo Lucyna Rajchel); b: Outcrops of Miocene rock salt dome near Praid, Transylvania, Romania (photo Krzysztof Bukowski); c: old workings in Wieliczka Salt Mine, Stanislav August Chamber, I level (photo Jerzy Przybyło); d: salt pool in Wieliczka Salt Mine, Gaisruck Chamber, III level (photo Jerzy Przybyło).

clay-salt rocks (the so-called zubers), salt clays, gypsum or anhydrites. In western Ukraine, potassium-magnesium salts (silvine, kainite, langbeinite etc.) also occur.

In the Miocene an epicontinental sea – the Paratethys, covered the fore-Carpathian area. The Paratethys developed during the late Eocene and early Oligocene (about 35 million years ago) as a northern branch of the western part of the Tethys Ocean. The Paratethys was an enclosed sea that consisted of an interrelated chain of basins of diverse tectonic origin (*Fig. 2.2*). These basins were covered most of the time by the same mass of water, and were linked between the Mediterranean and the Indo-Pacific areas (BÀLDI *et al.* 2002). During its maximum range, the sea expanded to the Rhône basin in France and as far as Central Asia (POPOV *et al.* 2004).

In the Lower and Middle Miocene, the marine connections of the Paratethys with the adjacent seas were temporarily closed and evaporates were deposited (*Fig. 2.2*). Finally, about 11.6 Ma, the Central Paratethys was isolated, creating first the Pannonian sea and later the Pannonian Lake. The present-day Black Sea is a remnant of the Paratethys.

The first climatic change which allowed for the development of evaporates in the Carpathian areas occurred in the Early Miocene (Egenburgian, *Fig. 2.3*).

The relevant salt-bearing series is called the Vorotyshcha Formation: 1100–2300 m thick, composed of clays with sandstones, potash and rock salts layers and lenses (PETRYCZENKO *et al.* 1994). All these beds have been subjected to intensive folding. The Vorotyshcha Formation is widely spread in Ukraine, and created potash salt deposits (e.g. at Stebnik). In the Polish part of the Carpathian Foredeep, the Vorotyshcha formation is found only over a small area south of Przemyśl, where zuber and salt clays have been identified (GARLICKI 1979).

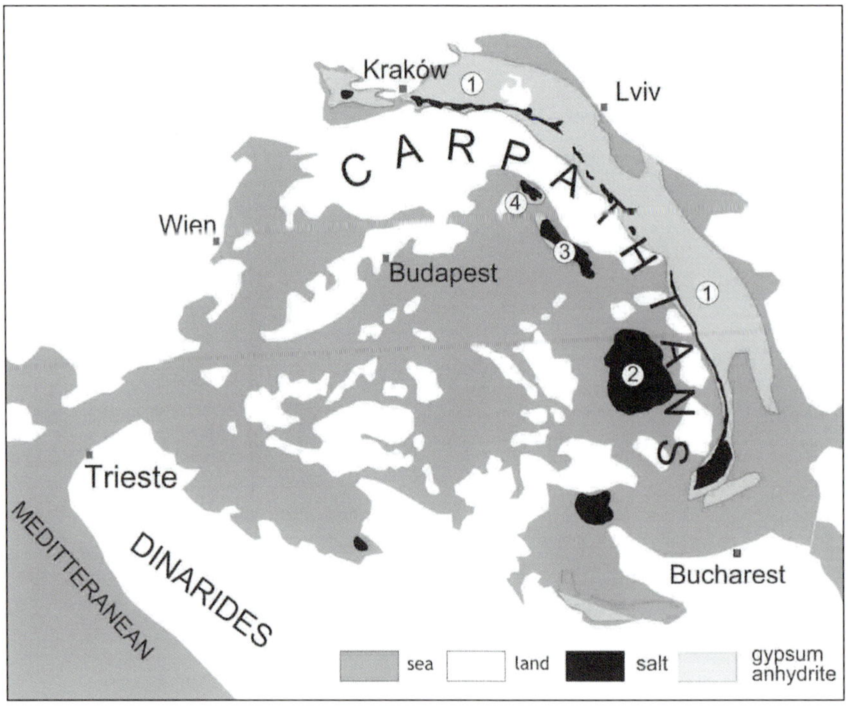

Fig. 2.2. Distribution of salts and sulphates (gypsum and anhydrites) in the Central Paratethys (after BĄBEL *2004) during the "Badenian salinity crisis", projected onto the palaeogeography of an Early Badenian Carpatho-Pannonian Basin complex at ~16.4 Ma (after* RÖGL *1998). 1: Carpathian Foredeep; 2: Transylvanian Basin; 3: Transcarpathian Basin; 4: East Slovakian Basin.*

AGE (Ma)	TETHYS (MEDITERRANEAN) STAGES		WESTERN PARATETHYS STAGES	UKRAINIAN CARPATHIAN FOREDEEP UNITS	POLISH CARPATHIAN FOREDEEP UNITS
12	MIDDLE MIOCENE	Serravalian	Sarmatian	Dashava	Machów
13			Badenian	Kosiv	
				Tyras	Wieliczka
14		Langhian		Bohorodchany	Skawina
15					
16					
17	EARLY MIOCENE	Burdigalian	Karpatian	Balych	MESOZOIC BASEMENT
18			Ottnangian	Stebnyk	
19			Eggenburgian	Vorotyshcha	
20				Polyanytsya	

Fig. 2.3. Simplified Miocene stratigraphy of Polish and Ukrainian Carpathian Foredeep (adapted from HRYNIV et al. *2007).*

One of the distinct changes of the palaeoenvironment in the past dozen million years was the Badenian Salinity Crisis (*Figs 2–3*). During that crisis a complex of climatic, environmental and geological conditions caused the occurrence of a continuous series of evaporite deposits (salts, gypsum and anhydrite) over a large area of central and southern Europe. Evaporites of Middle Badenian age have accumulated in the Carpathian Foredeep area, stretching from Upper Silesia in Poland (GARLICKI 1979) to the Doftana Valley in Romania (PERYT 2006).

Contemporaneous evaporites were also deposited in the Transylvanian Basin in Romania (BALINTONI – PETRESCU 2002), the Transcarpathian Basin of the Ukraine (KITYK *et al.* 1983), and the East Slovakian Basin (*Fig. 2.2*). Based on recent studies, the onset of evaporite deposition is dated at 13.81 Ma ± 0.076 Ma, and the entire event may have lasted ~400 000 years (DE LEEUW *et al.* 2010).

Carpathian Foredeep

In the Polish part of the Carpathian Foredeep Basin, evaporite deposits are included in two formations: the Krzyżanowice Formation and the Wieliczka Formation (ALEXANDROWICZ *et al.* 1981). The Krzyżanowice Formation comprises deposits of gypsum and anhydrites (up to 60 m thick). The Wieliczka Formation consists of rock salt (primary thickness 30–100 m). Rock salt occurs in Upper Silesia, then in the form of a narrow band extending along the Carpathian Overthrust from Kraków to Pilzno (BUKOWSKI, K. 2011; GARLICKI 1979), and further to the east where it occurs under the overthrusted Carpathian nappes (*Fig. 2.4*).

In the Ukrainian Carpathian Foredeep Basin, its foreland is much wider. The Badenian evaporites are included in the Tyras Formation (*Fig. 2.3*). The Tyras Formation achieves a thickness of 300–800 m and consists of salt breccias and rock and potash salts. Further out, the Carpathian Foredeep has an average thickness of 20–70 m and contains claystones, sandstones, carbonates, sulphates and rock salts. These beds are correlated with the Wieliczka Formation in the Polish part of the Carpathian Foredeep (ANDREYEVA-GRIGOROVICH *et al.* 2003). Within the Tyras Formation is located the Kalush potash salt deposit.

In the Romanian part of the Carpathian Foredeep, the evaporites (salt breccias, rock salt and gypsum) are included in the Cosmina Breccia and the Slănic Prahova evaporitic sequence. The Slănic Prahova Formation is ca. 120 m thick and is mostly halite with thin marl and anhydrite intercalations (GEORGESCU *et al.* 1994). The salt appears as a lens-like accumulation, with the widest thickness in its central area (~500m). In the central-north area, there is a small diapiric tendency, due to the lateral pressure developed by the Homorâci anticline. The salt is mined at Slănic, in the Victoria salt mine.

In north-west Bulgaria, Badenian Ca-sulfate evaporites occur, up to 40 m thick (TRASHLIEV 1969). They are mostly stromatolitic and laminated gypsum, included in the Deleina Formation (KOJUMDGIEVA – POPOV 1988). It is very probable that the same development characterizes the adjacent part of the Romanian Carpathian Foredeep (PERYT 2006).

Later, during the Sarmatian, tectonic processes associated with the remodelling of the Carpathian range (uplift and overthrusting of the Carpathians to its foreland) folded the Badenian salt-bearing deposits. At the cores of the strongly piled folds, the thickness of the originally thin salt deposits was multiplied. At the same time, the salt layers were stacked steeply and uplifted close to the surface. The narrow zone of such salt folds continues from Poland to Romania (*Fig. 2.4*). Prehistoric and contemporary salt springs often appear near outcrops of these folds.

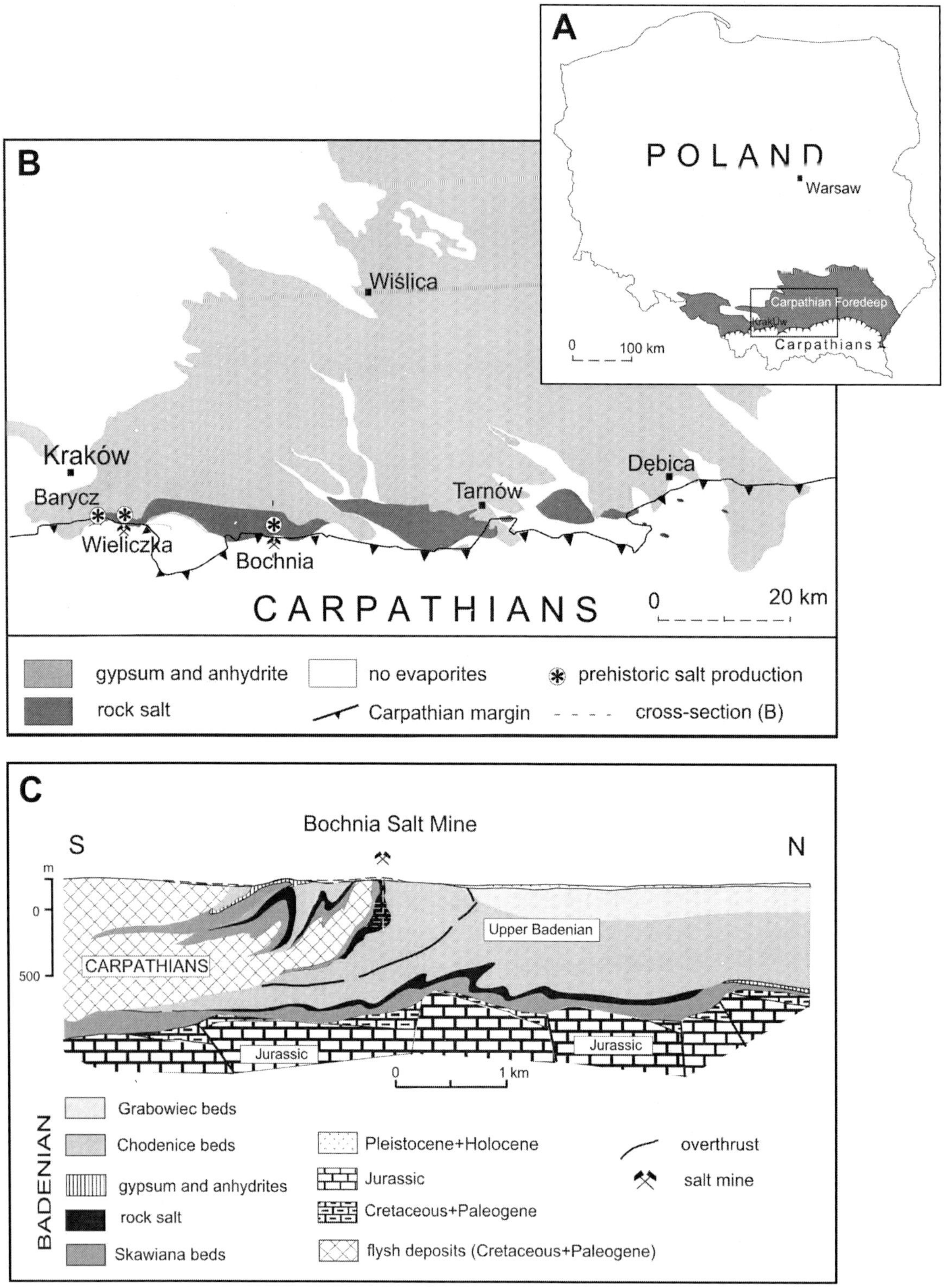

Fig. 2.4. Distribution of Miocene rock salt in the Polish part of the Carpathian Foredeep. A: Location of the Carpathian Foredeep in Poland; B: Occurrence of rock salts in the form of a narrow zone along the Carpathians (Poland, after BUKOWSKI *2011); C: Geological cross-section through the Bochnia rock salt deposit (after* CZAPOWSKI – BUKOWSKI *2009).*

Transylvanian Basin

In the Transylvanian Basin the Middle Miocene history begins in the Early Badenian with the deposition of acid pyroclastic rocks of the Dej Formation, covered by the Middle Badenian evaporites – salt and gypsum – of the Ocna Dejului Formation. The gypsum beds are a few tens of metres thick, and outcrop at the western border of the basin. Originally, the salt deposit thickness was around a few hundred metres, but the present distribution and thickness of the salt was controlled by post-depositional salt-tectonics (KRÉZSEK – BALLY 2006). The salt outcrops along the western, southern and eastern borders of the Transylvanian Basin are due to diapirism (*Fig. 2.5*). Diapirs commonly intrude vertically through more dense rocks because of buoyancy forces associated with relatively low-density rock salt, which form diapirs. By pushing upward and piercing overlying rock layers, diapirs can form salt pillows, salt domes and other structures. In the Transylvanian Basin many diapir structures are known (e.g. Dej, Turda, Ocna Mureş, Praid, Sovata). The salt outcrops associated with diapirs follow two major, roughly north-south oriented, lineaments situated near the western and eastern borders of the Transylvanian Basin (*Fig. 2.1B*, *Fig. 2.5*)

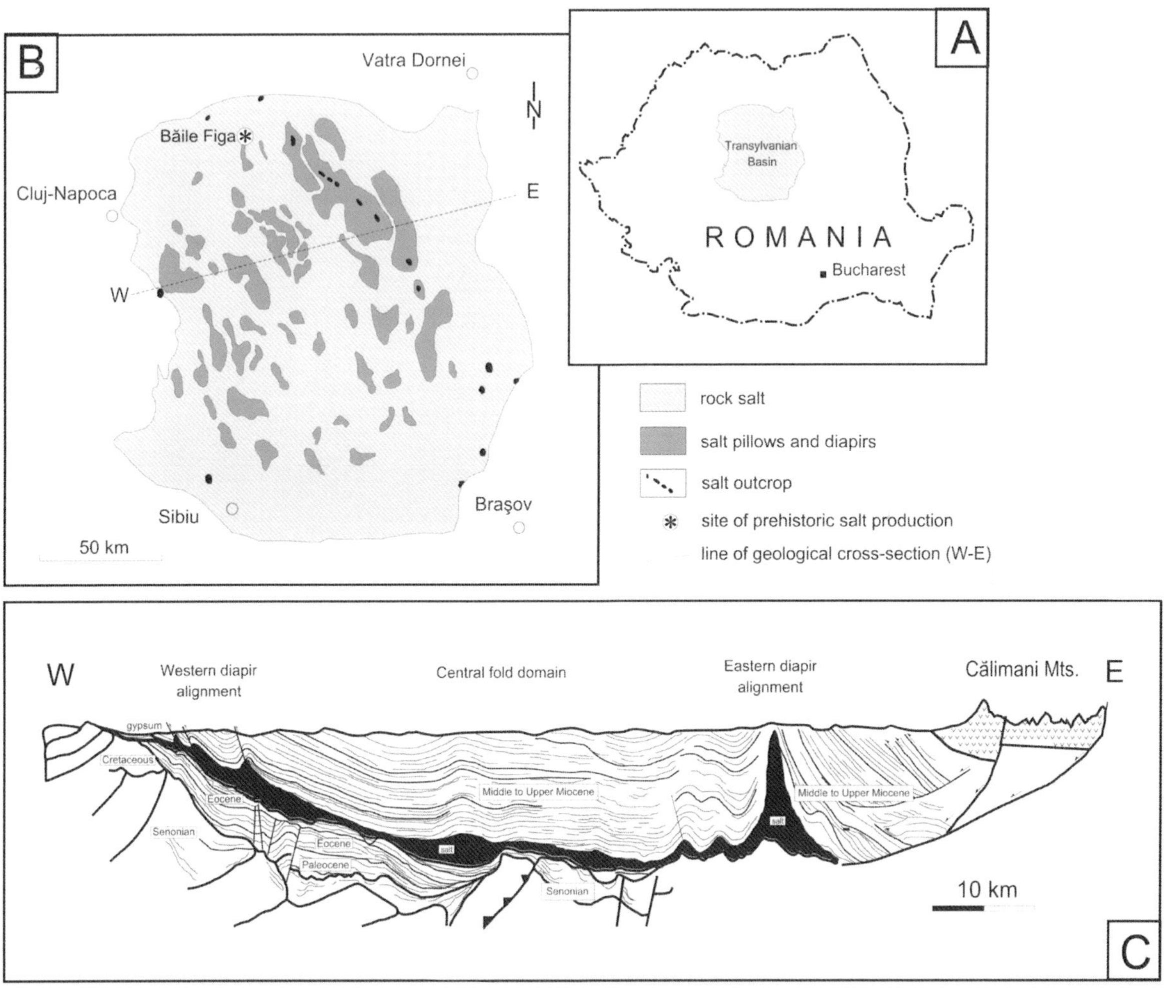

Fig. 2.5. Distribution of Miocene salt-bearing formation in the Transylvanian Basin (Romania). A: Location of Transylvanian Basin; B: Regional salt distribution (Middle Badenian) in the Transylvanian Basin (after KRÉZSEK – BALLY 2006); C: Regional cross-section of Transylvanian Basin showing salt pillows (thicker then 200 m), salt walls and outcropping salt diapirs (after KRÉZSEK – BALLY 2006).

East Slovakian Basin

Within the East Slovakian Basin (Inner Carpathians, in the area between Košice and Michalovce), two evaporite intervals dominated by rock salt with siliciclastic and anhydrite interbeds occur. The lower interval of the early Miocene age (the Lower Karpatian) is distinguished as the Solna Bania Formation and developed locally in the western part of the East Slovakian Basin (Solivar near Prešov). The upper interval of the middle Miocene (the Badenian) is defined as the Zbudza Formation, accumulated in local depressions of the eastern part of the East Slovakian Basin. Salt-bearing deposits contain pure or clayey halites interbedded with siliciclastics (laminated silistones and claystones, sandstones and conglomerates) and anhydrites. Lithological features of Zbudza salts are comparable with features of salts that make up the age-equivalent Wieliczka Formation in the Carpathian Foredeep (BUKOWSKI, K. *et al.* 2007).

Transcarpathian Basin

Evaporites of the Transcarpathian region occur only in the Badenian salt-bearing formation. There were formed sediments of the Tereblya beds, mostly consisting of evaporites, to various degrees contaminated with terrigenic materials. The Tereblya beds are divided into the lower (anhydrite-gypsum) and upper (salt-bearing) ones. In the Solotvino area salt tectonics has been active, resulting in ubiquitous salt diapirs (KITYK *et al.* 1983).

Chapter 3
Traditional salt exploitation in Transylvania

ANDREA CHIRICESCU

The ethnographic research campaigns initiated by the National Museum of the Eastern Carpathians and Exeter University, conducted in 2003–2005, were intended to establish the role, place and significance of salt in traditional village communities, in an attempt to outline a specific 'ethnographic scenery' created by the presence or absence of nearby salt sources. Another aim was to research and document the traditional exploitation methods of all types of salt source, in other words the processing and use of salt in the household in everyday life.

Comparable work has been carried out in Moldavia by a Romanian-French team, including an especial concentration on the health effects of halotherapy (among several similar contributions: ALEXIANU – WELLER 2007; ALEXIANU *et al.* 2007; SANDU *et al.* 2010). We have not attempted any detailed comparison with their results, but note many similarities.

Earlier work had established the existence of four large saliferous areas in eastern Transylvania, As a result, ethnographic research was conducted in four geographical areas:

- the Reghin – Gurghiu Valley area (localities Ideciu de Jos, Jabenița, Gurghiu, Orşova)
- the Homoroade Valley area (divided into Homorodul Mic – localities Lueta, Mereşti, Jimbor, Mercheaşa; Homorodul Mare – localities Comăneşti, Mărtiniş, Sânpaul; and the Homorod confluence – localities Rupea, Racoş)
- the Bistrița – Năsăud area (localities Blăjenii de Jos, Caila, Cepari, Dumitra, Figa, Mintiu, Sărățel, Săsarm,Țăure)
- the Praid – Sovata – Corund area (localities Corund, Praid, Ocna de Sus, Ocna de Jos, Sovata)

These geographical areas stood out because of the large number of salt sources, covering a wide range of source types: salt marshes, brine springs, wells, pools, lakes and even salt rock mountains. This raised a number of questions from an ethnographic point of view, related especially to traditional exploitation methods and the role that salt sources play in the everyday life of nearby communities.

Because of the large number and variety of salt sources, the methods used to exploit them are many and varied. However, a very important aspect is that they do not vary much from one area to another, thus confirming the unitary character of Romanian popular culture. This is why we consider that there are no significant differences in the traditional methods of exploitation and appearance of salt sources in Transylvania, or for that matter in Romania. Ongoing research will prove us right or wrong.

Identification and exploitation

In all the areas researched we found all the types of salt source mentioned above, the underground rock salt being quite close to the surface in many places. In each area we also had to take into account what people believed about the rock salt bed that is common to these regions. Although in some places rock salt was systematically exploited (Praid, Sânpaul etc.) in other places traditional, empirical, methods were used to extract it. These methods are as follows:

- 'Daytime exploitation', or extracting lumps of rock salt directly from the edges of the salt bed where it occurs close to the surface. This is one of the oldest exploitation methods, first practiced with rudimentary stone and wood tools, and later with iron tools. Thus, the stone pick-axe was

replaced by metal picks, the wooden wedge by iron wedges, the wooden mallet by iron sledge-hammers, and so on. Regardless of the tools used to practice it, this type of exploitation is still employed by villagers to extract salt for feeding animals. We found it at Orşova, Praid, Jabeniţa, Săsarm, Dumitra and elsewhere
- The use of brine (salt water), called 'murătoare', 'saramură', 'slatină' or 'sare lichidă' ('liquid salt') in the household, for cooking, feeding animals or treating diseases and injuries

  The 'in situ' use of the curative properties of brine and salty mud, by digging small pits near brine springs to collect water (as found at Comăneşti) or by covering the body with a layer of salty mud and then bathing in simple, unmanaged salt water ponds or lakes (as at Jabeniţa – La Gureţe)
- Grazing herds in salt marshes and other areas covered with halophile vegetation (CIOBANU 2003). Animals usually graze instinctively in these areas, without compulsion from the herdsman.
- Boiling and evaporating brine in order to obtain pure salt.

The peasants use their skills to identify salt-rich areas, by following and recognizing a range of specific indications. On a sunny day it is quite easy to distinguish a salty area, thanks to the salt crystals that form on the surface. There is also specific vegetation that grows on salty soils, plants that bear various names in popular tradition: the one found most often is called 'salt flower' (*Salicornia*), thanks to its salty taste.

Another way to recognize a salty area, or the presence of a brine spring close to the surface, is by following the reaction of grazing animals. They will instinctively search for salt. In many places, however, peasants avoid these areas since salt and brine can be harmful for animals if they consume it in large quantities. This is why locals avoid watering animals with brine, giving them rock salt instead. In some places, where rock salt is harder to obtain and brine is used for household purposes, they do water animals with brine, but closely control the amount consumed.

Although the locals know all these indications, they do not need to use them to identify salt sources. All the people we interviewed were born long after the sources were first discovered, inheriting the knowledge from their parents, grandparents and so on, as well as how to exploit and use the salt.

The use of brine in the household became a tradition, even a custom in most of the areas researched. There are many households in which powder salt is not used at all, even though it can be found in the shops. The elders of the village learned to appreciate the value of salt by hearing the stories told by their grandparents when they were children, about a period when access to salt sources was limited or prohibited entirely. They learned to use brine in daily household activities, including cooking, and regard brine as being a very precious foodstuff, a gift of nature, much better and healthier than the powder salt found in stores nowadays. Many cannot imagine the taste of food cooked with anything other than brine, or the taste of pickles prepared according to a recipe that does not include their own special brine.

Brine wells

Thanks to the importance brine has gained through history in many of the areas researched, we found carefully managed brine wells, some of them being locked and watched over in order to keep them safe and clean. In some places, such as Lueta, the house that covers the brine well is permanently locked, cleaned daily, and even blessed at Epiphany, just like the peasant houses.

We identified several types of brine well, grouping them according to their appearance: *unfurnished brine wells*, small brine pits, dug out by locals; *partially furnished brine wells*, that look like fresh water wells, being sometimes covered by simple, wooden structures; *furnished brine wells* or *house-covered brine wells*, These are real ethnographic monuments following archaic architectural techniques of the region, representing living proofs of cultural tradition and heritage.

A brine well is generally arranged just like a fresh water one. After establishing its position, according to the brine flow, a deep pit is dug out. In some places this pit is about 3 metres deep, since the people stop digging when they hit rock salt at the bottom (e.g. Ideciu de Jos, Mureş county). If they do not hit rock salt and the pit is deep enough, they place a layer of gravel on the bottom of the well, to filter the brine. The walls of the well are furnished with wooden boards, fixed at the corners into oak beams. The space between the vertical wooden walls and the soil is filled with gravel, to filter rainwater that might penetrate the well. Cristea Toader (b. 1946, Jabeniţa) told us about this technique, which he used to repair the well in Jabeniţa commune.

In some brine wells the salty water is caught in large oak trunks that were hollowed out and placed directly into the pits. This technique proves their considerable age (e.g. brine wells from the Homoroade Valley: Lueta, Mereşti, Mărtiniş).

Well-houses

The constructions that cover and protect the brine wells are real architectural monuments, since they follow the traditional building techniques that are used to construct the archaic single-unit (unicellular) houses. They constitute some of the few remaining indications of these techniques, since new houses have been built everywhere, replacing and thereby destroying the old ones. The house that covered the brine well had multiple roles: to protect the well, to keep it clean, to keep animals away, to avoid accidents, to keep them in good condition, as well as to limit people's access to brine during the Austrian period.

The foundation of the whole construction is supported on large boulders at all four corners. The sleepers are cross-bound. The walls are built of large wooden beams, placed horizontally upon each other, fixed at the corners with a technique traditionally called 'elder's key'. This is the oldest wall construction technique found so far. Some of the well-houses are built using newer techniques. The walls are made of wooden planks, cut even on all four sides, fixed together at the corners using the 'wolf's teeth' or 'swallow tail' technique. Two other types of well-house are found: houses made of wooden boards, and houses made of concrete or brick. Although they follow modern construction techniques, they were built on the site of earlier constructions.

The huts have shed, gable or hipped roofs, covered with shingle, tile, board, tar, corrugated iron or concrete.

The fact that the brine wells are protected by unicellular constructions resembling archaic houses demonstrates the importance they had in everyday life. They are individual constructions, part of the village or commune property from time immemorial. During the Austro-Hungarian period people's access to brine was limited, each family having the right to a certain amount of brine, based on the number of family members and the number of animals in the household.

The wells are locked and guarded, and were before and after the Austrian period as well, but for different purposes. A villager is in charge of keeping the well and its surroundings clean, opening the building by request or according to a certain schedule that is followed by all villagers. The oldest well-houses have wooden locks that open according to a specific code, known only to the well's caretaker (e.g. the well from Corund). Nowadays most well-houses are not locked.

The brine is drawn out with wooden buckets, with handles 2.5 m long, or attached to long wooden hooks. The brine has a higher density than fresh water, thus is found closer to the bottom of the well. The upper area of the well is filled with brackish water (brine mixed with infiltrated fresh water). Thus the water of the well has to be stirred before drawing it out, and the wooden handles and hooks served this purpose. Just like the house itself, all the objects found inside a well-house are made of wood, since salt is highly corrosive.

Brine is carried and stored in special wooden tubs. The tubs are made of wooden staves, bound together with two or three hazel hoops. Both the top and the bottom of the tub are closed. There is only one small opening in the top of the tub, through which brine is poured in and out. The tubs can hold up to 60 litres of brine, and have a special place in the house where they are kept.

The use of brine and rock salt in the household

From an ethnographical point of view, the most important use of salt and brine in the household is its role in nutrition. Its chemical and physical components are essential for survival, both for humans and for animals.

Although in most areas researched we found both brine wells and rock salt deposits, the rock salt lies quite deep beneath the surface and so is difficult to exploit using traditional methods. In the areas where the rock salt is close to the surface, it has not been exploited for many years, since this activity is strictly prohibited.

The use of brine in preparing food is a practice inherited generation after generation. It has become a tradition of the saliferous regions, still used nowadays. Brine is used to cook food, using a specific measure that may vary from one household to another: a wooden spoon, a ladleful, a scoop, a coffee spoon, etc. There are no specific recipes; the brine is added to a certain kind of dish according to its ingredients and the quantity to be cooked. There is one important criterion: the saltiness, which differs from place to place and from one source to another.

The brine is also used instead of powder salt to bake bread. In this case the baker adds a certain amount of brine to a certain amount of fresh water, in order to obtain the concentration he needs. In some areas 1 litre of brine is added to 10 litres of fresh water, but at Lueta, for example, they add 1 litre of brine to 4 litres of fresh water, or 100 ml of brine to one loaf of bread. In some localities there is a price one needs to pay for each litre of brine, especially if the person is a 'foreigner' or a business person who needs large quantities. Thus, there are some bakeries that prepare bread with brine, and instead of paying for it they trade a certain amount of bread to cover the price of the brine.

One of the most important aspects of salt is its preservative properties. This is why one of the most important uses of brine is to prepare pickles and and preserve cheese.

The highest consumption of brine, however, occurs in the winter, just before Christmas, when villagers kill off pigs. It is the time when each household needs a larger quantity, to clean, prepare and preserve meat, ham and bacon.

Brine was also used to obtain powder salt, by boiling and evaporating it. The powder salt thus obtained is very clean, white and pure. In earlier times, when the exploitation of rock salt was forbidden and the price of salt was very high, people used to exploit brine, boil and evaporate it, and trade the resulting salt for various products. Areas rich in salt tend to be poor in agriculture and fruit growing. Thus, the necessary amounts of grain and corn were obtained through exchange. In some areas, though, powder salt is much easier to obtain without it being necessary to boil the brine. At Lueta the brine is so concentrated that the salt crystallizes on the tubs in which the brine is kept. The locals scrape it off and consume it as table salt.

Animals also need a daily amount of salt in order to survive, which they usually get from their fodder. Sometimes the amount of salt that animals get in this way is insufficient. In most cases the owners put a piece of rock salt into the feeding trough, or on special wooden supports placed among the animals. Rock salt is very concentrated and the animals will consume only the amount of salt they need.

In areas where rock salt is difficult to find but there is a brine source, people use a variety of methods to feed it to animals, without exceeding the recommended daily amount. They sprinkle brine on hay and the aftergrass, thus keeping it fresh and mould-free. They also add a cup of brine to the pig trough.

The quantity of brine given to animals is closely supervised. Brine is not as concentrated as rock salt, so animals do not know when to stop consuming it. Excessive amounts of ingested salt can cause them considerable harm, or even kill them.

Another important property of brine is curative. People learned how to use brine to treat various diseases or injuries, in both humans and animals. In treating humans, the most important use is in healing sore throats, coughs and other cold symptoms. It is also used to treat rheumatism, to combat lassitude, and generally to strengthen the human organism, by adding a cup of brine to the water in which people bathe.

Brine and salty mud are also used to treat animals. To get rid of parasites or heal injuries peasants put compresses on the backs of animals, or on the injured area.

In most places, right next to the brine wells and springs, one can find natural or artificial pits that locals use to treat rheumatism. They use them empirically, as inherited from their predecessors, without consulting a doctor, judging by how they feel after following such a treatment. In some places the fame of the excellent curative properties of certain brine springs and salty mud has attracted many people, including doctors, and led to the development of spas. Today most of these places are in ruins, but locals keep using them, saying that a bath in a salty lake is much healthier than any medicine found in a pharmacy.

From an ethnographic point of view, we may conclude that all these salt sources were exploited empirically from ancient times, being inherited by today's people from their parents and grandparents, who inherited them from *their* parents and grandparents. We cannot establish for sure the age of most salt sources or the certain history of their exploitation. This is where archaeology can complete the picture, being able to confirm or deny the ancient use and exploitation of a salt source that is still in use today. Our goal is to find answers to all these questions concerning the start of salt exploitation in the region that we know today as Transylvania, based on the interpretation of archaeological finds through present-day ethnographic evidence. In this way we may eventually be able to reconstruct the 'salt route' in Romanian history.

Chapter 4
Transylvania

ANTHONY HARDING – VALERII KAVRUK

Introduction

Transylvania, as discussed above, contains many large salt deposits. Our work has concentrated on eastern and northern Transylvania (excluding the Maramureş), mainly Hargita and Covasna counties in the east and Bistriţa-Năsăud county in the north. We were aware of the discovery of wooden objects, including troughs, at Valea Florilor north of Turda in the 1930s, and of archaeological interest in the area near the saltworks at Ocna Mureş – indeed, Harding spent some days in that area in the late 1990s with Dr Horia Ciugudean (who has subsequently developed his own project). The area around the saltworks at Ocna Sibiului has also been mentioned in the literature, but as far as we know there has been no specific archaeological project investigating its potential as an ancient salt-working site.

The discoveries at Valea Florilor have been presented in the literature several times and do not need detailed discussion here (MAXIM 1971; WOLLMANN 1996, 246, Pl. CVI,3; WOLLMANN – CIUGUDEAN 2005). The finds were made in 1938 while enlarging the platform of the railway station at Valea Florilor, on the main line from Cluj to Turda and beyond. The area lies over a salt massif. The workmen came across a series of shafts, and in one of them a set of wooden objects at a depth of 8–10 m below the surface, mainly shovels and picks, but also a trough and parts of others. One trough was apparently well-preserved, 2 m long, 40 cm in diameter, with a line of fourteen holes in the base plugged with perforated tubular pegs of elder 3 cm in diameter and 15 cm apart from each other (*Fig. 4.1*). The trough was apparently complete, but the drawing made at the time looks rather schematic, so the details remain uncertain. Nothing was known about the other troughs until Wollmann found them, or parts of them, in Maxim's house in Cluj (WOLLMANN – CIUGUDEAN 2005, 98, Pl. III, 1–3); but because of the poor state of preservation little more can be said about their form.

The dating was naturally uncertain; Maxim suggested a date in the Dacian Iron Age. The rediscovery of the objects in Cluj and the obtaining of a radiocarbon date (3000 ± 80 BP), placing at least one of the troughs in the Bronze Age, was a crucial step in appreciating the true age of these extraordinary objects.

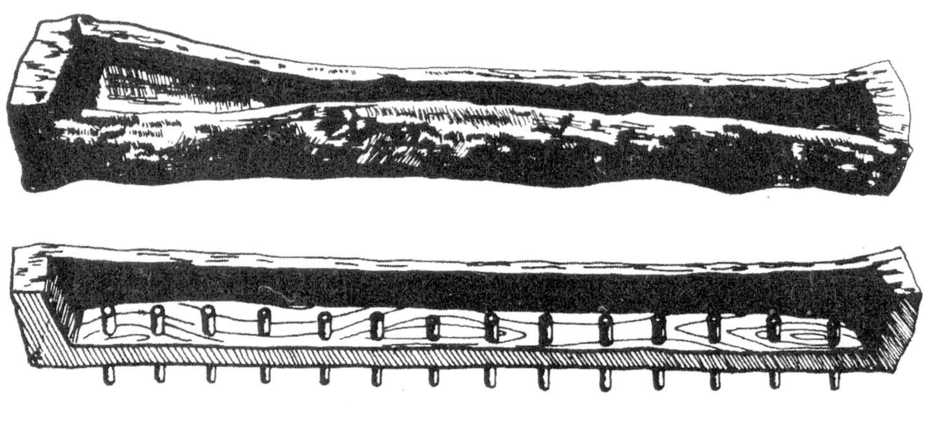

Fig. 4.1. Valea Florilor trough: drawing from MAXIM 1971.

The importance of the finds for the development of our understanding of salt exploitation in Transylvania is, however, impossible to overestimate, because they indicated the existence of a technology that was hitherto unknown – although the 19th century finds from Királyvölgy are in fact directly comparable (see Chapter 5). At the same time, there was clearly much field evidence on full view to those who had eyes to see it when we started our work; as far as we are aware, not one single site with evidence for prehistoric salt production in Romania had been recorded as such in the literature, let alone listed in the official register of sites and accorded legal protection (with the exception of the uncertain case of Valea Florilor).[1] Ioan Chintăuan had visited and described a number of salt springs and streams in Bistriţa-Năsăud county, in some cases with wooden piles visible (CHINTĂUAN – RUSU 1988), but their date was completely unknown and their potential for providing remarkable evidence for prehistoric salt production almost totally unappreciated, although Preisig had made perceptive remarks in this respect in the 19th century (PREISIG 1877). Until recently, few people had appreciated the significance of these finds – not surprising in the absence of datable associated artefacts, and prior to the possibilities afforded by radiocarbon dating.

As explained above (p. 22), our work began following the discovery of a mutual interest in ancient salt production. Visits to salt sources in various parts of Romania were supplemented by an extensive programme of ethnographic research in Transylvania.

Archaeologically speaking, our attention was initially attracted to brine springs in Harghita county where we could see the remains of above-ground installations (brine wells) and where there was historical evidence for salt exploitation. These visits and researches have been written up and published by Kavruk's team (e.g. CHIRICESCU 2006, among other places). Subsequently we extended our area of interest to Covasna and Bistriţa-Năsăud counties. We have not done archaeological fieldwork in Mureş county, though Kavruk's team has done a little ethnographic work there.

Fieldwork

In 2005 we obtained the funds to start work, which consisted of survey using differential GPS, and gradiometer survey using a Bartington series 601 machine. Survey traverses were at 1 m intervals, with sample readings 0.25 m apart. The grids (typically 30 m square) were laid out by tape and georeferenced by GPS.

Covasna county

Zoltan: This site, extensively excavated by Kavruk between 1996 and 2000, is known to have both domestic structures and an extensive ash deposit, resembling the *zolniki* found further east in Moldavia and beyond. The area north-east of the excavated part was surveyed by gradiometer, revealing a number of ditch and pit features (*Fig. 4.2*). In particular, the edge of the occupied area of the site was quite apparent, with a curving ditch running around the north-east and rectilinear features further south. Isolated anomalies, probably pits, lie in the field beyond the site.

Olteni: This site produced Late Bronze Age pottery that might possibly be considered briquetage in rescue excavations between 2001 and 2004; it has since been entirely quarried away. A small survey conducted a short distance from the excavated area, where the spread of sherds extended, produced no features. There is a salt spring a short distance away to the south, not far from a number of sites that span the whole period from Neolithic to early Medieval. The pottery in question does not in fact show the characteristic surface colours of authentic briquetage, and can be dismissed as such.

[1] This does not apply to the epigraphic information from Roman altars (see below).

Sites in the Homorod valley (Harghita county)

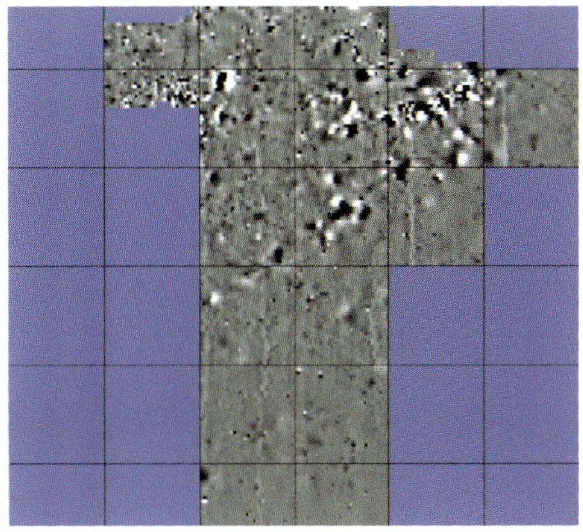

Fig. 4.2. Zoltan, Covasna county – geophysical survey plot.

The area around the village of Mărtiniş had been visited several times previously; many salt springs are known to lie in this valley (CHIRICESCU 2006, 165–6), including one in the centre of the village, now in a wooden well and hut (*Fig. 4.3*). Surveys were conducted at several sites in the vicinity.

Mărtiniş North: the field to the east of the salt stream was surveyed, revealing a number of features though no clear structural evidence (*Fig. 4.4*). The stream is rich in halophite vegetation (*Fig. 4.5*), and in dry periods the stream bed is white with salt crystals. No surface material is visible since the field is under pasture. Excavation would be needed to determine the nature of the site.

Mărtiniş-Bogozi: Part of the field, the site of an extensive Early Iron Age pottery scatter, was surveyed but most of the features recorded seem to be natural in origin.

Sânpaul: This locality is known as the site of a Roman fort and *vicus*, and as the findspot of a Roman altar bearing an inscription referring to one M. Caius Iulius Valentinus, who is described as a "*conductor salinarum*" (PISO 2004–2005 [2007]), one of several such inscriptions in Roman Dacia. The place is also known to have been the site of a salt mine in modern times, though no trace of this now remains. The fort lies around and under the church in the middle of the village, the *vicus* on the east side of the village.

Sânpaul lies in a plain where the Homorod Mare river has maeandered over a wide area. On all sides hills rise to around 600 m, with Sânpaul lying at 483 m (*Fig. 4.6*). The plain contains valuable wetlands which are protected as a natural reserve for halophite plants and birds. The lakes in the vicinity are home to a fish farm.

Along the road towards Ocland, 1.8 km south-east of the village and south of the road, lies the area investigated. We believe this was also the area of the former mine, but do not have specific information to confirm this. A low hill rises above the marsh and lake, and on it lies a typical salt well in its protecting

Fig. 4.3. Salt well at Mărtiniş, Harghita county.

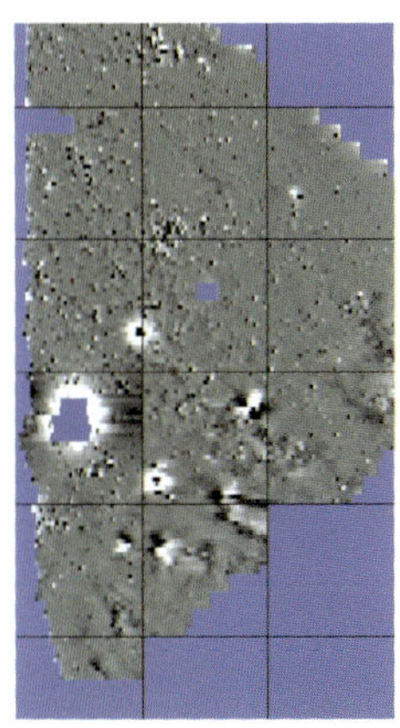

Fig. 4.4. Geophysical survey plot, Mărtiniş North.

Fig. 4.5. View of the salt stream at Mărtiniş North.

Fig. 4.6. The Sânpaul area viewed from the south-east.

Fig. 4.7. Salt well at Sânpaul, Harghita county.

Fig. 4.8. Salt stream at Sânpaul, showing timbers protruding from the stream bed.

hut (*Fig. 4.7*). A brine stream runs downhill towards the east from this hut and over a certain stretch wooden posts protrude from the stream bed (*Fig. 4.8*) (ANDRONIC *et al.* 2006, 80).[2] As the stream ebbs and flows in strength, more timbers became visible over the years that we visited it. Some medieval sherds were visible. The area was mapped by GPS and a small but inconclusive gradiometer survey was carried out. Samples from the timbers were taken and gave the following results:

Table 4.1. Radiocarbon dates from Sânpaul (calibration: OxCal v.4.1.7).

Lab number	Date BP	Cal date 1σ	Cal date 2σ
GrN-31522	405 ± 25	1444–1484 AD	1437–1618 AD
GrN-31523	370 ± 25	1458–1618 AD	1449–1632 AD
GrN-31524	375 ± 35	1451–1620 AD	1444–1634 AD
GrN-31525	2485 ± 30	759–541 BC	773–418 BC

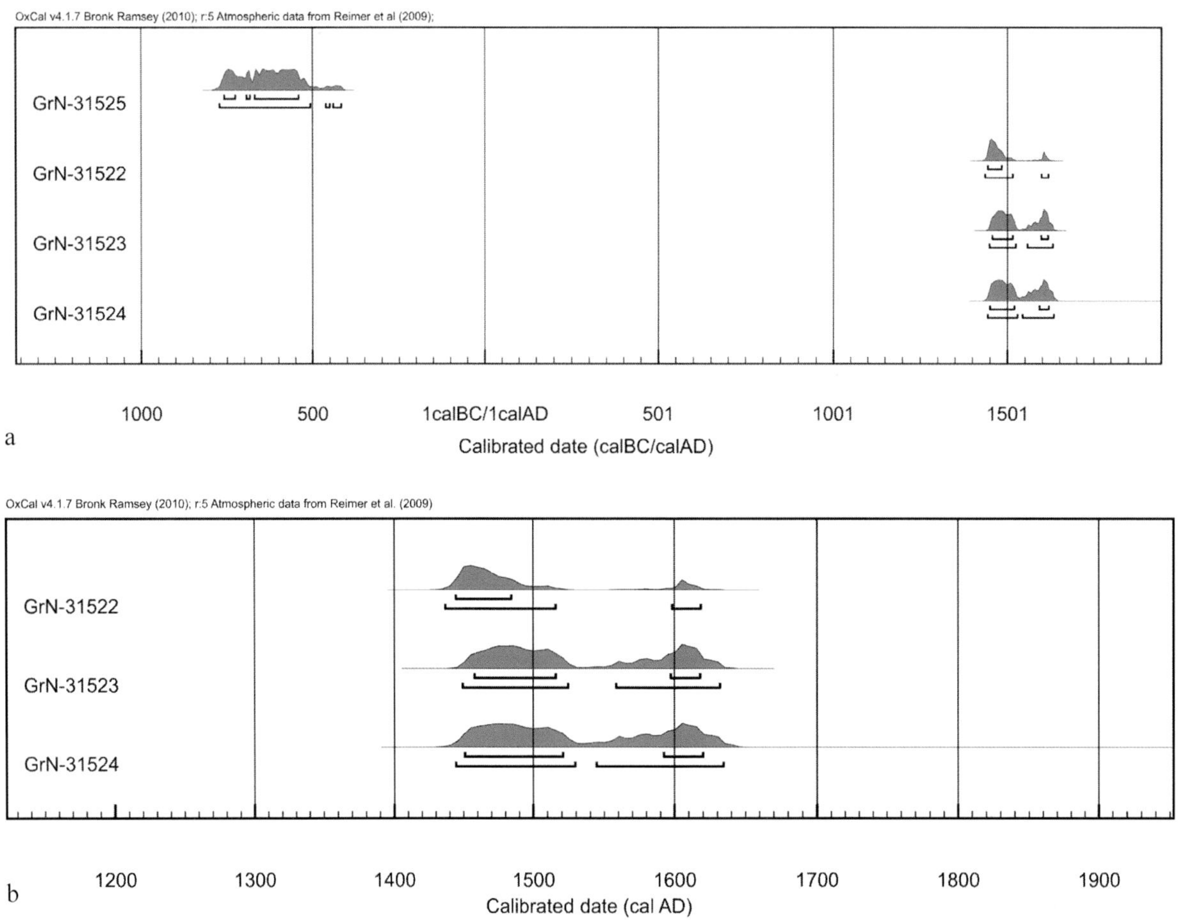

Fig. 4.9. Probability distribution of radiocarbon dates from Sânpaul. a: all dates; a: early modern dates.

[2] Photos of these were published by Kavruk in 2006, including one on the cover of the book *Sarea, Timpul şi Omul*, but no description was given.

The combined date range of the first three is AD 1453–1487 at 63.7% probability, or 1447 to 1514 at 81% probability, with a 14.5% probability that the range is 1601–1618, in other words a high probability that the date falls within a 75 year period around AD 1500 (*Fig. 4.9a–b*).

The range of the fourth date is relatively wide because the calibration curve is flat in this period, but there is no doubt that it falls in the Iron Age.

These dates are interesting for two reasons. First, none of them is modern, as those who have seen such timbers have tended to assume. All relate to the erection of wooden constructions a long time in the past. Second, the two periods covered are the period around AD 1500 (late medieval or early modern), and the Iron Age. Surprisingly, perhaps, none of the dates falls in the Roman period – which is the time for which there is independent evidence of salt working at Sânpaul. The fourth date perhaps provides some evidence for a pre-Roman exploitation of salt, which was then taken up in the Roman period. Further work on the site, and more dates, are needed. It is quite probable that the wooden constructions are more extensive than appears on the surface, to judge from our experience at Figa (see below).

At the time we worked at Sânpaul in 2005 we had no idea of the wealth of material that awaited us at Figa, which we visited after our work in the Homorod valley. In 2005 our work was cut short by floods and thunderstorms which made the use of our equipment impossible; there is clearly scope for more extensive work in the area.

Homorod valley: conclusions

At present, while we know of many salt wells and springs in the Homorod valley (Chiricescu 2006, 165–6), no places other than Sânpaul have wooden piles visible. In their absence, an archaeological project would be hard to set up, though we believe it is highly likely that other sites do or did once possess comparable installations, now destroyed or buried. In recent and presumably in historic times, the valley has been an extensive provider of brine, and if there was a significant mine at Sânpaul, presumably of rock salt as well.

Bistrița-Năsăud county

In 2005 we were informed about a discovery at Băile Figa, near Beclean in Bistrița-Năsăud county. The importance of the site was noted in 1977 by the geologist from Bistrița Museum, Dr Ioan Chintăuan, when he spotted many wooden piles in a brine stream named by the locals Pârâul Sărat ('salty stream') (Chintăuan 2005; Chintăuan – Rusu 1988); according to his account, a number of previous scholars had also noted the potential importance of the spot, starting with Ioan Al. Maxim, who had brought the Valea Florilor finds to the attention of the scientific world . Among these posts Chintăuan observed the broken end of a large wooden object, which turned out to be a trough. Chintăuan – Rusu (1988), relying on the similarity between this trough and the one found at Valea Florilor and published as dating to 'Dacian La Tène', assigned it to the second part of the Iron Age. In May 2005 Chintăuan removed the trough from the stream bed (*Fig. 4.10*) (unfortunately without proper recording and not accompanied by an archaeologist) and transported it to Bistrița Museum (Complexul Muzeal Județean Bistrița-Năsăud). Soon afterwards he published the trough, with the suggestion that it was used to obtain salt by evaporating brine (Chintăuan 2005). It is a remarkable object, with a row of perforations along its base, the holes plugged with pegs that are themselves perforated, twisted cord being present in at least one of the perforations (see below; *Fig. 4.11*).

In August 2005 we visited Bistrița Museum and saw the trough. We were then taken by Chintăuan to Băile Figa, where we immediately recognised the exceptional importance of the wooden remains in the stream. We took two samples of the trough for radiocarbon dating; the results, when they were

Fig. 4.10. Băile Figa, Bistrița-Năsăud county: Trough 1 in the process of removal, May 2005 (photos I. Chintăuan).

obtained in autumn 2005, indicated a Bronze Age date (the decades around 1000 cal BC; see below for a discussion of the radiocarbon dating). This compared with the date of 3000 ± 80 BP (1380–1120 cal BC at 68% probability) for one of the troughs from Valea Florilor (WOLLMANN – CIUGUDEAN 2005), and showed clearly that the troughs belonged to the Bronze Age, and that the Figa site deserved full investigation. Following a successful application for funds to the British Academy and the Ministry of Culture and Cults of Romania, work began in 2006 on a full surface survey at Figa, including a catalogue of all visible wood remains, and a partial survey at another site noted by Chintăuan at Săsarm (Valea Slatină, 'salt valley'), where a salt spring lies in a stream also full of wooden posts and other installations (see below).

Following the survey of 2006, and further successful funding applications, excavation began in April 2007. The situation was complicated by the fact that unknown to us at the time, permission had been given for the development of the area for a new spa and leisure complex, which, if executed according to the original plans, would have destroyed most of the northern half of the site. Additional funding became available from the developer and the local mayor's office, which enabled a larger area to be explored than would otherwise have been possible. Work in the northern half of the site thus took on a "rescue" character, and proceeded over many weeks in 2007 and 2008, and on a smaller scale in 2009, 2010 and 2011. Work in the southern part of the site was exclusively of a "research" nature, and was confined to short seasons in each of 2007, 2008 and 2009.

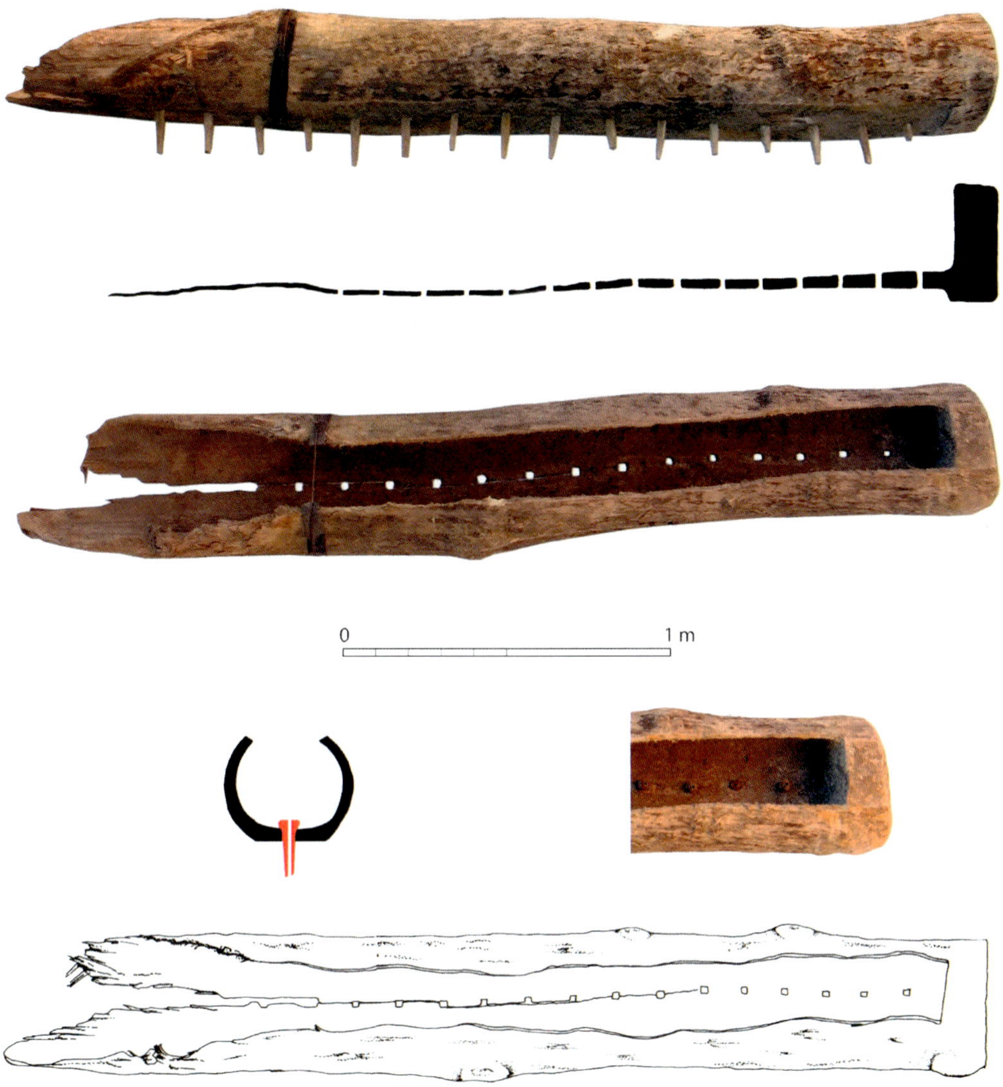

Fig. 4.11. Trough removed from Băile Figa by I. Chintăuan (Trough 1).

Băile Figa

Băile Figa (Figa Spa) is situated in the administrative territory of Beclean, about 2 km south-east of its eastern end, and about the same distance north of Figa village (*Fig. 4.12*). The area where salt occurs consists of a bowl-like depression about 290 m above sea level; this lies below a flat plateau to the north bordering the Someş valley; to the south the hills rise to around 350 m (*Fig. 4.13*).

The present morphology of the terrain has been much affected by human activities, mainly salt mining, from prehistory down to the present day. Thus there are many depressions and mounds on the site, resulting from digging for salt by local people over the decades (or centuries). Enquiries in Figa revealed that until the 1970s people from the neighbouring villages extracted rock salt from the site, using it mainly as a forage supplement. The locals say that when they dug for salt they often came across pieces of "black wood", which they attribute to the "Turks" (i.e. the Ottoman period), evidently

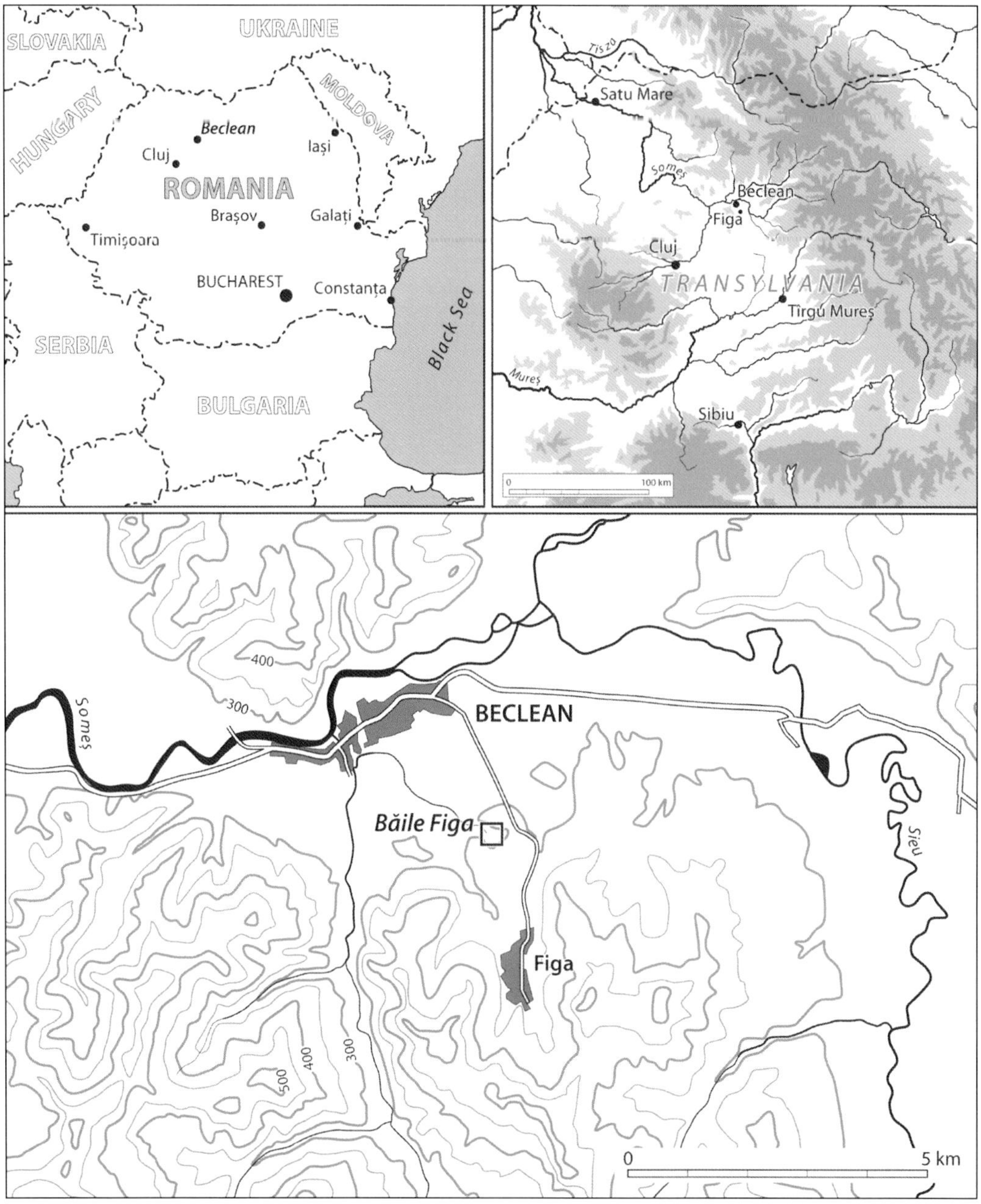

Fig. 4.12. Location map for Băile Figa.

the traces of earlier mining activities. There are a few brine springs in the main stream at Figa; some of them are still in use, others are blocked and abandoned. The locals use brine in order to preserve lard, meat, cheese and vegetables. At the same time, they use the salty mud and brine to cure rheumatic and blood circulation disorders in humans, and for disinfecting wounds on animals.

In the post-war period (1975 according to Chintăuan) the northern part of the site was partially developed as a spa and bathing complex. The remains of a dilapidated concrete bathing pool were present until 2009, along with other concrete foundations for buildings. At this period a number of chalets were built to the west, and an electricity supply installed. This no longer functions, and most of the houses are abandoned, though with the renewed interest in the area, new houses are being constructed, and the

a

b

Fig. 4.13. Băile Figa. a: general view of the area from the north; b: general view of the area from the south.

Fig. 4.14. Băile Figa: eastern side stream, after a period without rain (April 2007).

former rough road from Beclean is asphalted as far as the spa (though not the extra kilometre to Figa village – evidently not a priority for the authorities, whose main interest is in money-making ventures to help the local economy). By 2010 the spa was fully functioning and in the summer months very busy. While the spa buildings themselves have apparently not damaged archaeological remains, the surroundings of Trench III have been extensively altered. Each new season brings new threats, with new access roads being driven through. Potentially damaging also is the removal of soft mud from the stream bed for curative mud baths; such mud is often found around ancient wooden installations.

The site overlies a massive rock salt deposit (diapir) about 1000 x 1000 m wide and 1600 m thick. This deposit appears between 1.5 and 4 m from the present surface, at least in those spots where we have struck it. The site lies in a hemispherical depression about 600 m in diameter. The sediment that has accumulated in this depression has been transformed over time into a consistent layer of salty mud. The mud layer is overlain by a yellowish gravelly material, about 1 m thick and very hard to dig through. Underground water appears on the surface through several brine springs which, together with rainwater, create the salt stream. The main spring lies at the southern end of the depression. The stream flows from south to north across the depression, with one main and other small tributaries coming from other brine springs to the east. About 70 m further downstream the stream leaves the depression, turning to the north-west, and flows into the Valea Sărată stream, a tributary of the Meleş river, which in turn joins the Someşul Mare in Beclean. The volume of water in the Pârâul Sărat varies significantly, depending on the level of rainfall. In times of high rainfall, the depth of the stream rises to about 0.5 m and its width to about 3 m. Conversely, during droughts it is almost completely dry and the streambed is marked by a white strip of crystallised salt (*Fig. 4.14*).

Fieldwork 2006–2011

The 2006 survey

Survey work in 2006 consisted of a total station topographic survey of the whole exploitation area (formed by mounds and hollows), as well as the cataloguing and measuring in of all visible wooden pieces, of which some 460 were recorded. Most of these elements were observed within the stream bed or close to it, though they sometimes disappeared deep into the banks of the stream. Several timber structures were visible away from the stream bed, in pits dug in modern times. At first sight this appeared to suggest that there is no direct connection between the stream and the timbers, but further study showed that they are visible in the stream because that is where the overburden has been washed away.

From the survey it appeared that the area where timbers are present covers almost the entire depression, about 600 m (north-south) x 300 m (east-west). The principal elements other than posts were (proceeding from north to south): a "wall" of quartered timbers in close proximity to many posts, in the northernmost area, close to the bridge and concrete pool; what appeared to be a dam in the middle section of the stream (the "central dam"); another dam towards the southern end, formed of smaller roundwood poles; a large tree-trunk at the place where the salt and freshwater streams converged; and a possible trough in the stream close to the spring at the southern end. Many of the posts were close together and seemed to have formed constructions of some kind. It appeared likely that some of these represented the remains of shafts or adits going into the stream banks, similar to those that we saw at Săsarm, where some are probably of a fairly modern date (see below).

Subsequent survey at the site by total station has been conducted by S.C. Digital Domain srl and S.C. Vector Studio srl, whose staff have worked in close collaboration with our team.

At the northern end of the site, on both sides of the stream, a small number of prehistoric and some 19th century sherds were visible. The prehistoric sherds are not diagnostic but might be of Bronze Age date. More significant were finds made in the stream bed: a fragmentary stone shaft-hole axe, roughly shaped stone objects with a "waist", of the type known in English as "mining tools" (German *Rillenhämmer*), and a couple of wooden pegs or plugs of the type known from the trough extracted in 2005 (see below).

In 2006 a range of further samples for radiocarbon dating was collected, and six of them were submitted, from different parts of the site and different construction elements. The results indicated that four fall around 1000 cal BC, one around 250 cal BC and one around 450 cal AD (see below for a discussion of the dating). This confirmed the previous impression that the site included a strong Bronze Age component, but added the hitherto unknown information that some elements dated to the Iron Age and the late antique or early medieval period. This information made it imperative, from the point of view of understanding the technology of salt production on the site, to undertake excavation.

Excavation 2007–2011

Initially digging was done entirely by hand, but in view of the prospect that a large part of the site would be destroyed by the spa development, an earthmoving machine was brought in to speed up the work in the northern half of the site. Twenty-three trenches in total were opened across the area (*Fig. 4.15*) (mostly by machine, only five of them extensively investigated). Work concentrated in two areas: the extreme south of the site (Trenches I and XXI), and the northern area (Trenches III, VII and XXII); one further trench was hand-excavated in the central part of the site, where a trough was visible on the surface (Trench XV). Most of the other trenches were machine-cut simply to see if wooden structures

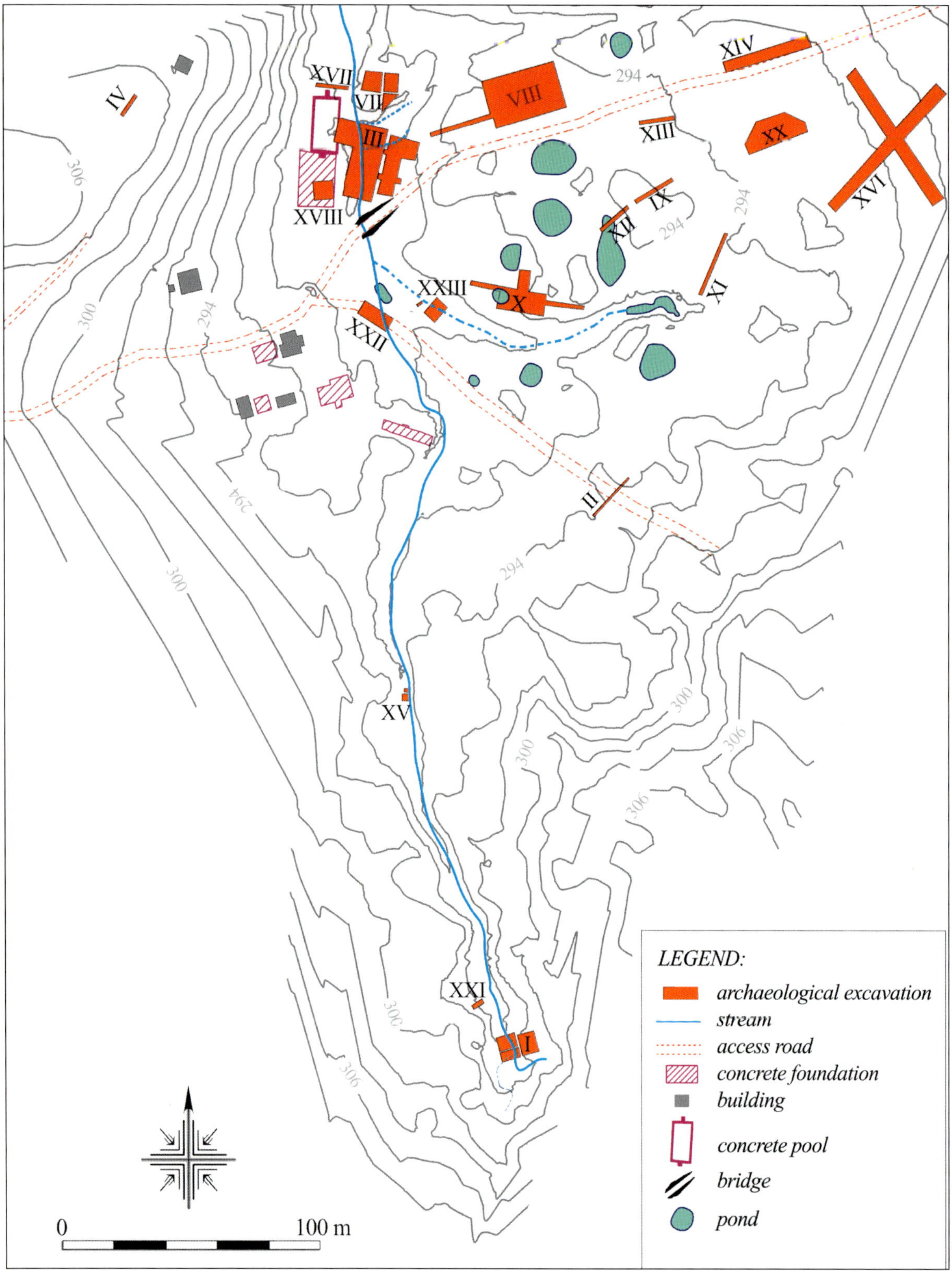

Fig. 4.15. Băile Figa: overall plan, showing the location of the trenches opened between 2007 and 2012.

were present, and if so, whether preservation might be necessary. Fortunately this was not the case outside the immediate vicinity of Trenches III and VII, so the development was allowed to proceed. It is hoped that most of the excavation area in the north will be preserved for future generations, or even that a part will be made visible to visitors to the spa complex.

Southern excavation area

Trench I (S.I)

Trench I was situated a short distance north of the salt spring at the extreme southern end of the site, and covered an area either side of the small stream that flows northwards from the spring (*Fig. 4.16.a–b*). Excavation took place in April 2007 under the direction of Harding with Dan Buzea, in August 2007 by Dan Buzea, in April and August 2008 under the direction of Harding, with Brynmor Morris and Adela Kovács, and in August 2009 under the direction of Harding. Because of the inaccessibility of the location and the steepness of the slopes either side of the trench it was not possible to get a mechanical digger to the site to assist with this process and all work was done by hand.

For convenience the trench is referred to in two parts, east of and west of the stream. Trench north was displaced by a few degrees from true north.

Initially in April 2007 an area 4 x 4 m was opened, crossing the stream at a point where large timbers projecting from the eastern bank indicated the possible presence of a trough. This area was subsequently extended to the north, first on the eastern side (6 x 2 m, with a 1 m balk left between this and the initial trench), and later to the western side. At this stage the trench measured a total of 10 x 8 m, the balk stretching across the whole trench (except over the stream itself) in an east-west direction. The only part of the trench which was excavated to a level below that of the topsoil, however, was the original south-east corner. Elsewhere the topsoil was removed and the exposed timbers cleaned up and planned. Already in April two new troughs, nos 2 and 3 (Nos. 1, *Pl. 9*; 7 *Pl. 10*), were exposed and removed, along with a number of other installations (posts joined with bindings, etc). In August most of the exposed timbers were long pieces lying more or less horizontally, or brushwood of modern origin.

Also in August 2007, an additional trench 4 x 2 m was excavated to the east (separated from the original one by a 1 m balk), deep into the material rising up from the stream. A number of large timbers were uncovered, lying in several orientations and slopes; but beyond planning their position no further work was done. Regrettably no profiles were drawn at this stage, and in the following winter the trench edges collapsed making further excavation dangerous and recording difficult. In the following seasons these edges were stabilised and the profiles (sections) drawn.

In April 2008, a further area was opened to the north-east, so that the entire eastern extension measured around 7 x 4 m (the trench edges were not aligned parallel to the east-west axis of the main trench). Because of poor weather conditions, and the use of untrained workmen for the removal of material, excavation was unusually difficult and conditions trying. Balk collapses, which had already occurred over the winter, continued with the constant tramp of feet on saturated surfaces; mud, to be expected in the trench itself because of the constant seepage of spring water, came to cover ever larger areas of the site. These conditions were eventually ameliorated in the drier weather of summers 2008 and 2009. In August 2008 all features in the western part of the trench were fully excavated after the removal of overlying timbers (see below). The eastern half of the trench was extended and the two parts joined up, exposing the rectangular box construction or shaft (see below). In August 2009 this eastern part was extended further, though the great depth of material as the trench went further into the rising ground meant that no great additional area could be exposed. The final dimensions of the trench at the

a

b

Fig. 4.16 a–b. Băile Figa, views of the area in part of Trench I prior to excavation.

close of excavations were approximately 10 x 14 m, though because of the terrain the trench edges in the eastern extension were not on quite the same alignment as in the western half of the trench.

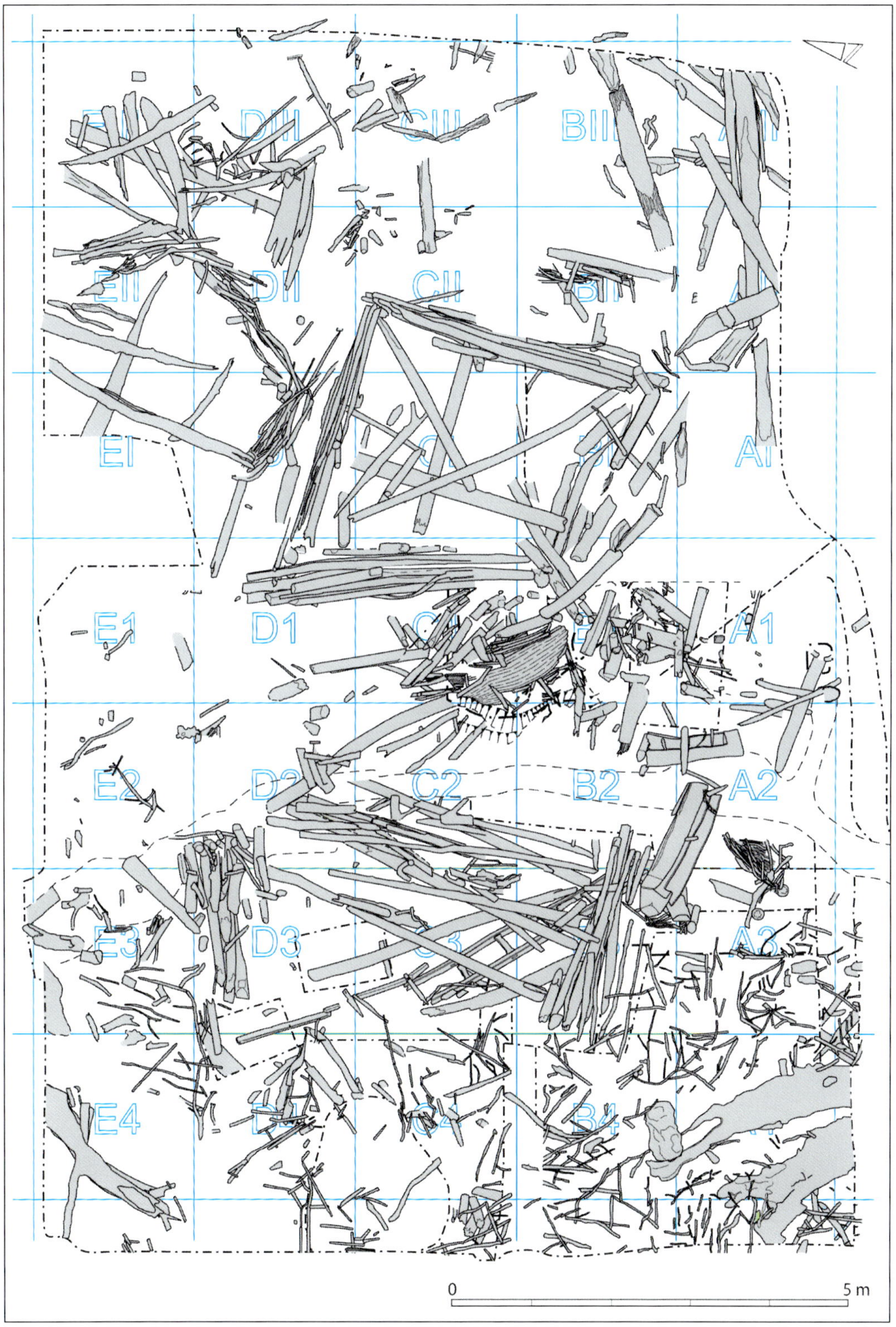

Fig. 4.17. Trench I, overall plan showing all wooden features and trench grid.

A grid of 2 m squares, denoted by an alphanumeric system, was used for recording purposes and is that shown in *Fig. 4.17*. Features are described according to the grid square or squares in which they lie.

The features recovered (Fig. 4.18)

Large numbers of wooden objects were present in all areas of the trench. While most emanated from activities presumed to relate to salt production, not all showed signs of working (though all were from cut branches or trunks). The situation was complex, and the situation was made more complex when the results of radiocarbon dating came through, since they showed that timbers at more or less the same depth could be of greatly different ages. Some elements, such as the ladder (below; No. 21, *Pl. 11*), lay

Fig. 4.18. Trench I, general view from the east, August 2008.

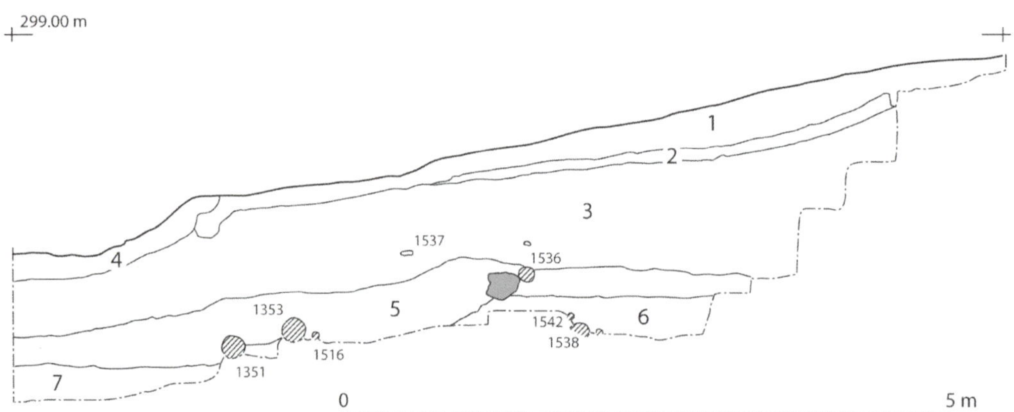

Fig. 4.19. Trench I, south-facing profile, north-eastern sector.
Explanation of layer numbers: 1: topsoil; 2: brown-grey silt; 3: brown-grey clay with grits; 4: redeposited clay; 5: variegated orange clay; 6: light grey clay; 7: grey clay.

in deep shafts which went far deeper than wattle fences of a much earlier date. The problem for the excavators was that no real stratigraphy was discernible in the muds and clays that filled the trench, even though the depths of deposit are considerable (*Fig. 4.19*); all were the result of inwashing over centuries of use and re-use. Thus the fact that one timber lay at a lower absolute level than another has little significance in the overall sequence of site use.

The description which follows proceeds from west to east. Numbers in square brackets are those from the field wood catalogue, numbers in round brackets are from the catalogue used in this book, with the associated plates.

A sizeable quantity of smaller branches and brushwood covered the upper levels of the extreme western part of the trench (squares A3/A4–E3/E4, up to 4 m from the western trench edge), along with a large tree-trunk, part of a felled tree lying in modern stream muds. A small quantity of modern pottery was found among these small branches, indicating that they and the felled tree are the result of recent activity (probably 20th century). A large roundish boulder was also present (C4), though there is no indication as to its function (if any). Profiles were cut through the deposit in two places (A2 and C4), but no earlier timbers or other material were evident and no further digging was done on this part of the trench.

In area B2–B3 a number of long posts were lying in an east-west direction, and in B2–D2 in a north-south direction, more or less horizontal. After cleaning it was evident that they overlay other features and after planning they were removed. Some had working on the ends which indicated they were part of a dismantled structure, probably a box like that found in the eastern half of the trench. Underlying them was a ladder (No. 21, *Pl. 11*), sloping steeply down into a large pit (*Fig. 4.20*). The ladder is described in the catalogue below; the largest part is 5.31 m long, but broken elements, including rungs, were present to the north-west. Two radiocarbon dates on ladder timbers indicate an Iron Age date, showing that the pit in which it lay must have been open at that period. The ladder was removed for conservation, but

Fig. 4.20. Trench I, ladder in western half of the trench, as exposed in the trench.

the pit could not be excavated to the bottom because when the ladder was fully exposed and removed excavation had already reached the water table.

A short length of wattle fence was present in A2, close to Trough 3 (No. 7, *Pl. 10*) in B2. Trough 3 was sloping steeply into the stream, and resting against a bank of material (B2/B3 boundary) with two large near-vertical timbers and some wattle remains (*Fig. 4.21*).

On the other side of the stream, immediately opposite Trough 3, Trough 2 (No.1, *Pl. 9*) was found (A1–B1) (*Fig. 4.22*). It lay nearly horizontal and pointing in a north-south direction. A length of twisted plant binding surrounded it (No. 6, *Pl. 9*), and originally fastened it to other wooden pieces, no longer *in situ* (it may also have held it together when it became broken). Trough 2 lay a little above the stream, not in it. To its south, in A1, were three oblique posts tied together with a plant binding, deeply embedded in the muds and going below the water table (*Fig. 4.23 a–b*). In spite of the depth involved, the radiocarbon dates obtained belong to the fifth-sixth centuries cal AD, indicating that activity was continuing on the site much later than the Bronze Age, to which the troughs belong (see below).

Further north and east, in B1, a collapsed wattle hurdle or fence [1156], lying horizontal, was found. It was unclear if it lay *in situ* (more or less), or had been uprooted from its original position and thrown to one side. In view of the situation with the box structure in the eastern half of the trench, the latter is probably more likely. Beside it was a bundle of poles [1157], apparently the uprights in wattle fences, perhaps in position ready to create such fences. Two radiocarbon dates fall in the 15th century cal BC.

A confused mass of timber lay to the east of this hurdle and Trough 2. The ground rises steeply at this point, so that the relationships between individual timbers was not clear. A line of superimposed timbers lay along the eastern edge of squares C1 and D1, perhaps forming part of the box structure or shaft in BI/BII–DI, though the alignment is different if the box was originally rectangular.

This box or shaft was composed of two walls, north and east, forming a right-angle junction (in CII), with the southern wall apparently destroyed or dislocated (*Fig. 4.24*). It was formed of split oak timbers lying one on top of another; at the deepest point examined, to the south-east, at least ten superimposed timbers were present in the east wall, and the wall thus formed continued below the water table. The recovered maximum height of the wall was around 1.35 m (in the sounding east of the east wall in BII); it is not known how deep it may have gone, but augering in this area by A.G. Brown showed that rock was not reached for a further 2.50 m, giving a possible overall depth of nearly 4 m (more if allowance is made for timbers no longer *in situ* and the additional height of the overlying soil is included). In the interior of the box several cross members at differing heights seem to have served as props to keep the walls vertical. One of these [1385] produced an Iron Age radiocarbon date. Vertical posts were placed at each corner and also in the middle of the preserved long walls. The horizontal members adjoined these verticals in an unusual and characteristic form of joint (see report by Richard Brunning, below), in which one side was composed of a concave hollowing out of the timber end, the other of a natural fork in the timber where a branch went off, the effect accentuated in some cases by a right-angle cut (*Fig. 4.25*). A radiocarbon date on a timber from this construction [1481] shows that it belongs to the Iron Age (4th–3rd centuries cal BC); this corresponds well with a small deposit of pottery found in the redeposited material to the south, and described as "Dacian" (see below).

Around the box construction, to south, east and north, many timbers were found, in a variety of positions and angles, but not forming any apparent construction, instead apparently collapsed or thrown aside from various installations that might have existed prior to the construction of the box. A stepped timber [1316] lying at an angle towards the east, and to the south-east of the shaft, also gave an Iron Age radiocarbon date. Possible indications of a line of posts running north-east – south-west through EIII–DII were tentatively suggested but could not be confirmed. The exception is formed by short lengths of wattle fence, in BII and CII, and a longer stretch forming a roughly circular arrangement in DI–DII. A date

Fig. 4.21. Trench I, Trough 3 as discovered.

Fig. 4.22. Trench I, Trough 2 as discovered.

a

b

Fig. 4.23. Post construction and binding beside Trough 2 (post-Roman date).

Fig. 4.24. Trench I, box structure in eastern sector, from the east, August 2009.

Fig. 4.25. Trench I, box structure, detail of joints around the upright post at the centre of the east side. Timber poles form the wall in the background.

on part of the wattle [1509] lies in the Bronze Age. This looks to be similar to the circular arrangements of wattle fence found in Trench III, where several radiocarbon dates have shown they belong to the Bronze Age (see below). If this is the case (so far these fences are undated in Trench I), it would appear that a series of Bronze Age constructions were cut into and broken up during the construction of the box during the Iron Age. On the other hand, two other dates on timbers lying to the north-east of the shaft fell in the Iron Age [1609] and the early medieval period [1336/1538].

Trench I: interpretation

Figs 4.26 and *4.27* show successive phases of the uncovering of datable features in Trench I. In *Fig. 4.26*, the brushwood and large tree-trunks believed to be modern have been removed, and the timbers of Iron Age date are shown coloured green. This includes the box structure to the east, and the piled up poles, probably emanating from it or another feature like it, and overlying the ladder, to the west. The troughs and wattle elements, known or believed to be of Bronze Age date, are shown in red. The pole construction of early medieval date is shown in blue. In *Fig. 4.27* the poles in the western half of the trench, and the early medieval construction, have been removed, showing the ladder and troughs more clearly. The wattle elements are evident in various places, notably the curving wall in the north-eastern sector. Timbers shaded grey are undated and their relationship to the dated elements is uncertain.

Trench I was evidently an important element in the overall scheme of salt production at Băile Figa, in view of the many and varied installations recovered within it. Of course it is not known how typical these were of the whole production site, and whether such elements were originally present more widely. The discovery of the Chintăuan trough (no. 1) some 130 m to the north, and the nearby existence of another (in Trench XV), may indicate that comparable constructions were originally more widely present on the site. No troughs have yet been found in the northern part of the site (Trench III and adjacent), but the wattle fences are similar to those in Trench I, if more extensively preserved.

To some extent the reconstruction of activities in Trench I depends on how the function of the troughs is interpreted (see below). Since they date to the Bronze Age, and the wattle fences and the bundle of poles [1157] are of the same date (like the comparable features in Trench III), it appears that the earliest elements consist of the troughs and wattle fences, most lying east of the stream but one trough and some wattle fragments to its west. At a later period (Iron Age), shafts were dug through the earlier elements, presumably to reach the natural rock salt directly. It has not been possible to expose this bedrock so far, though augering shows that it must be at a considerable depth, at least 1.5 m below the present stream bed. The wood-lined box or shaft was thus intended as a secure means of reaching the rock, and the simpler pit containing the ladder a means of access. The stepped timber [1316] was presumably a ladder for use in the shaft. It is important to note that Trough 3 in its present position lies over the bottom of the ladder pit, in other words it had been moved from its original position and only reached the spot where it was found much later on, after the ladder pit had been infilled. A further point worth noting is that the ladder pit and the box structure could not have reached down to rock salt in the presence of a stream flowing across the site in its present position; this strongly suggests that the springs and stream have shifted position over the centuries.

At a later date still, a post construction was placed in the southern end of the site near the stream and below the eastern trough (Trough 2). This occurred in the years around AD 500, some 2000 years after the troughs were first made. Given the relative location of troughs and posts, the people who inserted these posts and tied them together with withies must surely have seen the troughs; whether or not they simply pushed them to one side in order to continue with the work they were doing, or whether it is conceivable that the troughs were still usable so long after their creation is not a question that can be resolved at present. It does indicate, however, that digging in and around this very spot continued for many centuries. This is not so surprising when one considers that elderly villagers at Figa remember modern digging for salt in other parts of the area in the 20th century AD.

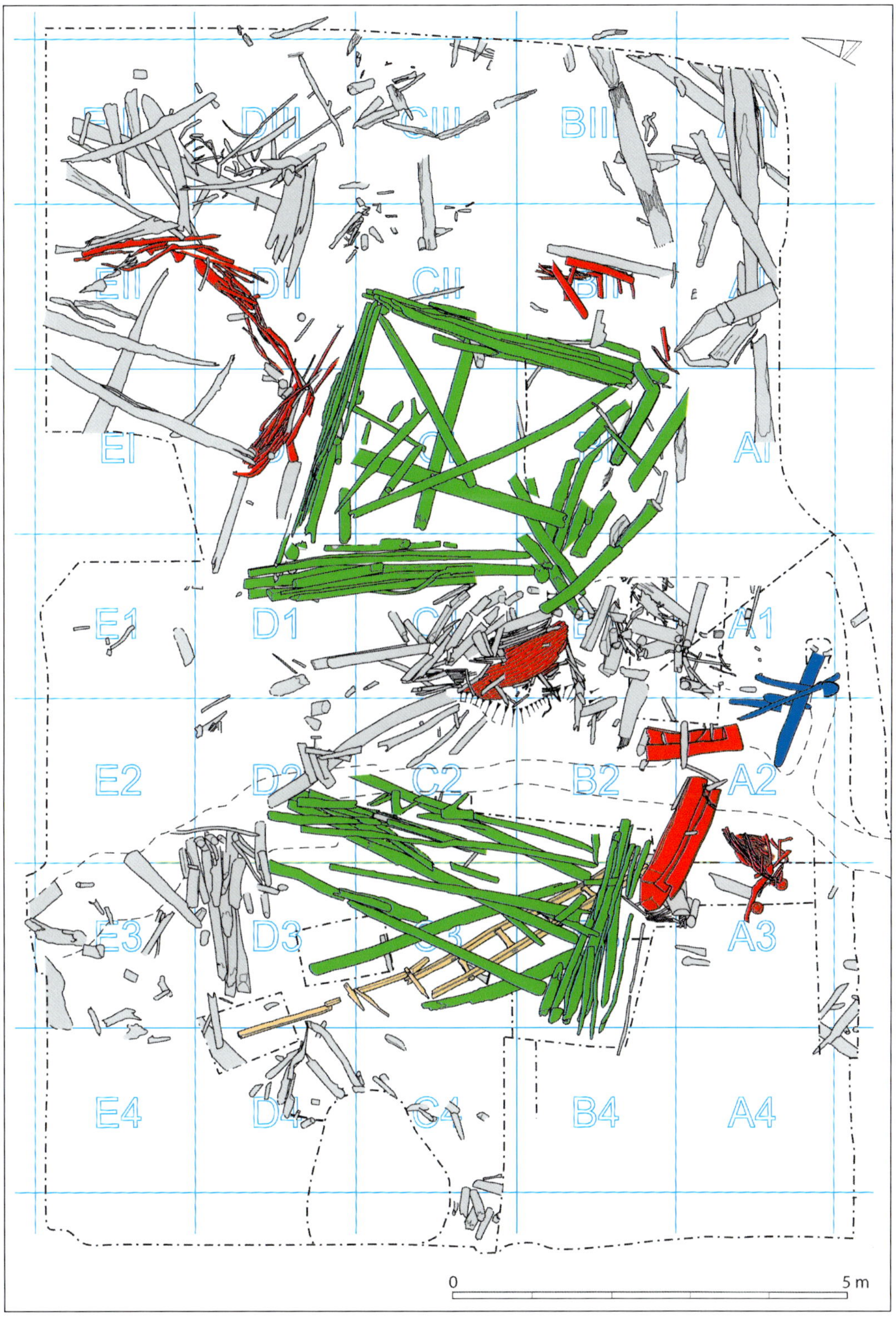

Fig. 4.26. Trench I, plan after removal of modern brushwood. Colour key to datable features: blue: early medieval; green: Iron Age; ochre: Iron Age ladder; pink: troughs 2 and 3; dark red: Bronze Age.

The date for timber 1538, to the north-east of the shaft, is the latest yet found in this southern end of the site, suggesting a period of use in the fifth-sixth centuries AD – a little later than the post construction to the south of Trough 2 (see above).

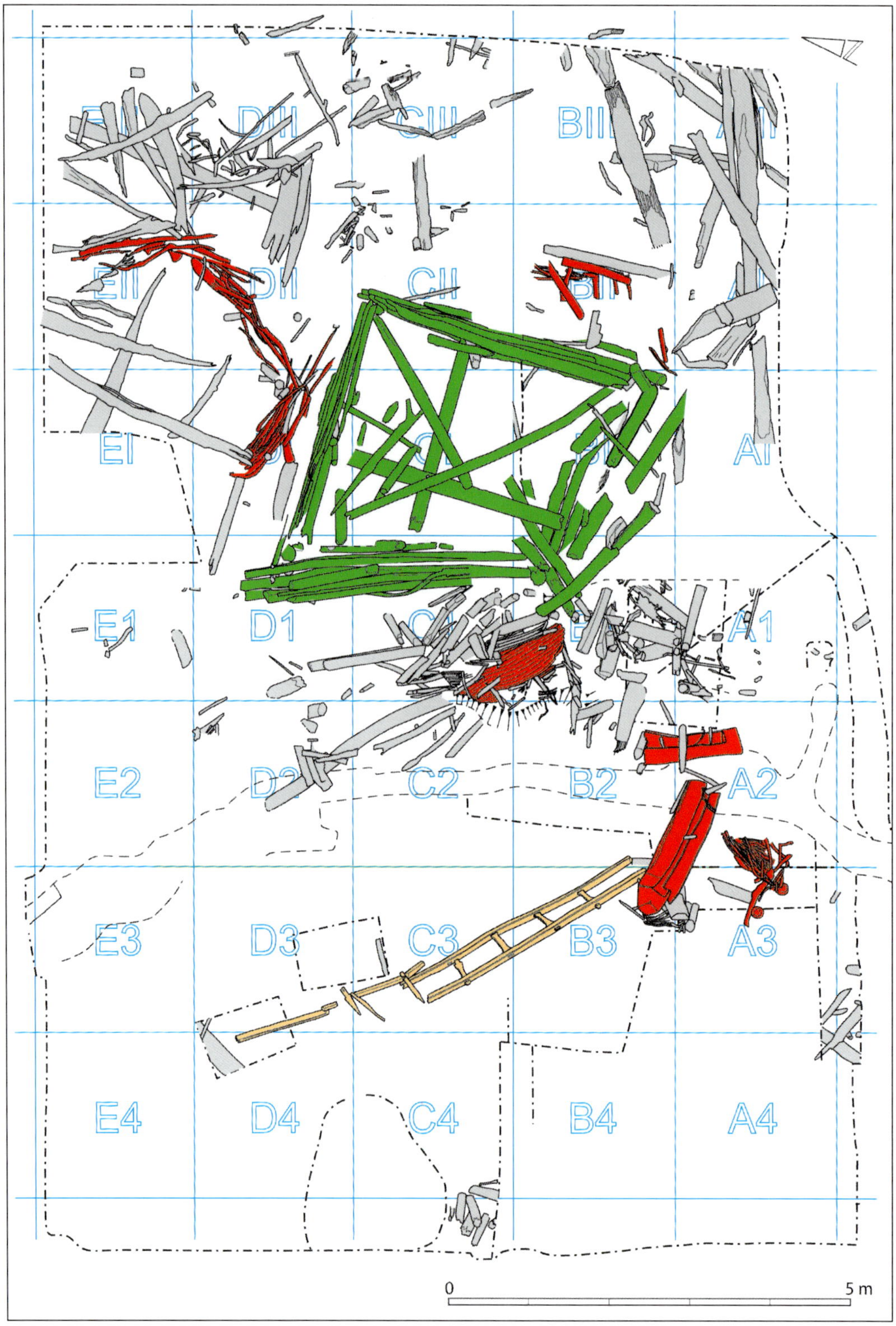

Fig. 4.27. Trench 1, plan with medieval and scattered Iron Age timbers removed. Colour key as for Fig. 4.26.

Trench XXI (S.XXI)

Trench XXI (*Fig. 4.28*) was excavated in August 2009 to investigate the line of timbers that runs obliquely across the stream, some 20 m north of Trench I. These timbers have a dam-like appearance, and five samples taken for radiocarbon dating in 2006 produced a consistent range of dates in the pre-Roman Iron Age (see below), between 350 and 180 cal BC. The trench, which measured 5 x 2 m, was intended to elucidate the context of these timbers in relation to the surrounding deposits.

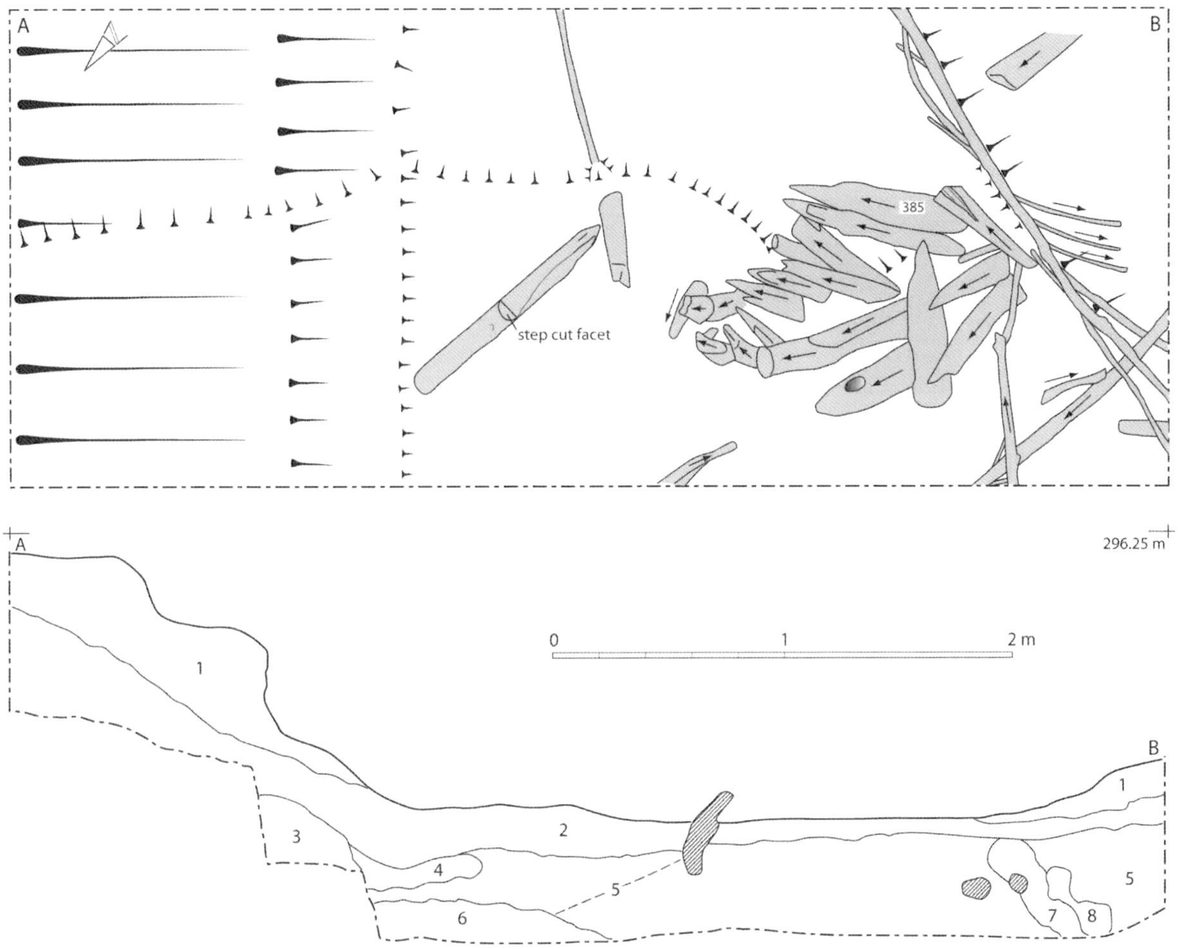

Fig. 4.28. Trench XXI, plan (above) and section (below).
Key to layer numbers: 1: topsoil; 2: mixed orange gravel; 3 and 8: dark humic material; 4: darker gravel; 5: grey or grey with yellow clay; 6: grey with yellow clay; 7: yellow clay.

Excavation was hindered by the presence of the stream running through the area, and the fact that its waters were swollen by the heavy rain experienced in the first two weeks of August. In addition, much of the upper deposit encountered was of a hard gravelly consistency, with compact grey and orange clays beneath. As a consequence, it became impossible to dig deeper than 1.20 m at the deepest point, at the south-western corner.

The eastern half of the trench was effectively barren of archaeological timber or other deposits. An oblique timber in the middle of the trench was notable for a marked step cut into its surface, with clearly visible facets from axing, though there was no indication of what it was intended to be joined with. The nine timbers that form the line across the stream slope towards the south-east and continued down into the muds; since they did not move when pushed, they presumably go on down some way. There are no signs of working on them. Behind them to the north-west is a group of some five other timbers, sloping in different directions. Further to the west was a group of small brushwood elements, which may be relatively recent in origin. No further structure was observed.

The excavation of this trench unfortunately provided no new evidence for the function of the line of posts in the stream bed, already noted and measured in the 2006 survey. No additional timbers were found on the east side, and nothing definitive on the west which would have suggested that further excavation would be worthwhile. All that can currently be said is that the structure dates to the Iron Age, and by analogy with other sites, notably Săsarm (see below), might represent the remains of an adit driven into the western stream bank.

Trench XV (S.XV)

Dan Buzea, with assistance from Björn Briewig

Trench XV was opened in the south-central sector of the site, very close to where I. Chintăuan had removed Trough 1 in 2005. During the 2006 survey, the upper part of a trough was noticed in the stream bed. In this area the valley narrows and has high banks, believed to be the result of human activities and not natural.

Excavation took place between 16 and 22 August 2008, in order to explore the context of the two troughs (*Fig. 4.29*). Initially, an area of 3 x 3 m was opened. As some wood pieces near the trough went into the trench edges, it was extended to the north with a box trench measuring 2 x 2 m, and a 50 cm balk in between the two areas (*Fig. 4.30*). The high ground on either side of the stream did not allow us to

Fig. 4.29. Trench XV, view from the east at conclusion of excavation, August 2008.

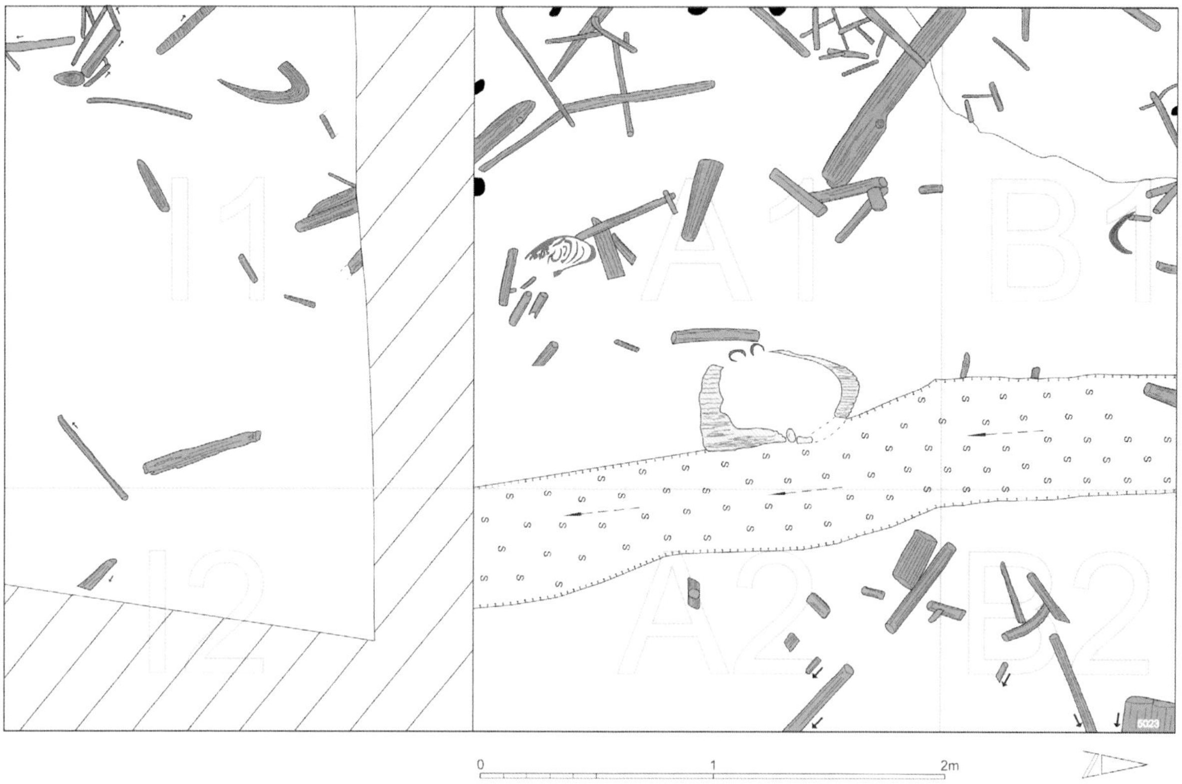

*Fig. 4.30. Trench XV, plan of the upper excavation level.
The trough shows as an oval in the right-hand part of the plan.*

extend east or west. The total area of the trench thus reached 5 m (N-S) x 3 m (E-W). The northern area was dug deeper in order to explore the trough and associated features (*Fig. 4.31*).

In the eastern section, along the line of quadrants A–B /1 (seen from the west), the following sequence was observed:

0–10 cm: gravel
10–30 cm: yellow clay mixed with small gravel stones
30–90 cm: grey mud

The natural subsoil was not reached.

Along the line of quadrants A–B / 2 (seen from the east), the sequence was:

0–20 cm: drift gravel
20–65 cm: a homogeneous layer of grey mud

All the wooden pieces (68 in total) were found in the grey mud layer. Most of them came from wooden structures, without context, highly fragmented and with wear traces. Apart from the trough, part of a ladder was noted, as well as a number of small wooden pieces and stone tools.

In order to understand the context of the trench and its contents fully, a significantly larger excavation would be needed. Time and money did not permit this in 2008, so the trench was covered and filled with salty mud to protect and preserve the wood, in the hope that in the future resources will become available for the purpose.

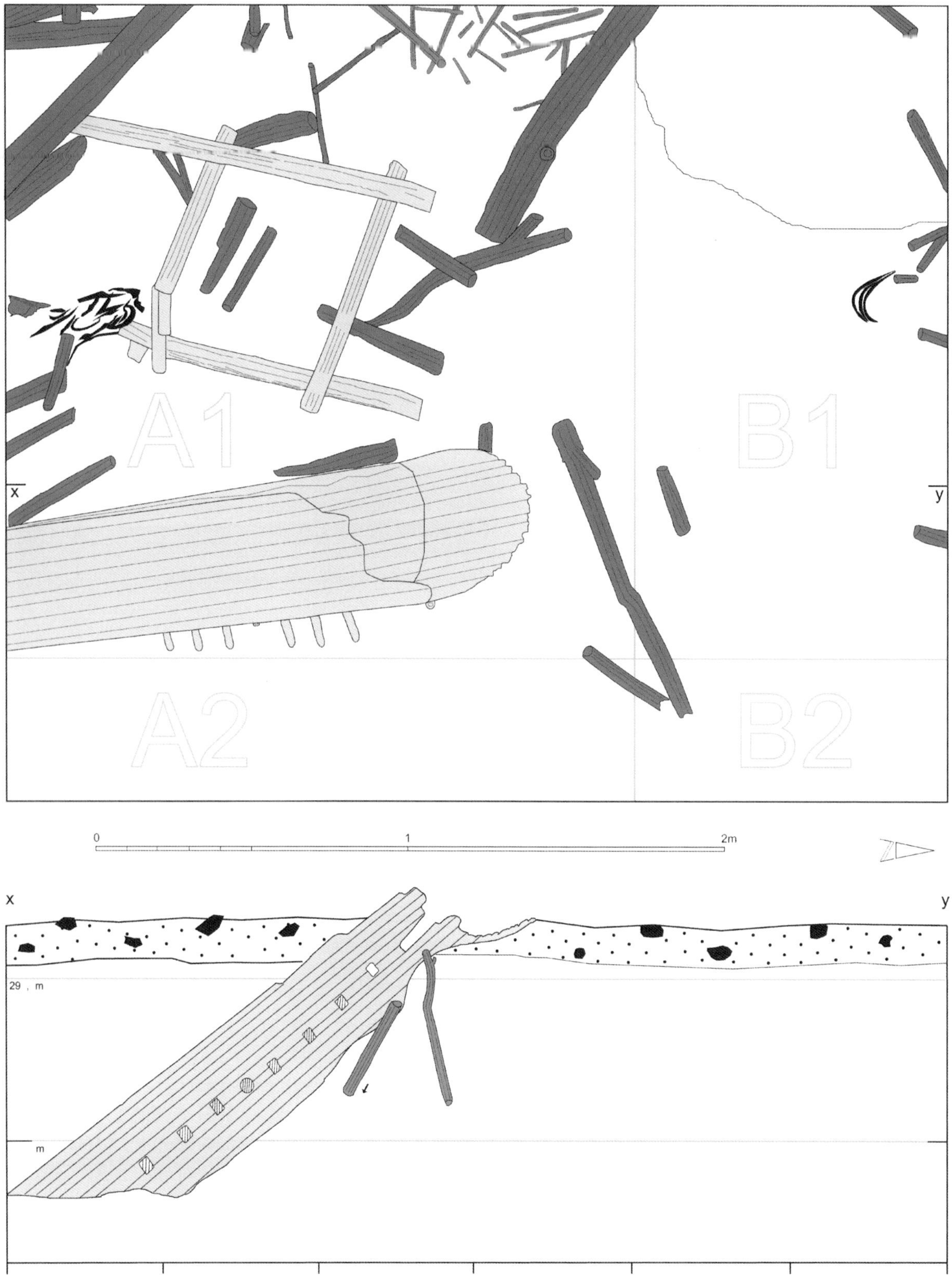

Fig. 4.31. Trench XV, plan and section of the lower excavation level (northern part only).

The trough (no. 4) lay in the central part of the trench. It sloped at an angle of about 45° in a N-S direction (high end to the north); the difference in height from top to bottom being 0.91 m. The upper end was broken, the other lying deep in the mud, and not reached by excavation so the overall length is unknown. As exposed it was 170 cm long, the maximum width 50 cm.

This trough was evidently large, very similar to Trough 1, which was found close by. The interior was filled with sand, gravel and mud. It was formed from a massive tree trunk (in the field thought to be probably oak), with a diameter of at least 50 cm. The base, as seen on the outside, was cut flat. On the axial line of the base, there is a row of rectangular holes (4 x 3 cm) at approximately equal intervals (about 10 cm) and one circular perforation. In six of these holes there was a wooden peg about 15 cm long, axially perforated and rectangular in section. These rectangular pegs were probably made of oak. The first three perforations from the top were empty, the pegs having been lost. Proceeding down, three perforations had pegs that were rectangular in section, followed by one circular in section, like those found in Troughs 2 and 3 in Trench I. This peg was probably made of elder, and it was shorter than the others, projecting 2 cm out of the base. In the next three perforations rectangular-section pegs were present. The interior was about 40 cm across, and there were signs of repeated chop marks (presumably from an adze).

The ladder was found in quadrant A1, in the layer of grey mud, at a depth of 0.9 m below the present ground level (absolute height: 292.33 m), about 20 cm east of the trough and partly under it. It was in a horizontal position and oriented NW-SE. A length of about 130 cm was exposed, both ends being visible (*Fig. 4.32*). The long sides had rectangular perforations in which the ends of the rungs were inserted. The width varies between 70 and 80 cm, and the distance between the rungs is about 50 cm.

Fig. 4.32. Trench XV, ladder with trough above.

A few other wooden objects were found, most of them small wedges (Nos. 122. 123; *Pl. 28*), a small splinter (No.124; *Pl. 28*), a piece of worked round wood (No.121; *Pl. 28*) and half a V-shaped object (Museum Inv. No. 15582). All the objects were made of oak.

Three stone tools were recovered: one hammer (No. 186; *Pl. 38*) and two possible whetstones (Nos. 191. 192; *Pl. 38*).

Northern excavation area

The largest excavation area was at the northern end of the site, on both sides of the Pârâul Sărat stream, in the area where many timbers were visible in the stream bed: vertical planks and split timber fences and/or walls, as well as many vertical poles and posts, some in groups and others apparently isolated. In this area, over a length of about 70 m (between the concrete pool and the northern edge of the site), Early Bronze Age and post-medieval sherds were seen on the surface, both in and on both sides of the stream bed. A sherd of porous clay fabric mixed with straw, resembling Neolithic fabrics, was also recovered.

Because most of this area was included in the spa project, excavation in this part of the site was of a rescue character, and the cost mainly covered by the local authority and the developer.

Four trenches were opened in this area: S.III, S.VII, S.XXII and S.XXIII. There were also two test trenches, one of them close to the southern edge of the concrete pool and another close to its northern end.

The excavations in the northern area were directed by Kavruk, assisted by Dan Buzea, Adela Kovács and Radu Zăgreanu. GPS and total station survey was conducted by Dan Ştefan, Maria Magdalena Ştefan and Cătălin Constantin.

Trench III (S.III)

Trench III was opened about 50 m south of the northern end of the site, close to the bridge crossing the Pârâul Sărat and the concrete pool, extending over both sides of the stream (*Fig. 4.33*). Excavation took place in April–May 2007, April–October 2008, April–September 2009, August–September 2010, August 2011, and September 2012.

Initially in April 2007 an area of 16 x 12 m was opened, extending over both sides of the stream. This area was subsequently extended to the north, south, east and west, so that by the end of September 2012 the trench measured about 600 m^2. In different parts excavation reached a depth between 0.7 and 3.6 m below the current ground surface. Only in the northern part of the trench did excavation reach the bedrock in the form of rock salt. The succession (from the surface downwards) of the layers in the trench was as follows:

- 0 – 0.4/0.8 m: yellow clay mixed with gravel, very poor in finds;
- 0.4/0.8 – 2.8/3.6 m: grey and blackish mud, rich in artifacts and wooden structures;
- 2.8/3.6 m: bedrock in the form of rock salt (reached so far only in the northern part of the trench)

Within Trench III a number of wooden structures were uncovered *in situ* (*Fig. 4.34*). Most of them, and the accompanying wooden artefacts, fall in the Late Bronze Age (12th–9th centuries cal BC) according to radiocarbon dating. Additionally, one structure, small and roundish (1.5 x 1.7 m), made of thin branches and twigs, connected with two larger wooden channelled pieces, produced a date in the 19th century AD. A few pieces are modern.

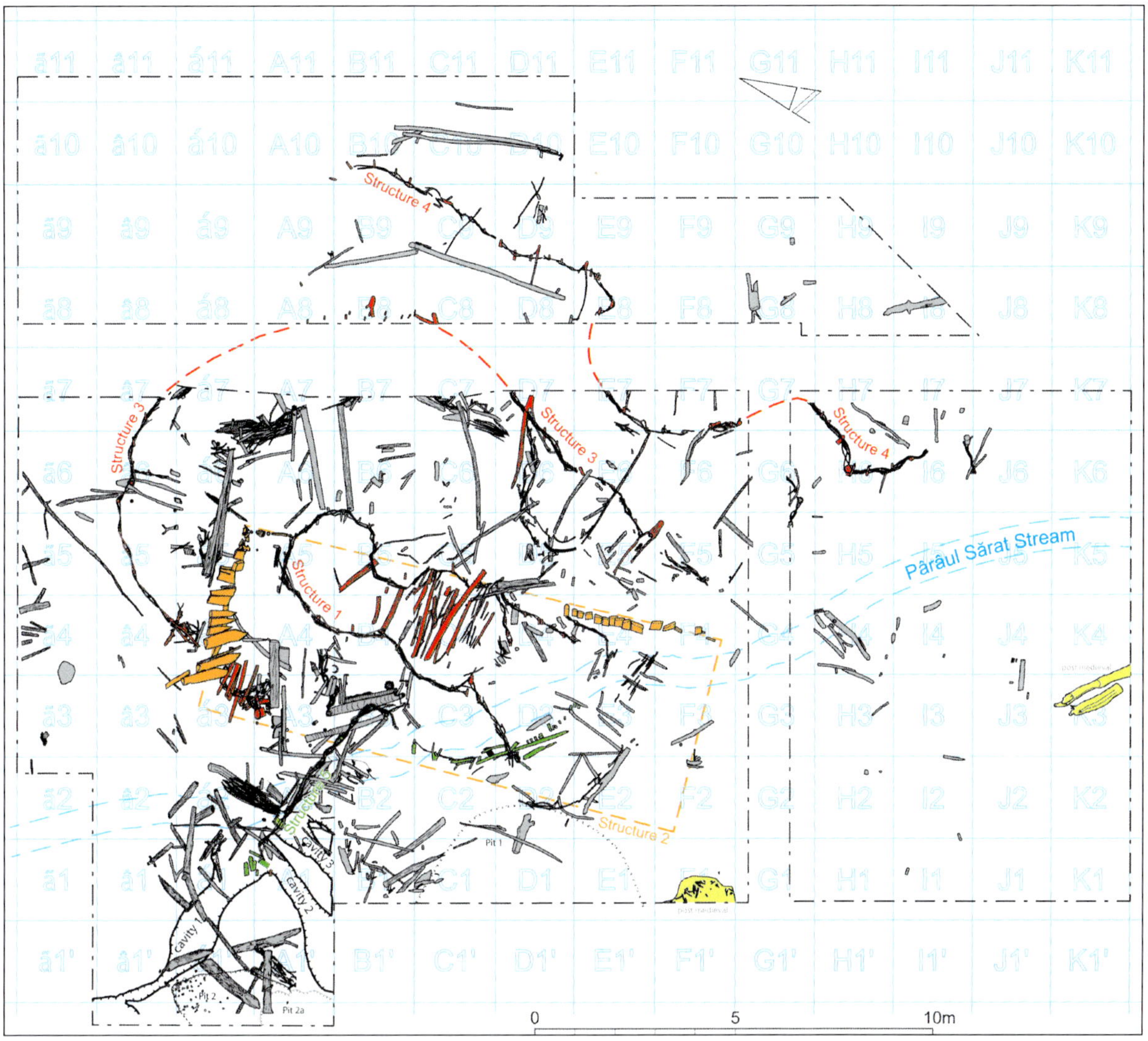

Fig. 4.33. Trench III, overall plan.

Structure 1

Structure 1 lies on the right (eastern) side of the Pârâul Sărat stream, its south-western end being found in the stream (*Fig. 4.35*). The length of the investigated part of this structure is 8 m, its width varying between 1.8 and 3.8 m. The total length of the surrounding wall is 16 m. The height of the wall (the excavation reached the bottom of the supporting poles only in the northern part of the Structure) seems to have been 1.5 m.

It was built of wattle (hazel branches), supported by vertical stakes and poles. Seen from above it has a U-shape with sinuous walls. The structure consists of three compartments, from south to north: a "corridor" with straight and more or less parallel walls, a narrow "neck", and a roundish "room". Its long axis is orientated north-south at an angle of about 15 degrees to the line of the stream. The southern end of the structure is open, and the distance between the two walls ends is 3.35 m. The southern end of the western wall lies in the stream, where it seems to be eroded by the flow of the stream. The northern end of Structure 1 lies about 2 m east of the stream, and it is closed by its roundish wall. In its north-central part the structure gets much narrower, creating a kind of neck (1.8 m wide and about 2 m long) between

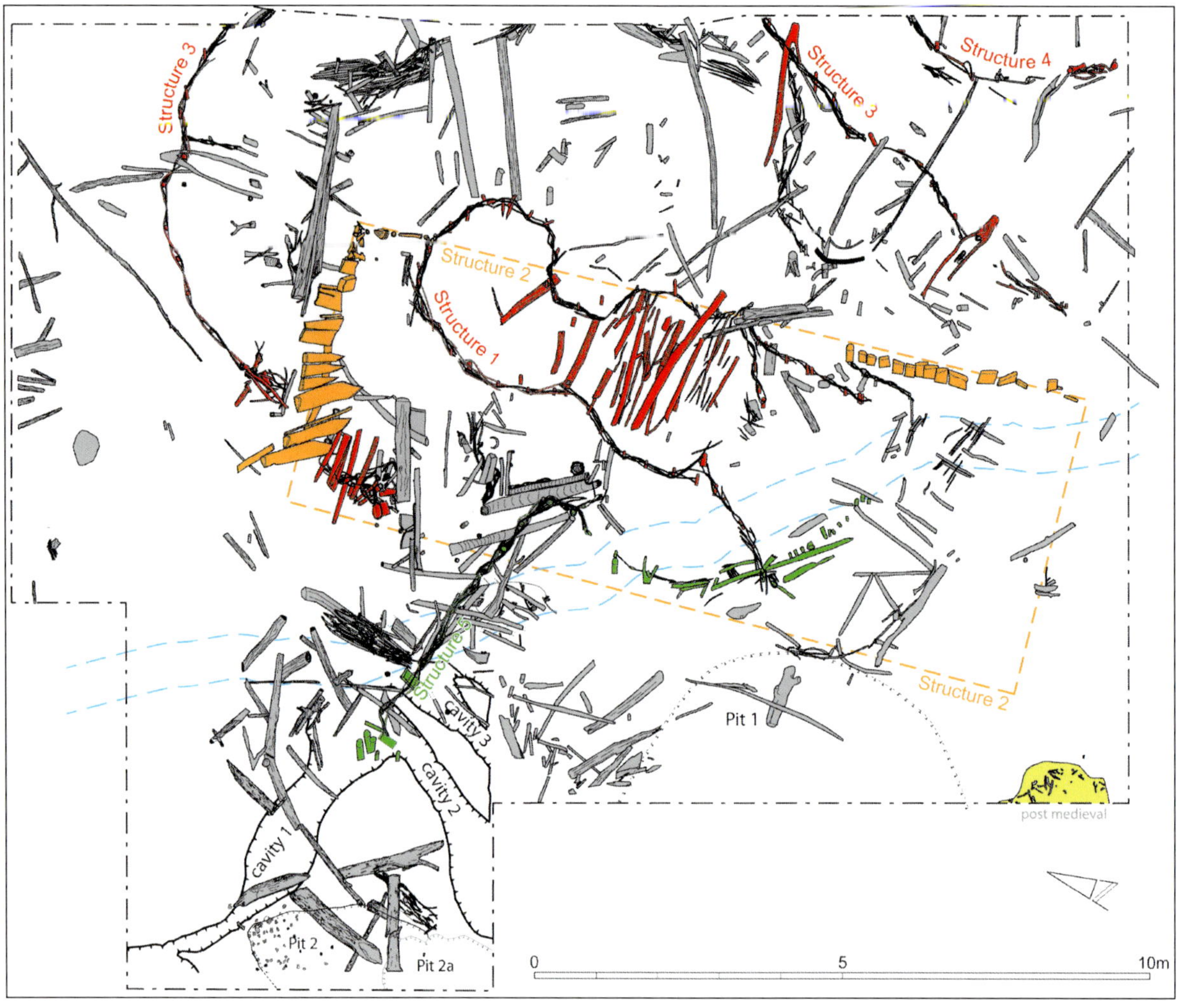

Fig. 4.34. Trench III, plan of central area.

the "corridor" and the roundish "room". North of this neck the walls curve and enclose a roundish space of about 2.5 x 3 m.

The central part of Structure 1 is covered by six massive beams and many hazel branches and twigs, some of them with leaves still present. The beams which cover this structure are rectangular in section and have square perforations at both ends, into which the upper ends of the supporting poles were inserted. Among the branches three hazelnuts were found (*Fig. 4.36*), which strongly suggests that this structure was built in late summer (when the nuts are ripe but have not yet fallen off the tree).

Chronology

In terms of relative chronology it is relevant that Structure 1 partially overlapped Structure 2 (below): the northern part of the Structure 1 fence cut through the plank-built wall of Structure 2, and the eastern wall of Structure 1 – about 7 m along – destroyed the plank-made wall of Structure 2. In addition, during the building of Structure 1 the planks from the eastern wall of Structure 2 had been removed. Two of these planks were found in a horizontal position about 2 m north of Structure 1, above the northern part of Structure 2. Although the south-western end of Structure 1 seems to intersect with the south-eastern

Fig. 4.35. Băile Figa 2008, central-western part of Trench III from the north-east showing about three-quarters of Structure 1; to the right the north-eastern corner of Structure 2; at the top the wavy fence of Structure 5.

Fig. 4.36. Băile Figa 2007, Trench III, hazel nuts found in the northern wall of Structure 1.

end of Structure 5 in the stream (below), at present we cannot define the chronological relationship between the two.

Two radiocarbon dates are available from wattle poles in Structure 1 (OxA-19277, 2858 ± 30 BP and OxA-19278, 2766 ± 28 BP). These fall some way apart from each other, but the overlap area (assuming they belong to the same building phase) indicates construction and/or use in the 10th century cal BC.

Structure 2

Structure 2 is situated on the right side of the Pârâul Sărat. Seen from above it has an L-shape. It is made of oak planks, pushed vertically deep into the mud (*Fig. 4.37*). Its total length is 17.4 m. It seems to be part of a rectangular structure, of which the western and southern walls had been removed in ensuing period(s) by newly built structures. If so, one might assume that this L-shape in fact represents the eastern and northern walls of a rectangular construction. The eastern wall runs at a 15 degree angle from the stream line, measuring 13.8 m. At 7 m along, the wall is destroyed by Structure 1. In its northern part, the wall turns sharply westwards, at a 100 degree angle. This part of the wall (perhaps the northern wall of a rectangular structure) is 3.6 m long.

The excavation reached the bottom of Structure 2 only in the western half of its northern wall, which included its last seven planks. All of these were in a slightly oblique position (about 75 degrees to the horizontal) which means that the wall had fallen slightly to the north (towards the outside). The bottom ends of the planks were sharpened in such a way as to produce a triangular shape. They go down to the bedrock (rock salt), which appears in this spot at a depth of 3.5 to 3.6 m from the current ground level.

Fig. 4.37. Băile Figa 2007, central part of Trench III showing the southern end of Structure 2 (vertical planks, some of them cut for dendrochronological sampling); in front of the planks the fallen part of a wattle fence (northern part of Structure 5) and the northern end of the western wall of Structure 1; behind Structure 2 (right upper corner of picture) the northern part of the eastern wall of Structure 3.

Fig. 4.38. Băile Figa 2009, Trench III, north-central part from the south, showing (lower left) the northern wall of Structure 2 and behind it the northern part of Structure 3.

The height of these planks varies between 1.80 and 2.00 m, their width varies between 20 and 24 cm, and their thickness between 4 and 10 cm.

Excavation in 2011 revealed that the western end of the northern wall of Structure 2 was cut by the curved wattle fence of Structure 3 (below) (*Fig. 4.38*). Thus, contrary to previous ideas, according to which this curved fence and the L-shaped aplank wall were part of the same structure (HARDING – KAVRUK 2010), we now know that the curved wattle fence (Structure 3) was built later than Structure 2 (L-shaped plank wall). It seems that during the building of Structure 3 the western part of the northern wall of Structure 2 had been removed. In the light of the recent excavation it appears that Structure 2 was part of a larger complex (rectangular?), of which the western and northern walls were removed in the following periods.

About 2.5 m north of the southern end in the eastern wall of Structure 2, a kind of niche was found. The niche was only partially preserved, because in this spot Structure 1 cut through the eastern wall of Structure 2. The existence of this niche was indicated by the fact that, in contrast to the rest of the wall where the planks were inserted into the mud to a depth of more then 2 m, here five planks were inserted to a depth of only 0.2–0.4 m. These short planks were found in a vertical position and were lined up with the planks that had been driven much deeper into the mud. The lower ends of these short planks were carefully worked: in two cases they were heart-shaped, and in three cases their ends were sharpened to form a broad chisel-like end. The bottom ends of the planks of the same structure uncovered in 2011 (northern wall) were different, triangular in shape (above).

The northern wall of Structure 2 is linked on its outer side to a massive horizontally placed timber, 3.6 m long and 12–16 cm thick. In turn, this timber is supported by vertical poles and planks. The eastern end of this timber was fastened cross-wise with a withy-like binding to one of the poles and one of the fence planks.

Chronology

In terms of relative chronology, Structure 2 was cut by both Structure 1 and Structure 3 and is therefore earlier than them (*Fig. 4.39*). Three radiocarbon dates are available from planks in the southern part of Structure 2: Hd-27309 (2896 ± 30 BP), OxA-19300 (2842 ± 27 BP) and OxA-19386 (2817 ± 31 BP).

Fig .4.39. Băile Figa 2009 Trench III from the north, showing Structures 1 (foreground), 2 (centre) and 3 (upper left).

These appear a little earlier, though not much, than those for Structure 1. The modelled dates (*Fig. 4.64*) suggest that while Structure 2 falls in the decades around 1000 cal BC, Structure 1 falls in the 10th century cal BC.

Structure 3

Structure 3 was uncovered at the eastern side of the Pârâul Sărat and in the stream-bed, both of its ends lying in the stream. This structure, seen from above, appears to be shoe-shaped. Its walls were made of wattle (hazel branches) supported by vertical poles pushed into the ground (*Figs 4.40*, *4.41*). Structure 3 encircles Structure 1 and cuts through the northern wall of Structure 2. It starts from the stream (square F5) and first goes north-eastwards, at an angle of about 45 degrees to the line of the stream, then gently turns to the north, towards square B8, and after that, in square A8, to the north-west and in square β7–8 to the west. In square α 4, the fence joins the stream, cuts through the western end of the northern wall of Structure 2 and gently turns to the south. Because the excavation is incomplete, it is impossible to say whether Structure 3 continues further to the south. The length of the uncovered part of the fence is about 26 m, and it comprises an area of 13 m (N-S) x 9.5 m (E-W).

At five more or less equidistant points Structure 3 has massive forked branches inserted into the fence at an angle of 90 degrees. These "forks" have short stems and long branches, and they are linked to the top part of the supporting poles of the fence so that the poles are placed between the stem and the

Fig. 4.40. Băile Figa 2009, north-eastern part of Trench III from south, showing the eastern part of Structure 3.

Fig. 4.41. Băile Figa 2010, Trench III, north-central part, view from north, showing (in front) the northern part of Structure 3; behind it in the centre of the picture Structure 2; to the right Structure 5.

forking branches; the stems are outside the fenced area while the forked branches are directed inwards. In some cases these forks had wattle above the junctions to reinforce the joint. One possibility is that these forks could serve as cross-beams for supporting some kind of roof above a fenced area.

Inside the fenced area of Structure 3 two fences were discovered joining the main one, which suggests that the structure was divided into compartments. One of these fences was uncovered in the central-southern part of Structure 3 (squares F6 and F7). It is straight, about 5 m long and oriented E-W (at an angle of about 40 degrees to the main fence line). Another similar wattle fence was uncovered in the northern part of Structure 3. It is straight, 1.2 m long, and joins the northern end of Structure 3 (square β6) at a right-angle, and is oriented N-S.

Within the northern part of Structure 3 part of a destroyed wattle fence was uncovered (square σA-7). Its preserved part is about 2 m long and oriented SE–NW, roughly parallel to the main fence. Apparently it comes from a structure of an earlier date than that of Structure 3.

In the northern part of Structure 3 (squares σ6 and σ5), two wooden upright stakes with forked top ends were found *in situ*, 2 m apart. These might have been used to support a cross-bar, from which items might have been suspended.

Chronology

No ^{14}C dates are available for this structure. In terms of relative chronology Structure 3 was built later than Structure 2.

Structure 4

On the eastern side of the stream, about 2–3 m east of Structure 3, another wattle fence was uncovered. It is oriented NE–SW (roughly parallel to the eastern fence of Structure 3), is wavy in shape, about 21 m long. It covers a distance of 20 m between square I6 and square B10. From square I6 it goes – for about 2 m – straight northwards to the middle of square H6; from that point it sharply turns to the north-east and goes to the eastern edge of square G7. From there it turns to north-west and after a further 2 m gently turns northwards and crosses the western edge of square F7. In square E7 it gently turns to the north-east and east, towards the northern part of square E8. From the middle of E8 it goes relatively straight for about 4 m to the north-east to reach the middle of square B10. From square B10 it turns to the north-west and after a further 3 m comes to square A9. No continuation to the NE is visible from that point.

Akin to Structure 3, Structure 4 was provided with forked branches linked to the upper parts of supporting poles; three such forks were preserved in the north-eastern half of Structure 4. Their long ends slant to the west-north-west, i.e. towards Structure 3.

Structure 4 might have been part of a larger structure. This could be elucidated by further research. Structure 4 appears not to be connected with other structures, but it might be in some way connected to Structure 3 in view of the fact that the forked branches forming the roof (?) of Structure 4 are directed towards Structure 3.

Chronology

There are no radiocarbon dates available for timbers in Structure 4. In terms of relative chronology it seems to be contemporaneous with Structure 3, but there is no firm proof of this.

Structure 5

On the left (western) side of the stream and in the stream bed another wattle fence was uncovered. It was made of wattle supported by vertical poles, stakes, split timbers and planks (*Fig. 4.42*). This fence is wavy in form and is oriented NW-SE covering a distance of 11.5 m from southern edge of square α1

Fig. 4.42. Băile Figa 2010, Trench III, north-western corner from the west, showing the western end of Structure 5; in front of it from north to south: the southern end of Cavity 1, Cavity 2, and Cavity 3; to the south-east (in the right of the picture) several timbers out of context, among them a massive 'forked' timber.

towards square E3. Its total length is about 15 m. It was perhaps part of larger structure, which future research may elucidate.

The fence starts in square α/A-1 and for about 5 m is straight and oriented NW-SE. After that, in square B3, it turns sharply (at an angle of about 90 degrees) to the south-west. Then, in squares C/D3 and D3, it gently curves towards the south-east to reach the intersection of squares D3 and E3. In square E4 it turns gently to the east-south-east. The south-eastern end of this fence was found between the southern ends of Structure 1. After that, about 2 m along, it is interrupted (i.e. destroyed) by the stream. However, it seems that in the area where the stream cuts through the fence, a little south from its line, there is part of the same (?) fence in a fallen position (south-western corner of the square E4), 2 m long. If this part was the continuation of Structure 5, the next stretch of wattle (1.2 m long), found *in situ* in the central/north-eastern part of square F4 and in the south-eastern part of square D4, may be seen as belonging to the same structure. Beyond that, we were unable to trace its course.

Like Structures 3 and 4, Structure 5 was provided with forked branches. Thus, a fragment of forked branch inserted into the fence and joined to the upper part of a supporting pole was found in the north-western corner of square C3.

Chronology

In terms of relative chronology, the position of Structure 5 in relation to Structures 3 and 1 is not completely clear, but it seems that it is later than Structure 3 (which in turn is later than Structure 2).

There are four radiocarbon dates on timbers in Structure 5. Two of them come from planks supporting the fence (OxA-24820: 2936 ± 27 BP, OxA-24821: 2921 ± 27 BP), the other two come from supporting poles (OxA-24841: 2773 ± 27 BP, OxA-24842: 2841 ± 25 BP). The first two (planks 158 and 161) come from long-lived trees, while the second two (timbers 154 and 155) were much younger when they were cut, which makes them more relevant for determining the date of Structure 5. The modelled dates (*Fig. 4.64*) suggest that Structure 5 lies a little earlier than Structure 2, at the start of the sequence.

Connections and chronological correlation between the structures uncovered in Trench III

The determination of physical connections between the structures uncovered in Trench III is important for an understanding of both their chronological correlation and their function. Since the excavation was incomplete, conclusions are provisional. Taking stratigraphy and horizontal position into account as well as the available ^{14}C dates, the following may be proposed.

The earliest timber structure in Trench III is Structure 2. At the beginning it was probably a rectangular structure (more than 16.5 x 3.6 m) built from closely spaced vertical oak planks driven deep into the ground. In its northern wall, this structure seems to have had a niche (the door?). Unfortunately no radiocarbon dates are available for this structure.

Later the area of Structure 2 was covered by a complex of four wooden structures (nos. 1, 3, 4 and 5). Approximately in the middle, Structure 1 was built. Its walls were made of wattle supported by vertical poles, and it was roofed by massive cross-beams, hazel branches and trusses of twigs. Structure 1 was surrounded by a shoe-shaped wattle fence – Structure 3. In turn, Structure 3 was flanked from east and west by two sinuous wattle fences, Structures 4 and 5. All the wattle fences surrounding Structure 1 were accompanied by large fork-shaped branches linked to the tops of supporting poles and directed inwards the fenced area. It seems quite probable that Structures 1, 3, 4, and 5 are physically connected to each other. If so, these structures were parts of the same functional system.

Fig. 4.64 shows the modelled dates for Structures 5, 2 and 1, with suggested boundaries between them. It should be noted, however, that the dates from the individual structures show divergence which makes this model uncertain. A larger number of dates, securely tied to individual structures, would be needed to resolve this problem.

The Early Bronze Age Pits

In the north-western part of the Trench III, on the western side of the stream, two or three pits were uncovered.

Pit 1

In squares CDE-1, 2 and underneath a group of unconnected timbers, at a depth of 0.6–0.8 m from the current ground surface, there were indications of a deepened area. The outline of this feature could not be defined precisely because of water seeping out of the bottom and sides of the excavated area. It appears to cover an area of about 2.5 m (N-S) x 1.5 m (E-W). The material filling the feature was mixed and flecked with small pieces of charcoal. A group of Early Bronze Age sherds was observed over an area of about 2.5 x 1.5 m (*Pl. 4.1. 2. 4–8*; *Pl. 5.4. 7. 13. 14*; *Pl .6, 12. 16*).

In total 1020 sherds were found in this spot weighing 12.147 kg. Many of them have a brushed surface, and one of them a textile imprinted surface; these elements are strikingly similar to pottery of the group called Iernut-Zoltan or Gornea-Foeni, which is assigned to the final phase of the Early Bronze Age (last centuries of 3rd millennium BC, see below).

This feature was not fully excavated and awaits further research.

Pit 2

In the north-western part of Trench III (squares α1 and β1) on the western side of the stream, an oval pit was uncovered and partially emptied. The investigated part of it measured 3.2 m (N-S) x 2 m (E-W).

The pit was oval in plan at 0.8 m below the current ground surface. The west side lay beyond the trench edge. Cutting the grey mud, the pit stood out because of its fill of blackish clay, with small fragments of charcoal mixed in. In the fill were 1722 sherds (total weight 23.656 kg) of Iernut-Zoltan/Gornea-Foeni type (final phase of the Early Bronze Age; *Pl. 5,3. 5. 6. 8–12*; *Pl. 6.1. 3. 4. 6–11. 13–15. 17. 18. 22*), including some with brushed surface and textile imprints (*Pl. 2.7*; *Pl. 3.1–7. 9–11*).

Pit 2a

After the filling of Pit 2 had been removed, rock salt was found beneath it. In the north-east corner the eastern side of an oval pit, dug into the rock, was found. The pit went under the west baulk of the trench. It lay 30 cm below the bottom of Pit 2. In the fill were found 32 sherds (total weight 0.585 kg) identical to those of Pit 2, including some with brushed surface and textile imprints (*Pl. 2.1–6. 8–12*; *Pl. 3.8*). At this stage of research, it is impossible to say with certainty whether or not Pit 2a belongs with Pit 2.

A piece of wood from this pit produced a radiocarbon date of 4435 ± 30 BP (OxA-24822), giving a huge calibrated range from 3329 to 2928 cal BC. As discussed below, this date is much earlier than that usually assigned to this type of pottery.

The bedrock

Excavation in 2010 and 2011 in the northern part of Trench III reached the rock salt bedrock in two places. It was so hard that it was impossible to crack it with iron picks and hammers. After the surface has been cleaned and washed, it was seen that in most of places it was white with grey stripes. In other words, the salt in the areas uncovered was mostly pure, with few impurities.

In most parts the rock surface is level. At the same time, in the north-western corner of Trench III (squares A/αβ-1 and β1), three deep interconnected cavities were uncovered in the bedrock. These cavities were found next to the area where Structure 5 stops abruptly, so the cavities were not overlapped by timber structures. Only a few unconnected timbers were found above them. No sure evidence for the nature and age of these cavities is yet available.

Cavity 1

Cavity 1 was uncovered about 4 m west of the stream, in squares α/β1 and β1. We managed to track it over a length of about 4 m. In the area uncovered the cavity looked like a straight narrow slot oriented SE–NW, more or less parallel to the stream. Its maximum width is 1 m, and its maximum depth from the bedrock surface 0.8 m. In section this cavity is V-shaped. Neither the filling nor the walls show any trace of human activity. All this may suggest that it is of natural origin. One possibility is that it was created by the former course of the stream.

Cavity 2

Cavity 2 joins Cavity 1 on the western side. From above it looks like a rectangular pit situated at right-angles to Cavity 1. We tracked this cavity for about 1.7 m but did not reach its western end. We emptied it to a depth of 1.6 m from the surface of the bedrock, but did not reach the bottom. In the exposed part Cavity 2 has vertical, straight-sided and parallel walls and is 1.4–1.5 m wide. No evidence of human activity was observed in its fill or on the walls.

Cavity 3

Cavity 3 is parallel to Cavity 2 and is situated at about 30 cm south of it. Only the eastern end was emptied. It was followed over a length of 1 m (E–W). The maximum width of the investigated part is 50 cm and its maximum depth is 0.6 m from the surface of the bedrock. In some places this cavity has small shallow holes which could be the result of the application of dripping water, but further research is needed to verify this possibility.

Structural elements in Trench III without context

Structural elements without context were recovered in several parts of Trench III. These include three short split logs (Nos. 116. 117; *Pl. 27*) [17699], one entire log (No. 109; *Pl. 25*), four massive planks (Nos. 118. 120; *Pl. 27*; and MNCR inv. nos. 17704. 17710), four massive stakes (Nos. 114. 115; *Pl. 26*) [17707. 17708], two thin stakes (No. 104; *Pl. 24*; MNCR inv. no. 17702), and a thin branch (MNCR inv. no. 17709). A triangular concave plank, which may come from a broken trough, later used as an element of a structure (No. 119; *Pl. 27*), was also found.

Finds

Fragments of 15 stone mining tools (Nos. 160–171; *Pl. 33–35*), 72 wooden artefacts (Nos. 48–120; *Pl.16–27*), and 2946 potsherds (weighing 42.969 kg) were recovered in Trench III. None had any clear context in terms of association with the structures identified. These are described in the Catalogue.

Excavation in Trench III in 2012

In order to obtain samples of Early Bronze Age pottery and rock salt needed for chemical analysis a small excavation (30 m^2) was carried out in the north-western corner of Trench III, where in previous years deep Early Bronze Age pits were found. The excavation was coordinated by Kavruk, assisted by Adela Kovács, over two weeks in September 2012.

During excavation in square β/ 2'at a depth of 1.65 m from the present-day surface, an iron pick, with part of a wooden haft inserted in the shaft-hole, was found in the mixed soil (see the Finds Catalogue). At different depths a small number of Early Bronze Age sherds were found. Rock salt was reached towards the end of the excavation in sqare β/1' at a depth of 2.80 m from the present-day surface.

Overall interpretation of Trench III

At present, the function of the structures uncovered in Trench III is unclear. On the one hand, they are not associated with settlement traces; on the other hand, close to these structures there is a deposit of rock salt. It is therefore likely that all these structures were related, in one way or another, to the exploitation of the rock salt. But how did they function?

Based on the fact that Trench III is crossed by the salt stream, we suggested previously that the structures were designed to collect brine (Harding – Kavruk 2010). Since then, the situation has changed: on the one hand, there are indications that the Pârâul Sărat stream followed a somewhat different course during the Bronze Age; on the other, excavation has now reached rock salt close to the timber structures. We therefore need to consider the possibility that these structures are related to the extraction of rock salt. One possibility would be that they were used for storing the extracted salt, which would have needed protection from the elements until it was taken from the site and transported elsewhere.

At present, there is insufficient evidence to allow a more detailed interpretation. Until the relation between the cavities in the rock salt and the surface complexes is established, through the completion of excavation in Trench III, interpretation will remain problematical.

Trench VII (S. VII)

Trench VII was opened 10 m north of Trench III. It included the high eastern bank of the Pârâul Sărat and the streambed. It measured 20 x 8 m (160 m^2). In this trench two built structures were discovered: Structure 6 and Pit 3. In 2010 this part of the site was destroyed by the extension of the spa.

Structure 6
In the stream bed half of a circular fence, 2.5 m in diameter, was uncovered (it seems that the other half was destroyed), made of wattle supported on poles driven into the ground.

Pit 3
On top of the right bank of the Pârâul Sărat a shallow depression was uncovered. It was oval (5 x 3 m) and shallow, its maximum depth being 25 cm from the level of the natural yellow clay. A cluster of stones covering an area of about 4 x 2 m was uncovered in the middle of the pit. The fill consisted of blackish/greyish earth and contained a few Early Bronze Age and a larger number of Late Bronze Age sherds (*Pl. 7.1–9*).

Interpretation
Trench VII covered a fairly sizeable area, but produced few features. Although it lies close to Trench III there is little sign that the structures found there continued to the north into Trench VII. Its interpretation therefore remains uncertain.

Trench XXII (S. XXII)

Trench XXII, measuring 12 m x 6 m, was opened approximately 25 m south of Trench III and extended over both sides of the Pârâul Sărat stream (*Figs 4.43, 4.44*).

The archaeological remains in this area were very vulnerable, because at this point the stream was periodically crossed by heavy vehicles. It was here that the Beclean authorities intended to build a bridge to access the private chalets on the western bank of the Pârâul Sărat. Following the signing of a new contract, rescue excavations were carried out in August–September 2009 under the direction of V. Kavruk, assisted by Adela Kovács and Radu Zăgreanu.

Examination of the balks and sections showed that the top layer consisted of gravel mixed with sand, and just below it, from a depth of 30 cm downwards, a layer of compact grey mud. In the eastern half of the trench well-preserved archaeological remains were found, consisting of planks and split trunks, as well as branches, none of them *in situ*, however.

There were no traces of buildings, most archaeological material having been eroded by the stream. However, near the southern wall of the section traces of a wattle fence were found, supported by oblique/vertical poles, built in the same technique as the wattle structures of Trench III.

A number of objects were found in Trench XXII:

- a wooden massive mallet made of a trunk cut at both ends (the part used for hammering) with a side branch forming the handle (No.131; *Pl. 29*);
- wooden shovel (No. 132; *Pl. 29*);
- 5 wooden pegs (Nos.126–130; *Pl. 28*) found in the upper layer of the stream bed, and presumably carried downstream from further south. The pegs are of oak and are square in section.

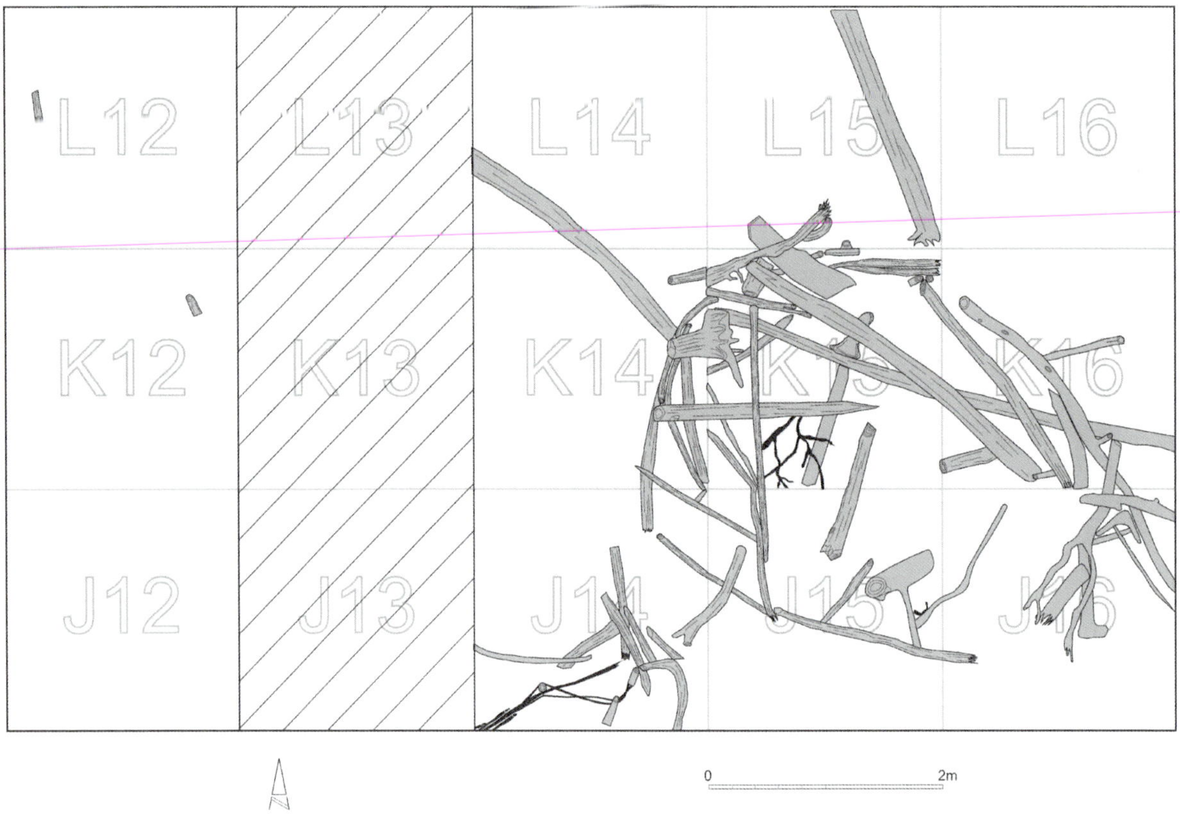

Fig. 4.43. Băile Figa 2009, Trench XXII, plan.

Fig. 4.44. Băile Figa 2009, Trench XXII, view from the west.

Trench XXIII (S. XXIII)

The main eastern tributary of the Pârâul Sărat stream is rich in material: many poles, planks and wattle remains are still visible in the streambed. Unfortunately, a large (ca 5 x 5 m) and deep (ca 2 m) pit dug in the 1970s near the tributary spring, as well as much human activity around this pool, has badly damaged most of ancient structures in the area. However, in the western part of the stream many timbers are still preserved *in situ*, often grouped in compact clusters in the streambed. A control trench (Trench X) dug in 2008 showed that timber structures are present 15 to 25 m north of the tributary. The radiocarbon dates indicate that the timbers here fall in the Bronze Age and the post-Roman / Early Medieval period (see below).

In 2011 the eastern half of the stream was incorporated into the Băile Figa spa. Although the management was told that that digging in the area was strictly forbidden (the site now being a protected monument), employees of the spa periodically dug pits into the streambed in search of the salty mud that is used for therapeutic purposes. Such diggings in the summer of 2012 uncovered a cluster of poles, planks and wattle, all of them seemingly *in situ*. The area affected by this work measured about 4 x 5 m, and achieved a maximum depth of about 0.8 m from the current ground surface. Arriving on the site on 11 August 2012, Kavruk requested the administration of the spa to stop the work, and conducted a rescue excavation in the affected area. After the necessary preparation, Trench XXIII was opened on 25 August. Also, during the widening and cleaning of the streambed in order to facilitate the removal of water from the trench, a cluster of wooden artefacts was uncovered 2.5 m north of its north-western corner. In order to research and rescue them, a small excavation area (1.5 x 2 m) was opened as part of Trench XXIII. The surface area of the trench excavated in 2012 thus eventually extended to 51 m^2. The excavation was coordinated by Kavruk assisted by Adela Kovács, and latterly by Dan Buzea.

Trench XXIII was placed 38 m south-east of the confluence of the main course of the Pârâul Sărat stream and its main tributary (*Fig. 4.15*). The trench measured 8 m (NE–SW) x 6 m (SE–NW), its north-eastern and south-western sides roughly parallel to the course of the tributary stream. The larger part of the trench lay on the right side of the stream, the maximum distance from the streambed to the north-eastern edge of the trench being 7 m. At the same time the trench extended over the streambed and onto its left side, so that the distance between the south-western wall of the trench and the streambed was 1.75 m. The trench was divided into squares of 2 x 2 m. The squares were marked in a grid system: from south-east to north-west by capital Latin letters, and from south-west to north-east by Arabic numerals (*Fig. 4.45*).

The excavation was done manually using picks, shovels and trowels. In order to stop the penetration of water from the stream into the trench, two narrow strips of earth were left unexcavated along its banks. Water which sprang from the walls and bottom of the trench was initially evacuated by directing it to the stream using shovels, and as the trench got deeper by means of buckets and a pump. The soil was removed progressively in layers of about 10–15 cm. The maximum depth reached was 1.7 m from the current ground surface. All *in situ* elements were left in place, cleaned, and periodically photographed. The data on every timber found *in situ* (position, dimensions, description etc.) was recorded in the Timber Catalogue. Finds not *in situ* were first put onto the plan, photographed, and the context described in the field diary, then removed and kept in black plastic bags together with the original mud, and placed in a cool, dark and damp cellar prior to transportation to Suceava for conservation.

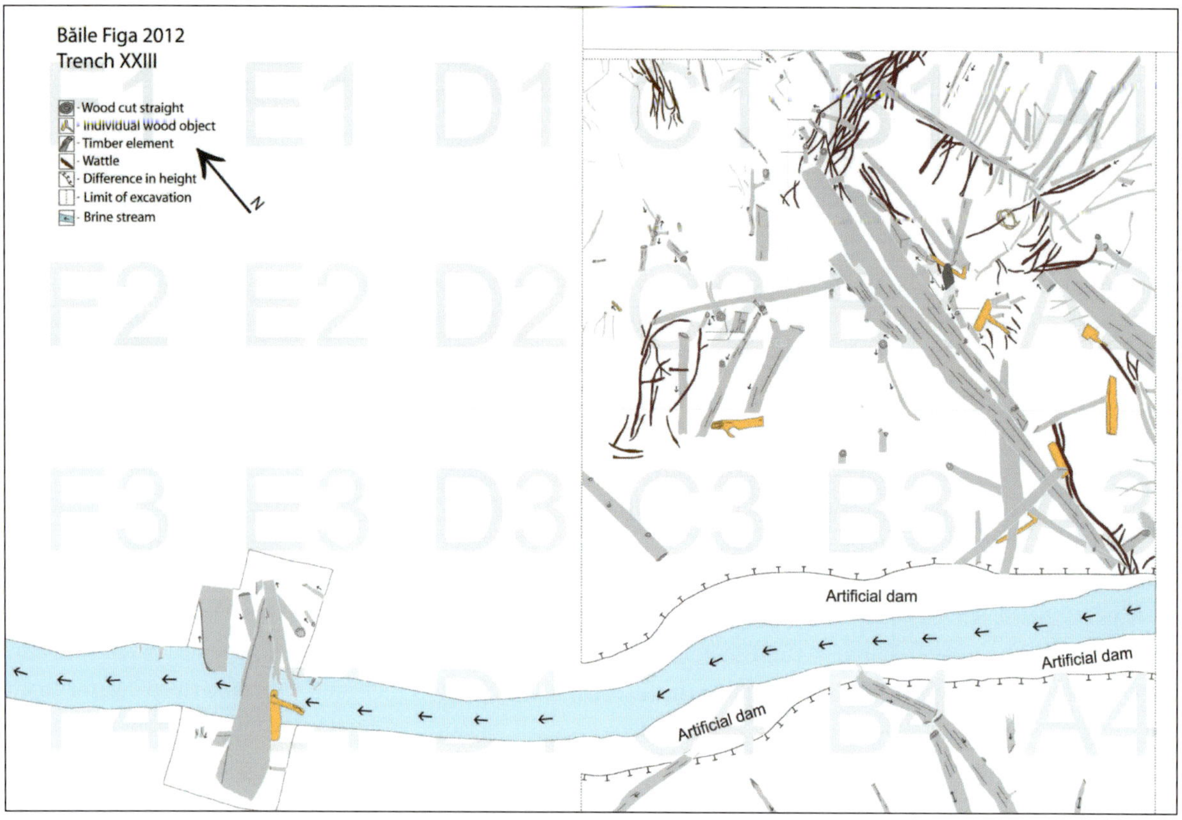

Fig. 4.45. Băile Figa 2012. Trench XXIII, plan.

Stratigraphy and its interpretation

The stratigraphy of Trench XXIII was recovered mainly by studying the vertical sections of three trench sides.

1. North-western section

- a. 0–0.35 m: yellowish soil, mixed with sand and gravel; no relevant artifacts;
- b. 0.35–0.85 m: orange-yellowish sandy clay, mixed with small and medium stones; no relevant artifacts;
- c. 0.85–1.60 m: grey and blackish mud mixed with medium and large stones, containing ancient wood pieces

2. South-eastern section

- a. 0–0.40 m: yellowish clay mixed with sand and gravel; no relevant artifacts;
- b. 0.40–0.85 m: grey and blackish mud, containing ancient wood pieces and Bronze Age pottery

3. North-eastern section

- a. 0–0.45 m: yellowish sandy clay mixed with gravel; no relevant artifacts;
- b. 0.45–1.20 m: grey and blackish mud, containing ancient wood pieces and Bronze Age pottery.

It is expected that with further excavation the rock salt will be reached under this mud layer.
The stratigraphy in Trench XXIII is similar to that of Trenches III and XXII. The upper layers, consisting of yellowish soil, poor in artifacts, were formed mainly as a result of natural alluvial processes. By contrast, the mud beneath is rich in artifacts and was formed after ancient salt production was abandoned.

In origin it was a fine humic soil, fertile and rich in organic material. It was carried to the bottom of the Băile Figa depression naturally by wind and water from the eastern slopes of the depression. After it reached the bottom of the hollow – which is rich in brine – it gradually turned into salty mud. Both types of soil contributed to the preservation of ancient timber. The wet conditions promoted by the mud ensured that the wood fibres remained elastic, and the liquid salt in the timber prevented the formation of microorganisms. At the same time, the upper deposits, rich in clay, gravel and sand, protected the mud against evaporation, which would have created conditions in which the salt would have crystallized. Rain and snow falling onto the mud thus contributed to the maintenance of wet conditions in the lower layers.

Finds

Trench XXIII produced many finds: wooden structures *in situ*, parts of structures no longer *in situ*, wooden artifacts (Nos. 135–147; *Pl. 30. 31*) and pottery sherds. Some of the *in situ* structures and elements were clearly interconnected, while other elements seem to have had a connection to them but without further indications of its nature. This suggests that all were part of the same building, which was labelled Complex 1. The sides of the building were made of upright posts and planks set fast in the ground, and of a wattle fence supported by upright posts. Its roof seems to have been made of horizontally positioned beams and stakes. The trench covered just the western part of this building, its eastern part (apparently the larger one) being situated east and south-east of the trench edge.

The uncovered part of the building has two walls placed at right-angles to each other (*Fig. 4.46*). The northern wall was made of wattle supported by vertical poles, and the western wall of a range of upright planks. The wattle wall was uncovered in the eastern part of the trench (square B-1). The trench uncovered only part of it, its eastern part continuing eastwards beyond the trench edge. The uncovered

Fig. 4.46. Băile Figa 2012. Trench XXIII from the south-east, showing Complex 1 with the north-south plank wall (centre) and part of the east-west wattle wall (left).

part of the wall is slightly curved, oriented west-east; it is 1.40 m long and 0.90 m high (*Fig. 4.47*). Five vertical posts supporting the wattle were visible in it, their uncovered height being 0.73 to 0.93 m. The western end of the wattle wall was connected to the northern end of a massive beam 17 cm thick and 4.70 m long placed horizontally, oriented south-north (square B-1, 2, A-3). The beam and wattle wall were connected by means of the insertion of the top of the supporting pole at the western end of the wattle wall into the rectangular perforation (8 x 10 cm) made in the north-eastern end of the beam. 1.6 m from its north-eastern end, at the beam's outer (western) side, a cavity was cut. In front of this cavity the massive post was uncovered in an oblique position, but it seems that initially it was placed vertically. All of this suggests that the beam was linked to the post, and the cavity cut in the beam was meant to keep the joint stable. The opposite (southern) end of the beam is not perforated and shows no connection with other elements. Close to that beam and alongside it lay another beam, apparently not *in situ*. It was 3.04 m long and 0.20 m thick. It was not linked to any other element, and did not show any traces of working apart from cuts at each end.

The wall made of vertical planks and posts set fast in the ground lay about 30 cm east of the *in situ* beam and close to the beam not *in situ* (square B-2) (*Fig. 4.48*). About 2.5 m of it were visible. The planks were made by splitting tree trunks, 25 to 29 cm wide and 6 to 8 cm thick. Their uncovered height was 0.88 to 0.45 m. The distance between planks and posts was between 15 and 20 cm. In the northern part of this wall a few poorly preserved fragments of twigs were visible, which might suggest that horizontally placed twigs were inserted between planks and posts to close the gaps between them.

Many other wooden elements were found inside the space delimited by the walls of the building. A group of 14 straight sticks (6–3 cm thick and 1.98–0.50 m long) were found in the eastern corner of the

Fig. 4.47. Băile Figa 2012. Trench XXIII, detail of wattle wall and its junction with the plank wall, from the south-west.

Fig. 4.48. Băile Figa 2012. Trench XXIII, plank wall of Complex 1 seen from the north-east.

Fig. 4.49. Băile Figa 2012. Trench XXIII from the east, showing the interior of Complex 1, and group of parallel poles on the right.

trench (square A-1); they were parallel to each other and to the eastern wall and seem to have been part of the roof of the building (*Fig. 4.49*). Attached to one of the constructional elements in the northern corner of square A-2 there was a binding made of twisted twigs.

Several wooden artifacts were found inside the building: two socketed axe hafts (Nos.140.141; *Pl. 30*), a fragment of what seems to have been a trough (No.135; *Pl. 30*), a perforated peg coming from a trough (No.136; *Pl. 30*), three mallets, of which two were made of T-shaped branches (Nos. 139. 142; *Pl. 30*), while the third has a haft inserted in the opening of the active part of the mallet (No. 138; *Pl. 30*). A few wedges found inside the building (No. 143; *Pl.31*) were probably used for splitting massive timbers or making cracks in the rock salt (cf Chapter 8). More details about the finds in Trench XXIII are in the catalogue (Nos. 135–147; *Pl. 30. 31*).

Wooden objects found outside Complex 1

In the northern part of Trench XXIII (squares C-1, 2, 3; B-1, 2) several wooden elements were recovered. Some – fragments of wattle in square C-1, 2, 3 and posts in square C-2 – seem to be component parts removed from destroyed structures. Two complete timbers consisting of thick branches with short cut smaller branches, which seem to be parts of built structures, are unique; at the mid-point of their wider parts they have perpendicular and oblique cuts, in steps and at different angles. These objects seem to have been parts of composite devices or wooden structures.

Finds in square D, E-3

During cleaning of the streambed 2.5 m north-west of Trench XXIII in square D, E-3, a group of wooden artifacts was uncovered (*Fig. 4.50*). This group consisted of three oblique posts, the eastern ends lying in the eastern bank of the stream. Under them lay a massive wooden channel laid in a horizontal position

Fig. 4.50. Băile Figa 2012. Trench XXIII, north-western part, showing the group of timbers and wooden channel, viewed from the south.

perpendicular to the course of the stream. The channel was made from a hollowed-out tree trunk. As found, it was 1.88 m long, both of its ends being under the stream banks. Its maximum width was 0.5 m. Above the channel part of a device consisting of an arc-shaped wooden plank (broken, and part missing) with a round perforation at the end and two rectangular perforations at the middle part, was found. It is similar to the two arc-shaped objects found in Trench III (Nos. 88. 89; *Pl. 28*). In one of the rectangular perforations a plank was inserted (described in more detail in the Finds Catalogue). The area was protected for future excavation.

At the end of the excavation of Trench XXIII the banks of the trench were reinforced with planks, and the trench was backfilled with plastic containers, plastic sheet and mud. The mayor of Beclean undertook to protect the area by means of a fence and to create suitable conditions for the preservation of the site.

North-Western excavation area
George G. Marinescu with Valerii Kavruk

In order to check for settlement(s) related to prehistoric salt production at Băile Figa, we examined surface features all round the valley sides. Apart from the many earth mounds and depressions, the result of modern digging for salt, no sign of human habitation was apparent anywhere on the valley slopes. If any ancient settlement had existed, the modern digging would have disturbed it, so that some indications would most likely have been visible on the surface (charcoal, ash, daub, pottery etc); none was found.

The northern part of the western slope (the north-western sector of the site) lies outside the area of the rock salt deposit, and no mounds and depressions are visible. Most of the area is today covered by woodland. This area seems to offer good conditions for settlement: it is dry, has good visibility over the whole valley; and there are fresh water springs nearby. For this reason small test excavations were carried out in May 2007 to determine if any settlement traces might be present. The work was conducted by George G. Marinescu and Adela Kovács, under the general direction of Kavruk. Four trenches were opened: IV, V, VI and XIX.

Trench IV (S. IV)

Trench IV (10 x 1.5 m) was opened on the plateau 100 m west of the Pârâul Sărat stream, at the edge of the forest opposite Trench III. The plateau seemed to offer suitable conditions for settlement, but excavation showed that it had been created artificially in modern times. There were no archaeological features.

Trench V (S. V)

Trench V was opened about 100 m north of the northernmost chalet on the north side of the forest track. This was the sequence:

- 0 to 15 cm: blackish humic material typical of woodland
- 15 to 30 cm: greyish soil with a few post-medieval/modern wheel-made sherds in yellow fabric
- 30 to 50 cm: greyish-yellowish soil with a few small Bronze Age sherds
- 50 cm: natural (yellow clay)

There were no traces of construction in this trench. In the northern profile 6.5 m from eastern edge of the trench, a small spot of burnt earth was observed. In order to find whether this was part of a larger feature, a 2 x 2 m extension of the trench was dug, but no features were found. Among the Bronze Age sherds, close to the burnt earth, a vertical handle, oval in section, with a button on top (*Pl. 8.9*), was found (see below).

Trench VI (S. VI)

Trench VI (10 x 1.5 m) was opened 12.5 m east of Trench V. It was oriented N–S. Its stratigraphy was similar to that of Trench V. A few Bronze Age sherds were recovered over the whole trench up to a depth of 50 cm.

In the western half of the trench there was a shallow depression down to 65 cm below the present-day surface. A group of Bronze Age sherds was found in this area. In the western profile of the trench, the contours of a small storage pit (40 cm wide) and a stake-hole were noted. To investigate these features two small extensions (C2: 3 x 2.5 m, and C3: 2 x 2 m) were opened, but the outline of the depression could not be clearly delimited. Its fill consisted of ashy soil mixed with a few Bronze Age sherds, and a few pieces of daub.

In C3 the storage pit (Pit 1/Gr.1) was investigated (diameter 40 cm, depth from the surface 65 cm). In its fill a few Bronze Age sherds with brushed surface (*Pl. 8*) and a flint scraper were recovered (see below).

Trench XIX (S. XIX)

In order to find the western edge of the settlement area, Trench XIX (8 x 1 m) was opened 50 m north of Trench VI. No evidence for occupation was found in this trench. This suggests that the western edge of settlement lies somewhere to the east, between trenches VI and XIX.

North-western excavation area: interpretation

The evidence recovered from the northern part of the western valley slope is slight. It consists of a few shallow depressions associated with stake-holes, ashy soil and single sherds. Features in Trenches V and VI, perhaps the remains of shelters for temporary habitation, were barely visible. In these trenches some 60 sherds were recovered, most dating to the Bronze Age (with the exception of a few sherds coming from post-medieval/modern wheel-made vessels). While most are undiagnostic, a few features suggest a Late Bronze Age date (see below for discussion of the dating).

Geophysical investigations at the prehistoric salt production site at Băile Figa

DAN ŞTEFAN

Introduction

Geophysical applications in archaeology have a long tradition, the first attempts being documented as early as the end of the 19th century. However, systematic research dedicated to the establishment of geophysical methods expressly designed to be applied to the recognition and description of archaeological structures and vestiges really only developed after the Second World War. Since then, instruments, acquisition methods and processing algorithms have been adjusted to better suit the particularities of detecting and characterizing buried objects and structures in a wide variety of archaeological contexts. Above all, there is agreement that the customary practice of employing a combination of geophysical methods – magnetic and electrical, occasionally electromagnetic (ground-penetrating radar, electromagnetic induction methods) – has produced good results for a varied set of archaeological issues.

The unusual geology, local geomorphology and archaeological context represented by the prehistoric salt production site at Băile Figa challenged both archaeologists and geophysicists to select and amend their investigation methods to the site's particular characteristics. Geophysical investigations applied in relatively similar, though not identical, conditions are described by Kayt Armstrong (2010, with literature). In this regard, the geophysical methodology applied at Băile Figa may be seen as significant not only for the help it has given to the archaeological investigation, but also for its expected contribution to future geophysical investigation on other sites located in similar environmental conditions.

This report presents the main results of several preliminary geophysical investigations carried out on the salt production site at Băile Figa. The survey was designed to reveal the stratigraphic connection between the salt rock-bed, buried archaeological structures, and the present-day ground surface.

Survey design and data acquisition

The petrographic features of the minerals, soil and archaeological remains located beneath the Băile Figa basin appear, at first glance, unfavourable to the geophysical exploration of the site. Structures that have undergone thermal treatment are absent, as are those of stone. Virtually all of the archaeological structures were made of timber, while most of the tools were made of wood, with the notable exception of stone hammers. Wood by itself is non-magnetic, and so is salt, whether rock or in solution. As a consequence, the magnetic response of the archaeological structures at Figa was expected to range over low values.

At the same time, the overall level of humidity in the site area is typically high, but the high concentration of dissolved salts in soil moisture leads to a very high electrical conductivity. This can be a serious issue for any electrical method used for site surveying, since a very high conductivity in the soil produces low electrical contrasts. The irregular nature of the ground surface and the very high conductivity of the soil are also severe obstacles to survey by ground-penetrating radar (GPR).

Given these difficulties, it was clear that only a careful survey design would reveal useful underground geophysical contrasts. An initial assessment of the magnetic properties of soil and river terrace gravels was made on the basis of magnetic susceptibility measurements carried on with a ZH Instruments SM30 Resonance k-Meter. A major part of the stones which make up the geological horizon are of igneous and metamorphic origin; as a consequence, their magnetic susceptibility is very high (24 to 28 $\times 10^{-3}$ SI). The soils over the whole area were formed as a consequence of the disaggregation process of the stones making up this river terrace. Consequently, a high level of magnetic susceptibility was transferred into the soil matrix (2 to 5 x10^{-3} SI). At the same time, the timber structures (more specifically timbers soaked in saline solution) show a very low magnetic susceptibility (0.05 x10^{-3} SI or even lower). Therefore, a measurable magnetic contrast between the archaeological structures (made of timber) and soil was expected to stand out. Unfortunately, the significant modern man-made disturbances in the topsoil were reflected in the measured magnetic signal. Given this, it is more likely that a low to medium magnetic contrast between buried timber structures and soil matrix would be obtained. In order to reveal such magnetic contrasts by regular magnetic survey, one should avoid and filter out any other high magnetic influences of non-archaeological origin that might overwhelm the measured signal.

The magnetic survey was applied over a rectangular area of 1200 m^2 located in the central part of the site (*Figs 4.51*, *4.52*). The area was selected due to its relatively even, less disturbed, surface. The occurrence of some timbers in the river bed was also taken into consideration. The magnetic grid measured 40 x 30 m and was divided into 30 traverses at 1 m intervals. The instrument used was a Bartington Grad 601–2 dual fluxgate gradiometer with a measured density of four readings per metre along traverses.

It was assumed that the uneven nature of the site surface, rather inappropriate for magnetic or electromagnetic study, could have responded better to geoelectrical methods. This is why we selected electrical resistivity tomography (ERT) as the method best suited to the 2D imaging of the subsurface along stratigraphic profiles. Given the very high level of conductivity in the area, the *pole-pole* configuration of the electrodes was employed. Given the large distance between the P1 and P2 (the two potential electrodes of the *pole-pole* array, one of them being placed remotely), the measured values of the electrical potential were maximized. In practice, the level of the measured telluric noise, the other consequence of the *pole-pole* configuration of electrodes, was low enough not to significantly influence the data. What is more, the *pole-pole* array is known to have the widest horizontal coverage and reach the lowest depths of all the regular electrode arrays used in geoelectrical surveys. In total, geoelectrical measurements were completed along four profiles, using a multi-node resistivity meter in a 36-electrode configuration. The position and direction of the ERT profiles is represented in *Fig. 4.51*. The distance between electrodes was set to 1 m, thus establishing the length of any individual ERT profile at 35 m.

Apart from magnetic and geoelectrical prospection, a GPR survey was tested along five profiles of various lengths in central areas of the site, mainly along the access road and river. In order to maximize the coupling factor between the electromagnetic antenna and topsoil, winter was selected as the best period for carrying out the survey. Although the thin layer of snow was useful in leveling the topsoil irregularities, thus favouring a better coupling factor for the antenna, the general level of high conductivity in the area caused significant attenuation of the electromagnetic signal. Since the profiles obtained do not add useful information, they are not discussed in this report.

Data processing, results and interpretation

Magnetic data equalization between survey lines using a zero mean traverse algorithm and certain filtering procedures (applied in order to remove spurious measured spikes) were the only adjustments required in the magnetic data processing flow. The dynamic range was restricted to an interval of ±60 nT/m for enhanced visualization of the magnetic data. In addition, the grey scale used was inverted in order to emphasize the negative magnetic anomalies that we anticipated in the timber structures.

A moderate match between magnetic data and areas with a higher density of buried structures, as revealed by excavation in Trench XXIII (see above), may be seen in *Fig 4.52*. However, the high amplitudes of the vertical magnetic gradients, ±60 nT/m and above, rather suggest that morphological

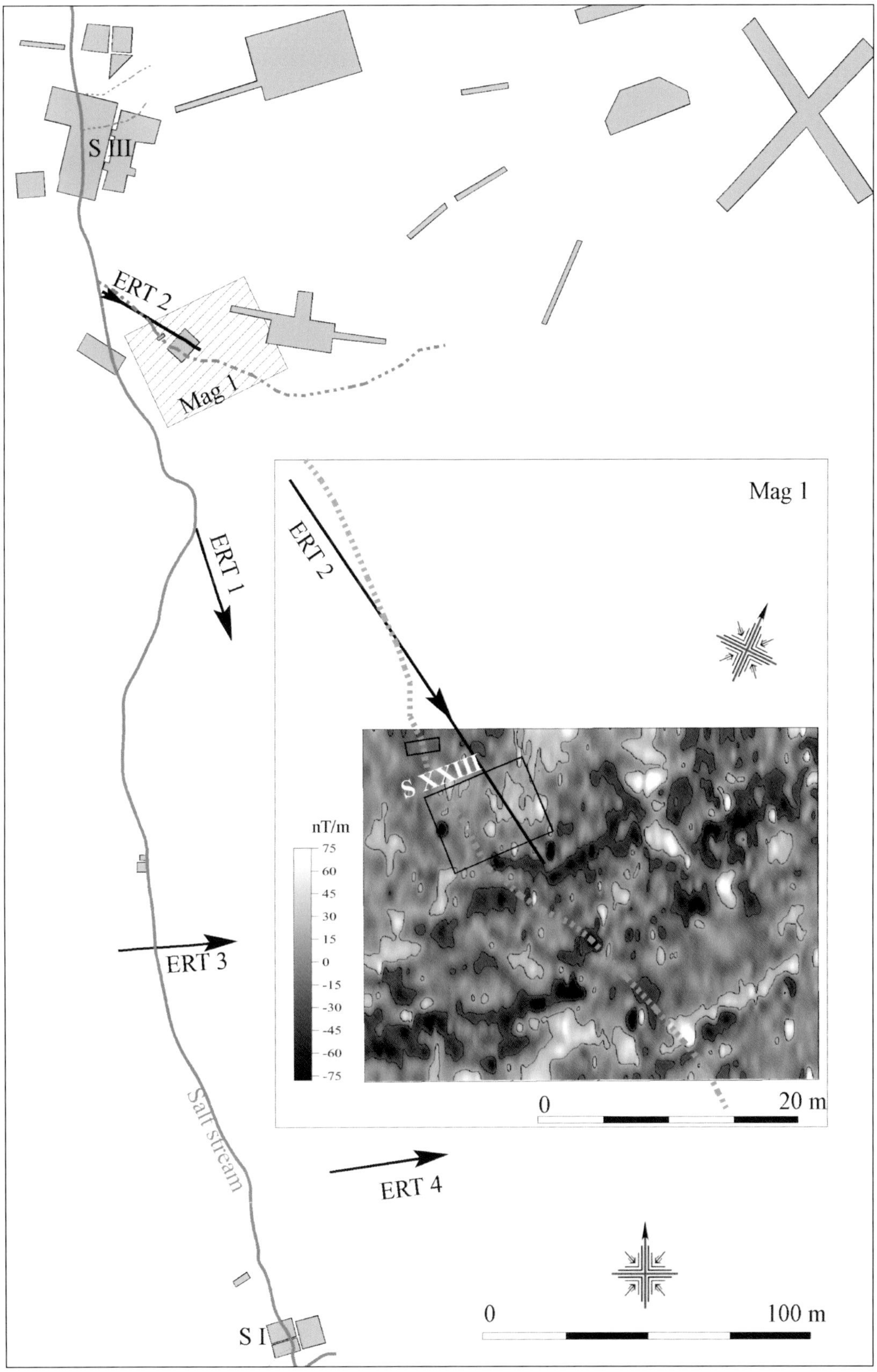

Fig. 4.51. Băile Figa. General map, showing the area investigated with magnetic methods (Mag 1 and inset) and ERT profiles.

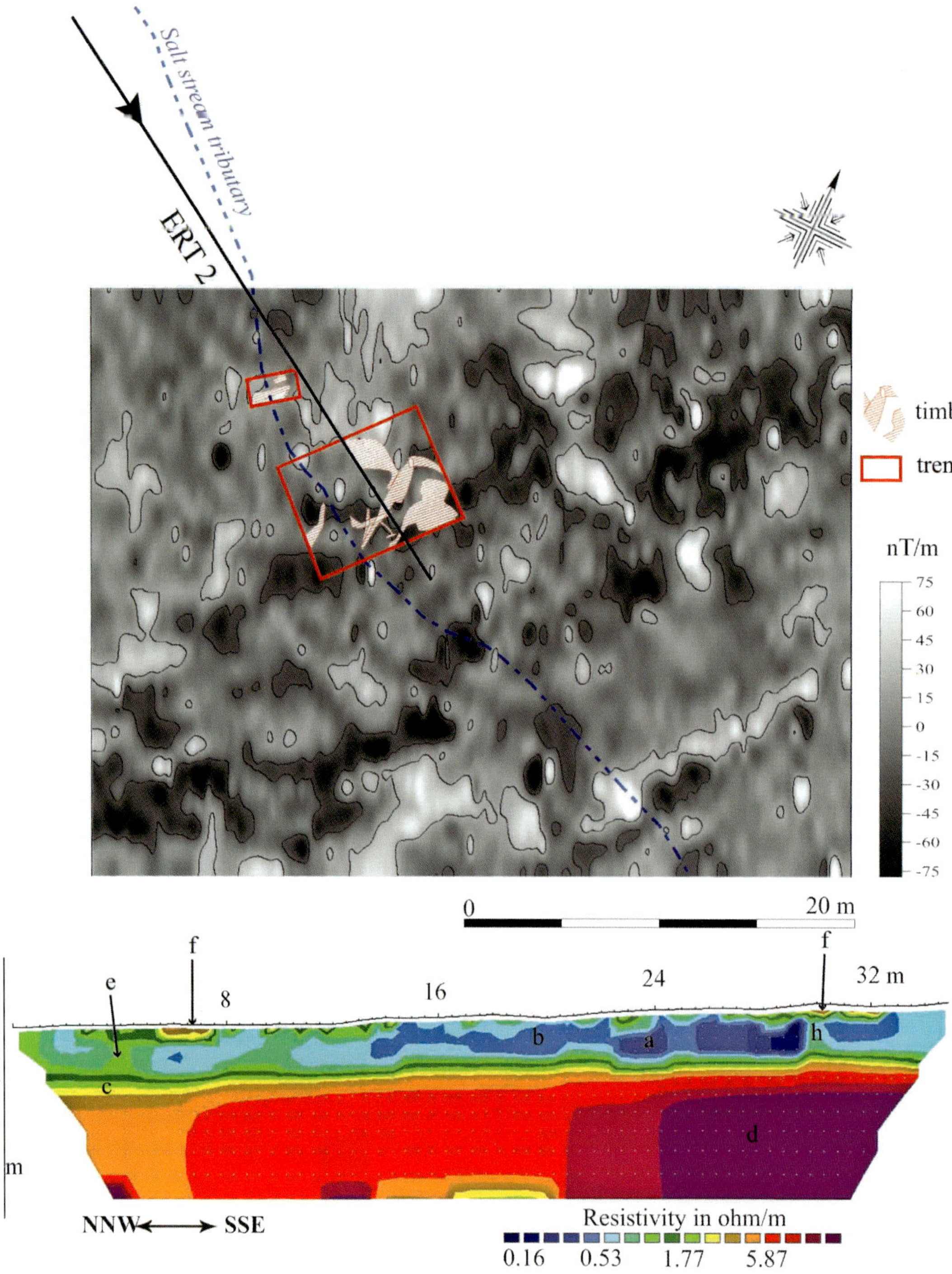

Fig. 4.52. Area investigated with magnetic methods (upper); ERT profile 2 (lower).

irregularities in the soil matrix were the real contrast generators, since the contrasts produced by archaeological structures should be much weaker (in any event less than ±10 nT/m).

Significantly more useful information was extracted from geoelectrical prospection after processing of the apparent resistivity pseudo-sections, and obtaining two-dimensional resistivity distributions

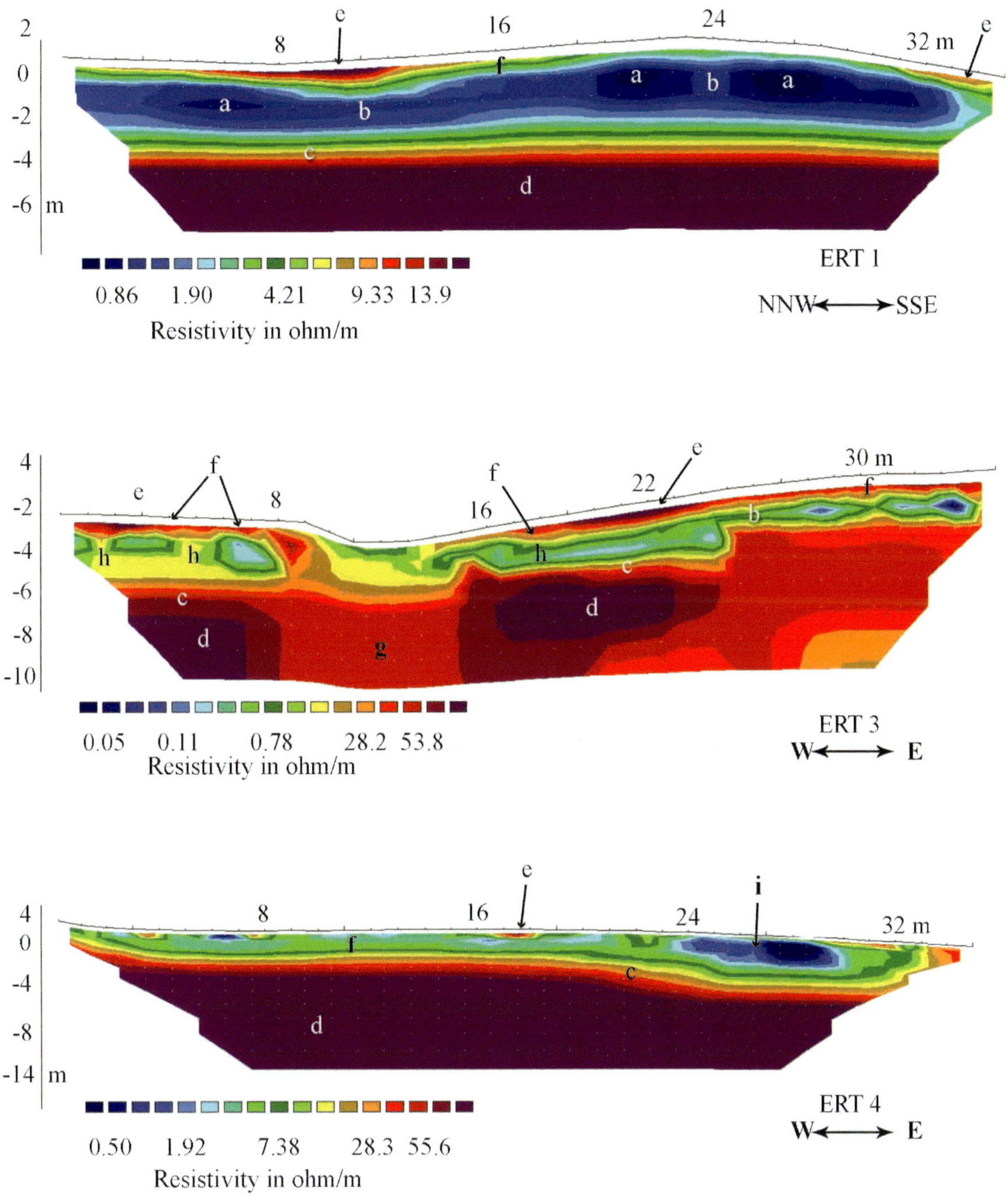

Fig. 4.53. ERT profile 1 (top); ERT profile 3 (middle); ERT profile 4 (bottom).

(LOKE – BARKER 1996). The resulting geoelectrical tomographic profiles (*Fig. 4.52* lower; *Fig. 4.53*) show an extremely large dynamic range, from fractions of an ohm per metre up to 10 000 ohm/m. This is not a picture of a common soil profile but follows quite well a stratigraphic model consisting of a resistive basal level covered with an uneven soil layer in which fluids contain dissolved salts.

An overall assessment of all the ERT profiles indicates that the depth of the rock-salt is on average 4 m. There is also a good match between the observed topographic profile of the topsoil and the basal level. Other details can be seen in each of the ERT profiles, and are interpreted as follows:

a. soil saturated with saline solution; extremely low resistivity values;
b. soil rich in accumulated moisture and soluble salts; saline waterlogged sediments with very low resistivity;
c. transition area from soil rich in moisture and soluble salts to the rock bed (solid salt); the upper and altered part of the salt rock massif which interleaved with soil and/or alluvial deposits;
d. salt massif; very high resistivity values;
e. dry soil and/or soil with gravel;
f. Solonchak[3] – visible in all ERT profiles in thin layers in unforested areas of the basin and slightly thicker in the forest (see ERT 4); these are the only continuous soil areas, the result of local pedogenesis, often separated by alluvial intervening strips; resistivity values can vary over a wide range, typically 3 to 30 ohm/m, depending on the soil moisture level;
g. disaggregation area in the salt massif; small rifts and alteration zones allowing infiltration of water, thus lowering the resistivity to the 15 to 40 ohm/m range (especially near or under the salt stream bed, see ERT 2 and ERT 3);
h. possible anthropic structures; abrupt variations of the resistivity values (see ERT 2) or unusual cuts in the layer (see ERT 3); maybe of archaeological origin;
i. clay (mud) filling with low resistivity values (see ERT 4).

The ERT 4 profile is particularly interesting, especially the section between 24 and 32 m. Here, the ERT profile went over a modern excavation made by local people to access the rock salt. This is just one of many similar depressions in the Băile Figa basin. We assume that the salt massif was reached in the excavation; soon afterwards, the spot would have been abandoned. The image of the ERT 4 profile clearly shows the process of natural, progressive, filling up of the excavation with clay sediment (mud). This is due to the combined action of the rainfall/evaporation cycle on the abandoned pit (and on the others of this sort).

Overall, the low values of the apparent resistivity, and resistivity calculated by inversion in the upper layers (a, b), can only be explained as having been caused by the high percentage of saline solution in the soil. This configuration of geophysical parameters would not support any other geological model than one with a natural rock salt basal layer (d).

Conclusion

The prehistoric salt production centre at Băile Figa is certainly exceptional in archaeology. It is similarly unusual as far as geophysical investigation is concerned. Conventional near-surface geophysical techniques, as widely applied in archaeology, encountered serious challenges in the particular environment of Băile Figa. Nevertheless, non-invasive studies should be undertaken in all circumstances, even when the result is indirect and/or partial.

The preliminary attempt to explore the Băile Figa basin by geophysical methods described here was a logical step in the extensive research programme conducted at the site. In the end, it became clear that the methods applied (gradiometry, ERT, GPR) could not precisely detect and characterize archaeological features buried in alluvial deposits. However, we were able to obtain important information regarding

[3] Solonchak – soil with high soluble salt accumulation formed from saline parent material as a result of moisture evaporation.

the close relation between the salt massif, alluvial deposits in which timber structures are found, and the topographic profile of the topsoil.

Future geophysical investigation at the site should bear in mind lessons from our present experience. GPR should be used at lower frequencies (100 MHz or even lower) in order to reduce signal loss generated by the saline solution accumulated in the soil. ERT survey should have a smaller minimum distance between nodes (smaller than 1 m). Interesting results can also be expected from 3D ERT surveys.

Magnetometry should be done with a total field magnetometer, driven by optical pumping or Overhauser sensors. The need here is not for better sensitivity but for the complete characterization of the earth's magnetic field in order to discriminate between the magnetic response of buried structures and other magnetic influences of non-archaeological origin. Qualitatively speaking, fluxgate gradiometers are only able to provide very limited magnetic information; this is the gradient of the vertical component of the earth's magnetic field, while classical processing algorithms require magnetic total field datasets; these can be upward/downward continuation, pole reduction, Euler deconvolution and other inverse and direct modelling techniques (Blakely 1995; Telford *et al.* 1990).

Acknowledgements

The author would like to thank Constantin Ştefan, Cătălin Constantin and Maria-Magdalena Ştefan for their significant contribution during the fieldwork. Many thanks go to Maria-Magdalena Ştefan for her help throughout the survey and for the preparation of the plans.

The finds

Pottery

Adela Kovács

General overview

Work at Băile Figa between 2005 and 2011 recovered 4095 sherds weighing 56.172 kg in the different parts of the site. Most was recovered during research excavations in 2007–2011. Pottery was found only in trenches I, III, V, VI and VII (in 5 of the 22 trenches excavated).

Of the total quantity of pottery recovered, 3622 sherds (50.076 kg) can be attributed to prehistoric periods (mainly Bronze and Iron Age), and 473 sherds (6.096 kg) to the post-medieval or modern periods (19th–20th centuries AD).

The distribution of the pottery recovered on the site is quite uneven, as follows:

Trench I: 359 sherds (5.605 kg), of which:

17 (0.606 kg) are Bronze Age
40 (1.096 kg) Iron Age
302 (3.903 kg) post-medieval/modern

Trench III: 3116 sherds (42.969 kg), of which:

1 Neolithic (?)
2946 (40.810 kg) Bronze Age (predominantly Early Bronze Age)
170 (2.159 kg) post-medieval/modern

Trenches V and VI: 60 sherds, of which 50 were Bronze Age (mostly Late Bronze Age) and 10 post-medieval or modern

Trench VII: 611 sherds (7.444 kg), of which:

513 Bronze Age (mostly Late Bronze Age)
98 undatable (small and undistinctive)

In the following discussion, the pottery assemblage is divided by vessel part, fabric (based on colour, inclusions and temper, wall thickness and surface treatment), and decoration.

Trench I

Bronze Age

The 359 sherds of Bronze Age pottery were mainly found in the earth covering the Iron Age box-like timber construction, and consist of:

Rims (23.5 %)
Body sherds (58.8 %)
Bases (11.8 %)
Handles (5.9 %)

The fabrics are divided between coarse (29.41 %) and medium (70.59%).

Inclusions

Sand, pebbles and grog (58.82%)
Grog and sand (29.41%)
Sand and pebbles (5.88%)
Sand and grog (5.88%)

Because of the high degree of fragmentation it was impossible to determine vessel shapes. The only shape that could be reconstructed was a medium fabric pot with slightly splayed rim and a strap handle on the shoulder (*Pl. 1.2*).

All sherds have a smooth surface. The range of colours of the outer surface is

Brick-red with spots (41.18%)
Brick-red (17.65%)
Greyish (17.65%)
Blackish-greyish, light brown, brown with spots and blackish-greyish with spots: 5.88% each

58.82% of the sherds come from vessels fired in an oxidising and 41.18% from vessels fired in a reducing atmosphere.

Iron Age (Dacian)

40 sherds from two vessels of Dacian type were found in Trench I, most found together in a small area close to the southern wall of the box-shaped timber structure. Most of the sherds are of medium fabric, made by hand from clay mixed with grog and pebbles, and have smoothed surfaces. Both vessels were burnt in a reducing atmosphere, their predominant colour being dark brown with blackish spots. Reconstructions are shown in *Pl. 1.1, 3, 4*

Vessel 1: sack-shaped pot with slightly splayed rim and rounded rim, short neck, sack-shaped body and flat base. On the shoulder cylindrical button(s) with deep finger impressions. Decorated with a raised cordon covered with finger impressions. Height approx. 28 cm, mouth diameter approx. 21 cm, body maximum diameter approx. 23 cm; base diameter approx. 12 cm. MNCR inv. no. 16640 (*Pl. 1.1*)

Vessel 2: conical with flat base, slightly rounded walls, straight rim with splayed edge, and tab-like handles on the shoulder. Height approx. 9 cm, mouth diameter approx. 29 cm, maximum diameter approx. 30 cm; base diameter approx. 19 cm. MNCR inv. no. 16641 (*Pl. 1.3–4*)

Trench III

Spatial distribution:

Pit 1 (Gr.1): 1020 sherds (12.147 kg)
Pit 2 (Gr.2): 1722 sherds (23.656 kg)
Pit 2a (Gr.2a): 32 sherds (0.585 kg)
Without context: 342 sherds (6.298 kg)

2460 sherds of the 3116 recovered in Trench III) have been analyzed. 486 badly eroded Bronze Age sherds, as well as all the 170 post-medieval/modern sherds, have not been analyzed.

Vessel parts

Rims (11.26%)
Body sherds (85.48%)
Bases (2.7%)
Handles (1%)

Fabric classes

Fine: a few sherds only (0.85%)
Medium (80.52%)
Coarse (18.24%)

Inclusions and temper

Sand and pebbles (47.6%)
Sand, pebbles and grog (25.7%)
Coarse-grained sand (11.8%).
Other around 10%.

Shapes

Coarse fabric:

Since about 90% of the coarse sherds are body sherds, and only 10% are rims or bases, vessel form is hard to assess. The following forms are present:

Large dish (mouth diameter approx. 40 cm), conical with wide mouth, straight neck and rounded rim. Width much larger than height (*Pl. 4.9*)

Jars – height greater than width

Conical vessels with slightly curved walls, mouth diameter significantly greater than that of the base, but a little smaller than that of the body; rim decorated with raised cordon (*Pl. 4.7*)

Carinated vessels, with short neck, upright rim with the edge cut off straight, sometimes decorated with raised vertical and horizontal cordons (*Pl. 6.1*)

Vessels with flat base and carinated body (*Pl. 6.21*)

Sack-shaped vessels with slight rounded rim, sometimes with horizontal raised cordon (*Pl. 5.13*)

Vessels with straight rim whose top surface is cut flat. Under the rim a vertical strap handle, the ends attached under the rim and shoulder (*Pl. 6.3*)

Vessels with everted and rounded rim, short neck and pronounced shoulder. Immediately under the rim a strap handle, the ends attached to rim and shoulder (*Pl. 6.4*)

Vessels with carinated body and vertical oval-section handles attached to the widest part of the pot (*Pl. 6.14*).

Medium fabric:

This pottery is highly fragmented. About 85% of the sherds are body-sherds and undiagnostic as to shape. Only about 15% of the sherds are rims or bases. The following forms can be recognised:

Bowls: vessels whose maximum width is at the upper part, their maximum diameter is larger than their height and larger than their base. Two forms can be distinguished: conical, with slightly rounded upper part, straight rim with rounded edge (*Pl. 4.3–6*); and with spherical body and rim with rounded edge slightly inturned (*Pl. 4.2*)

Jars: rather large vessels, the height greater than the width

Conical vessels with wide mouth and straight walls and no neck (*Pl. 4.1*)

Sack-shaped vessels with outturned rim, short neck, and slightly profiled walls. The maximum diameter is usually two-thirds of the way up the vessel (*Pl. 5.1–14*)

Surface treatment

Brushed (1331 sherds, 59.79 %)
Textile-imprinted (50 sherds, 2.25%)
Polished (401 sherds, 18%)
Smoothed on both surfaces (444 sherds, 19.95%)

Textile impressions are always on the outer surface of medium fabric pottery. In a few cases there are imitations of textile impressions produced by a sharp-ended tool (*Pl. 2.1–12*; *Pl. 3.1–11*).

The brushing technique shows some variation. Most sherds have *deep brushing* (69.87%) and less than a third *superficial brushing* (30.13%). In the overwhelming majority of cases it was the outer surface which was brushed (*Pl. 5.3, 5, 9–11, 13*; *Pl. 6.13, 16–18, 22, 23*). A few sherds also have brushing on their inner side (*Pl. 4.1, 4–9*; *Pl. 5.14*).

Decoration

181 sherds from Trench III are decorated. The techniques consist of the application of clay to the surface, and incision or pricking.

Applied clay: 60.89%
Pricking: 31.25%
Incision: 4.47%

Applied clay:

Bands. In most cases the bands were attached to the rim (*Pl. 5.1, 6, 11, 14*) or immediately under the rim (*Pl. 5.5, 9*), on the neck (*Pl. 5.6*), rarely on the upper part of the body (*Pl. 6.8*). The bands are finger-impressed (*Pl. 5.1, 14*) or notched (*Pl. 5.5, 9, 11*). In two cases the bands are vertical, placed on the upper part of the vessel (*Pl. 6.1, 2*).
Tabs may be attached to the vessel body (*Pl. 6.16*)
Buttons are usually attached to the vessel bodies. Some of them are conical in shape without other elements (*Pl. 6.12*) or are finger impressed on the top (*Pl. 5.3, 10*). The only double button was attached to the vessel under the rim (*Pl. 4.3*), and probably served as a handle.
Cordons, usually nail pricked or vertically incised (*Pl. 4.7*)
Incision: vertical or oblique lines (*Pl. 6.9*), as well as horizontal parallel lines (*Pl. 6.10*)

Pricking: only horizontal motifs were used. Three shapes of prick marks could be distinguished (determined by the shape of the tool used): round (*Pl. 6.11*), triangular (*Pl. 4.4, 8*) and rectangular (*Pl. 4.1*; *Pl. 5.2*). These kinds of decoration were placed at the outer edge of the vessel rim (*Pl. 4.1, 4, 8*), immediately under the rim (*Pl. 5.8*), or at the shoulder (*Pl. 5.2*).

Firing

Reduced: 84.34%
Oxidised: 14.44%
Refired: 1.22%

Colour range

The colour range is quite varied. Most sherds are greyish, black-greyish or brown.

Trench VII

611 sherds of Bronze Age date (7.444 kg) were found in Trench VII. All come from Pit 3. 403 sherds were analyzed; 98 are very small and totally undiagnostic, 110 come from the same vessel.

Intact and reconstructable vessels

The only intact vessel on the entire site came from Trench VII. It is a small cup in medium fabric made of clay mixed with coarse-grained sand (*Pl. 7.1*). It has a wide mouth, slightly curved walls and a flat base. Height 6.2 cm, mouth diameter 6.2 cm; base diameter 4.1 cm. Both surfaces are yellowish-brown. It was found in the upper part of the fill of Pit 3 (MNCR, inv. no. 15272).

Two other vessels could be reconstructed. One of them (110 sherds) is in medium fabric, made of clay mixed with fine sand. The greatest width is at mid height. The walls are rounded, the rim straight and the base flat. Two vertical handles are on the shoulder (*Pl. 7.9*). Both surfaces and the section are black and dark grey. Height 36 cm, mouth diameter 22.5 cm; maximum body diameter 34 cm, base diameter 16.5 cm, wall thickness varies from 0.8 cm to 0.6 cm (MNCR, inv. no. 15295).

The second vessel (7 sherds) is a bowl in medium fabric made of clay mixed with sand. Wide mouth, the diameter twice as large as its height. Walls and rim straight, base flat. A small handle under the rim (*Pl. 7.5*). At the same height, decoration of two horizontal roughly straight lines of prick marks. Height 4.8 cm; mouth diameter 14 cm, base diameter 8 cm, wall thickness 0.4–1.1 cm (MNCR, inv. no. 15278).

Vessel parts

Rims 9.9%
Body sherds 85.6%
Bases 1.8%
Handles 2%

Fabrics

Coarse 19.1 %
Medium 80 %
Fine 1 %

Inclusions

Sand and grog 35.24%
Sand, pebbles and grog 22.3%
Sand and pebbles 20.3%
Coarse-grained sand 16.6%
Other mixtures: less than 5%

Surface treatment

Smoothing 96.98%
Polishing 1.76%
Other treatment types: less than 1%

Surface features

Brushed: 21.6 % (*Pl. 7.3, 4, 7*)
Textile-impressed: two sherds (*Pl. 7.6*)
Brushing with horizontal lines: two sherds (*Pl. 7.3, 7*)
Brushing with oblique lines: two sherds (*Pl. 7.4*)

Decoration

Only 3.72% of the 403 sherds found in Trench VII are decorated, as follows:

Horizontal bands: 8 sherds (53.3%). Two variants: single vertical band (*Pl. 7.8*), and double band (*Pl. 7.7*)
Oval pricks: 7 sherds (46.7%) (*Pl. 7.5*)

On 60% of the decorated sherds the decoration covers the whole body, and on 40% only the rim area.

Firing

Oxidised: 74.9%
Reduced: 24.6%
Refired: less than 1%

Colour range

Bricky: 23.1%
Bricky with spots: 22.8%
Light brown: 11.2%
Other colours: less than 10%

North-Western sector (Trenches V and VI)

A few undiagnostic Bronze Age sherds were recovered from Trench V. A few Bronze Age sherds were recovered across trench VI up to a depth of 50 cm. A group of sherds came from a shallow depression. In the storage pit in C3 (Pit 1/Gr.1) a few Bronze Age sherds with brushed surface were recovered (*Pl. 8.1, 2, 6*).

The recovered sherds come mostly from vessels made in coarse and intermediate fabric. Most come from pots with a more or less conical body, flat bottom and a slightly splayed rim. Sometimes these pots had a brushed surface. Some pots have small conical or hemispherical buttons under their rim (*Pl. 8.8, 11, 12*). Most come from well fired pots, and only a few sherds from poorly fired vessels. The predominant fabric colour is reddish, yellowish and brownish on the outer surface, and brownish and/or greyish inside. A few sherds combine black and brownish spots on their surface. The great majority of these sherds are undiagnostic in cultural terms.

Of the 60 sherds found in Trenches V and VI only eleven are significant:

1. Handle with cylindrical button on the top; medium fabric, clay mixed with fine sand. The piece might come from a vessel with two upright handles. The predominant colour is reddish. Height 5.6 cm, thickness 1.9 cm, button diameter 2.1–1.9 cm. Trench V, depth 45 cm. CMBN inv. 23431 (*Pl. 8.9*) In Transylvania this kind of handle is specific to the Late Bronze Age (Reinecke Bz D/HaA) and usually appears in Noua and late Wietenberg contexts.

2. Rim from a sack-shaped pot (?), coarse fabric (clay mixed with fine sand and small pebbles). Under the rim oblique parallel lines made by a comb (*Pl. 8.6*). Inner surface polished. The predominant colour of the sherd is blackish. Height 3.5 cm, thickness 1.1 cm. Trench VI, 40 cm below the modern surface. CMBN inv. 23436.

3. Flat base, coarse fabric pot of clay mixed with sand and small pebbles. Exterior surface rough, inner side smooth. Predominantly blackish-brown and brown. Length 5 cm, thickness 1 cm. Trench VI, 40 cm from the modern surface. CMBN inv. 23438 (*Pl. 8.4*).

4. Rim fragment, coarse fabric, clay mixed with sand and small pebbles. Exterior surface brushed, inner smooth. Greyish-brown. Height 10 cm, thickness 1 cm. Trench VI, 40 cm below surface. CMBN inv. 23439 (*Pl. 8.14*).

5. Rim fragment, coarse fabric, clay mixed with sand and small pebbles. Exterior surface rough, inner side was smooth and slightly polished. Brown-yellowish and brown-reddish. Height 8.2 cm, thickness 1.4 cm. S. VI at a depth of 40 cm. CMBN inv. 23440 (*Pl. 8.7*)

6. Rim fragment, sack-shaped vessel, coarse fabric, clay mixed with fine sand and small pebbles. Conical button under the rim. Yellowish and yellowish-brown. Height 6 cm, thickness 1.1 cm. Trench VI. CMBN inv. no. 23441 (*Pl. 8.11*).

7. Rim fragment, sack-shaped pot, medium fabric, clay mixed with fine sand. Conical button under the rim. Exterior surface smooth, inner side slightly polished. On both surfaces the sherd is brick-coloured. Height 4.2 cm, thickness 0.8 cm. Trench VI at a depth of 30 cm. CMBN inv. no. 23442 (*Pl. 8.12*).

8. Rim fragment, sack-shaped pot, medium fabric, clay mixed with fine sand. Conical button under the rim. Brick-coloured with a black line on the rim. Height 4.5 cm, thickness 0.8 cm. Trench VI at a depth of 30 cm. CMBN inv. no. 23443 (*Pl. 8.8*).

9. Rim fragment, sack-shaped pot, medium fabric, clay mixed with fine sand. Both surfaces of the sherd smooth and brick-coloured. Height 6.8 cm, thickness 0.8 cm. Trench VI, square C2 at a depth of 50 cm. CMBN inv. no. 23446 (*Pl. 8.10*).

10. Rim fragment, sack-shaped pot, coarse fabric, made of clay mixed with sand and small pebbles. Both surfaces are brick-coloured. Height 5.7 cm, thickness 1.1 cm. Trench VI, square C2 at a depth of 50 cm. CMBN inv. no. 23447 (*Pl. 8.5*).

11. Body sherd, medium fabric, clay mixed with fine sand. Outer surface shows oblique straight lines made by comb. Both surfaces are brick-coloured. Height 5.8 cm, thickness 1 cm. Trench VI, Pit 1 (Gr.1) at a depth of 60 cm. CMBN inv. no. 23445 (*Pl. 8.3*).

Stone

Adela Kovács

Forty stone objects collected from various parts of the Figa site have been catalogued. They are illustrated in *Plates 33–38*.

Two main types of stone object are present: grooved hammers (sometimes called "mining tools" in English; German *Rillenhämmer*); and shaft-hole axes. A smaller category consists of possible whetstones.

The grooved hammers have a waist around the mid-point for the attachment of rope or plant fibre binding material, probably leading to a wooden handle attachment to enable the object to be used as a sledge-hammer with which to hit the rock surface or crush lumps of rock into smaller fragments.

The shaft-hole axes are fragmentary and presumably date to a period earlier than the Middle and Late Bronze Age – perhaps even the Neolithic. This raises the possibility that they were used in earlier times for the same purpose.

Prehistoric woodworking at Băile Figa

Richard Brunning

The recording and assessment of the Figa wood was carried out in 2009 on material undergoing conservation in Suceava Museum, and on objects and structural timbers from Trench I and three pieces from Trench XV. The material being conserved was mainly from Trench III with a smaller amount from Trench I. The structures in the process of excavation in Trench III were briefly inspected but no recording took place.

The wood is catalogued in the Wood Catalogue (Appendix 1) and illustrated in *Plates 9–32*.

Condition

The observed condition of the wooden remains was variable. Virtually all the material being conserved was in excellent condition with very good surface detail remaining. Some items showed signs of wear, but this was from their active use rather than from post-depositional damage or decay. The material being excavated in Trench III appeared to be in a good condition, although in general only the upper parts of the posts were visible. In Trench I the condition of the material was variable. The condition of the upper parts of the square box structure had suffered very significantly from desiccation, no doubt greatly exacerbated by evaporation drawing out the salt water to the surface of the wood. Possibly as a consequence the surface condition of these timbers was extremely poor and, although the wood was very solid, it showed evidence of much shrinkage and cracking. Some freshly exposed material outside the square box structure in Trench I was in a poor condition, which supports the probability that the observed decay occurred in the distant past, possibly before the wood finally entered the burial environment. In contrast the freshly exposed ends of timbers composing the structure were in a visually excellent condition, as were several newly uncovered artefacts both inside and outside the structure. Where desiccation has not occurred the preservation of the wooden structures and artefacts is exceptionally good. It is clear that the combination of high salt concentrations and wet or waterlogged deposits are maintaining a superb burial environment for organic materials. This suggests that the potential for preservation in the unexcavated areas remains very high.

Structural timbers and joinery

The main structure in Trench I was a large square box structure that had partially slumped on its southern and western sides (*Fig. 4.24*). The walls of the structure were composed of half split logs that retained their bark. The majority of the wall timbers had suffered very badly from desiccation. Where surface condition allowed definite species identification, the logs appeared to be made from young oak trees, under 50 years of age. In addition to being split in half, the logs often had flat tapering scarfed ends to allow them to fit snugly behind retaining uprights.

In the south-west corner of the structure a jumble of half split oak logs was present. These retained their bark and were derived from trees generally under 30 years of age. They all had scarfed ends at their west ends. The similarity with the other walling timbers suggests that they were probably part of the collapsed southern wall.

Eight vertical posts retained the walls, one in each corner and in the centre of each wall. Two of these uprights on the southern side had slumped to the south-west, along with most of the wall planks. The uprights in the centres of each wall were braced against each other by a pair of large oak logs [1367 and 1385] that formed a cross in the centre of the structure (*Fig. 4.54*). These braces were angled slightly downwards from the south to the north and from the west to the east. Additional bracing was provided by another oak log [1435] that joined the corner timbers [1126 and 1361] on the northern side and also helped to support the central upright [1360] on that face. On the eastern side of the structure the central post [1363] was braced against the corner posts to the north by timber 1431 and to the south by 1430. The top of an angled timber [1388] seen on the southern side of the structure displayed a double spur joint and may represent a bracing timber that has been moved out of position by the collapse on the southern side.

The junctions of the bracing timbers were uncovered and examined on the east side of the structure in the centre and at the northern corner. Around the central upright [1363] the northern [1431] and southern [1430] braces were at the same height with the east-west internal brace [1367] below them. In the northern corner the north-south brace [1431] was still in place but the brace along the northern wall [1435] had slipped out of contact with the corner upright [1361].

All the upright posts and all the bracing timbers were made of young oak trees. The bracing timbers retained their bark and were only modified at their ends where they joined other timbers. The upright posts were squared off and given chamfered edges where the bracing timbers joined them in order to provide a more secure fixing.

Each bracing timber required a joint at either end. These joints involved a central flattened or slightly curved area, with projecting spurs on either side that wrapped round the sides of the uprights (*Fig. 4.25*). When the structure was under tension this would have provided a simple but secure join. Several of the joints were no longer in place because of the slumping on the southern side.

Two variations in joint design were consistently used at the top and bottom of the trunk. The end that was from the base of the tree was invariably carved into shape with the projecting spurs carved from the normal girth of the tree. At the upper end of the trunk the presence of a side branch was utilised to create one of the protruding spurs. This was obviously a very deliberate choice of the builders, presumably because it was a slightly easier way to make the joint and because it utilised the natural strength of the tree in its design. In most cases considerable effort had been taken in producing a regular tapering design

Fig. 4.54. Băile Figa 2009. Trench I, showing the box structure with bracing timbers crossing in the middle (background); wattle fence (foreground left).

to the projecting spurs. This may have helped to ease their placement through sediment. The central north-south bracing timber [1385] was an exception to the general pattern as its northern end had only one projecting spur. The end of the spur was still covered by sediment but it may have gone through a gap in the wall plates.

Numerous structural timbers had been excavated in previous years and stored on site in water. Many of these were incomplete, having one broken end. The complete ends exhibited a range of joints similar to the bracing timbers still *in situ* in the square structure, but with some significant variations (*Fig. 4.55*).

Non-structural material

A wide range of non-structural wooden components was recovered from Trenches I and III. Many of these are without obvious parallels elsewhere in prehistoric Europe. These are described more fully in the Wood Catalogue. The large troughs and their associated perforated pegs are exceptionally interesting pieces. The hollowing out of logs and creation of perforations in their base is a common practice on prehistoric logboats (e.g. McGrail 1978; Mowat 1996), but the combination of square holes and lines of pegs acting as nozzles is unique. The pegs or nozzles would have projected from the base of the

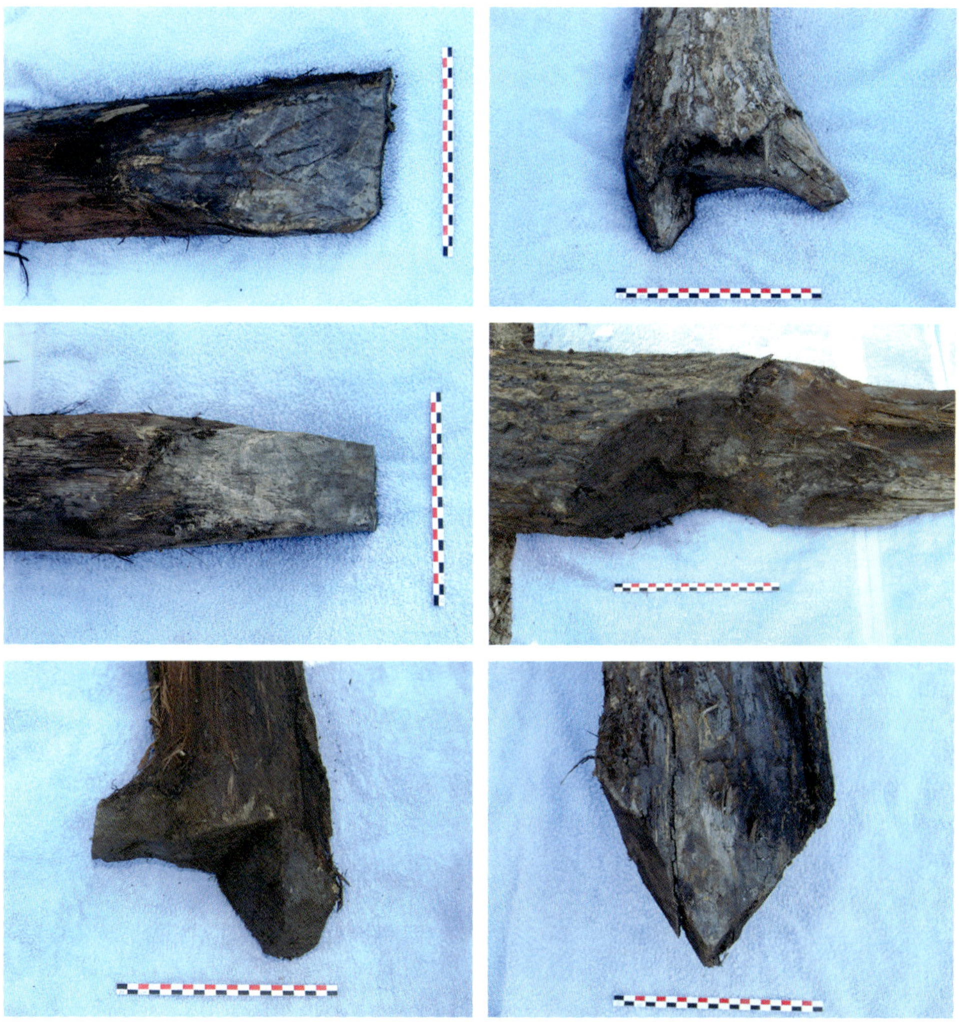

Fig. 4.55. Băile Figa 2008. Trench I. Worked ends (part of joints similar to those on the box structure) on timbers removed from the west part of the trench.

troughs so they must have been supported at either end, possibly by withy ties going round the deep incisions near the end. Fragments of twisted hazel ties (14846 and 15258, *Pl. 22.82*) were recovered from Trench III, and others from Trench I. The style of hollow pegs varied significantly in the two trenches.

The use of water in the salt production process is further attested in Trench III by a possible funnel (14729, *Pl. 18.53*) and several small bowls or troughs (14839, *Pl. 17.51*; 14972, *Pl. 18.55*; 15263, *Pl. 17.49* and 15264, *Pl. 17.50*). All these items were found in a broken and fragmentary condition, possibly reflecting hard use. They are similar in design to other prehistoric examples of carved vessels (EARWOOD 1993). but were not very finely finished. This suggests that they were used in the salt extraction process rather than being 'domestic' vessels of the workforce.

Wedges for splitting wood or salt blocks were recovered from Trench I (14845, *Pl. 16.45*; 15597, *Pl. 14.35*; 15581, *Pl. 28.122*; 15584, *Pl. 28.123* and two unnumbered examples). These were all made of oak heartwood with the narrow end aligned along the rays of the wood forming a ca.0.4 cm wide tip. In addition there was an oak object [1158] with a similar splitting edge but with a handle at the other end.

No wedges were recorded from Trench III but there was a tangentially split object which had a curving narrow blade and a crude handle at one end (14841, *Pl. 23.93*). This could possibly have been used to help chop up blocks of salt. A more robust item from the same trench was a piece of radially split oak (14847) with a handle at one end. The other end showed wear on one side where it seems to have been used as a mallet or pounder. This could have been to drive in wedges, but as none was found in that trench it may have been used to break up the salt. Two chunkier pounders (14844 and 15578, *Pl. 13.29–30*), which used side branches for handles, were found in Trench I. These were probably used for crushing the salt.

Several oak and beech shovels were recovered from Trench III (14840, 15259, 15262, *Pl. 21.74–76* and 15261, *Pl. 23.95*). Two of these had handles that gradually curved into the blade (15259 and 15262) while the others had straight shoulders between the handle and the blade. One of the latter had the handle on the left side (14840).

A tangentially split oak timber from Trench III (14973, *Pl. 23.85*) had been carved into a rough semicircle. It had broken across the wider side through a central hole 3.4 cm across. Its unbroken edge is cut flat so it cannot have functioned as any kind of wheel. Before it broke it may originally have been a roughly circular lid of a large tub with a central hole, like the lid of a butter churn. Perhaps salt could have been broken up in such a tub with a crusher on a round handle passing through the hole.

Material may have been carried around the site on wooden 'stretchers'. The evidence for this comes from Trench III from two naturally curving oak branches that had multiple perforations. One (15266, *Pl. 23.88*) was broken across the holes that consisted of a circular hole (5.7 cm and 6.1 cm diameter) near either end and three rectangular holes in between. The other (15267, *Pl. 23.89*) had two round holes of similar dimensions near the ends and two rectangular holes in between. The round holes could have accommodated round timbers used as handles, while the rectangular holes would have contained split timbers forming the base of the cradle that would have been 71.9 to 79 cm wide.

A quarter split oak timber from Trench III (14843, *Pl. 23.84*) was also from some sort of rectilinear framework or structure. It was 84.5 cm long and broken at one end, with three oval perforations spaced unevenly along its length.

An artefact type that also appears to be unique to the site is a series of perforated 'hooks' 19.8–22.0 cm long, with diameters of 3.5–4.3 cm. Three (14963, 14964 and 15225, *Pl. 22.78–80*) were found in Trench III and one (15582, *Pl. 28.121*) in Trench XV. Two of them are oak, one beech and one unidentified. They were made from naturally curving stems, although the outside of the curve was enhanced in two of the cases. Each had a rectangular perforation, 5.9–6.1 cm from the straight end, going through in the same plane as the curved end. The largest hole was 3.7 x 2.2 cm and the smallest 2.6 x 1.4 cm. One of the holes (in 14964) was cut through at a significant angle with one entry beginning 6.1 cm from the end and the other 5.2 cm. In two cases a split timber, broken at both ends, was seen to penetrate the hole. In both cases the split fragment was significantly smaller than the hole, suggesting that a tight fit was not required. A U-shaped object from Trench III (16663, *Pl. 22.81*) may represent the complete form of these hooks. It was made from a forked branch with opposing rectangular holes in each arm. With a piece of wood through these holes it could have operated as a suspension loop, holding perhaps the end of withy ties. The hooks may represent such artifacts broken across the apex and or the holes.

Several oak blocks were recorded that had been cut across in opposing directions at two ends at angles of 35 to 50 degrees. One of these was from Trench III (15332, *Pl. 25.109*) and the others from Trench I (15577, *Pl. 15.37* and an unnumbered piece). These were of a very deliberate manufacture and may have been used as props to keep separate two larger timbers that had curving or rounded bases. Two short timbers from Trench XV (15585, *Pl. 28.124* and 1395) had been cut across in the same direction at two ends and may also have functioned as a from of wedge or spacer keeping other timbers in place or apart.

Evidence for the methods of climbing in and out of the excavations came in the form of a log ladder and a ladder with rungs like modern ones. The log ladder [1316] was not fully excavated when examined (*Fig. 4.56*). It was made of a half split oak log, over 94 cm long, that had two steps, and a third step at the top, cut spirally into it. The steps were therefore triangular in shape and 6–7 cm wide at the widest point. The steps were 15.9 cm and 23.2 cm apart. Many prehistoric log ladders have been found with angled steps along one side but this spiral design appears unique. The advantage of the spiral design over the normal type was that it could be set vertically into the ground whereas the more common log ladders are set at an angle. This may have been more convenient in a cramped excavation.

The ladder with rungs (15580, *Pl. 11.21*) was incomplete but was over 5 m high. The sides were boxed trunks, neatly chamfered at the edges, with a gently curving side profiles. There were at least

Fig. 4.56. Băile Figa 2009. Trench I from the east, showing the box structure behind with the stepped log ladder (foreground left).

seven rungs on the most complete side, which penetrated rectangular holes in the side timbers. The rungs were very carefully fashioned, with a gracefully carved taper towards the middle and well finished tenons which were pegged in place on the outer side of the mortise holes through a square hole at either end. A parallel is represented by an example from the Iron Age settlement at Glastonbury Lake Village, UK (BULLEID – GRAY 1911).

Tools, toolmarks and woodworking

The preservation of toolmarks was very good on many of the structural timbers and artefacts. A wide range of different facet shapes was noted from concave, to slightly concave, to flat in profile across their width. This variation in facet shape, taken together with the facet widths, lengths and angles of cut suggests that both bronze and iron axes were being used on the site . This is significant because it suggests that activity on the site spans the transition from bronze to iron.

There was considerable variation in the size and edge profile of the axes being used. The blade profiles were often visible as stop lines, or 'jam curves', in the wood where a blow had been delivered at a steep angle and had come to rest in the wood. The width of the axes used was also reflected in the facet width where this had not been subsequently reduced by other blows. This information could be used to build up data about the blade profiles of axes, which could then be compared to the profiles from the axes themselves (BRUNNING 2007; SANDS 1997).

Bronze axes were being used to work wood in both Trench I and Trench III, leaving distinctive facets that were concave in profile across their width. In some cases adzes were used to hollow and finish the large troughs, the shovels and the smaller bowls and troughs. Two tool handles were found in Trench III. One of these (14837, *Pl. 20.71*) is designed to accommodate a socketed axe. The other handle (14838, *Pl. 20.72*) had a rounded projection at a more oblique angle and was probably designed to accommodate a blade used as an adze.

The blade profiles of bronze axes and adzes were recorded on seven pieces of wood. The blade profiles were not always complete but together with the facet dimensions suggest the use of tools with blades between 3.5 and 5.0 cm wide with the majority being less than 4.4 cm wide. Some blade impressions appear to represent the full blade widths of 2.9 cm, 4.0 cm, 4.2 cm, 4.4 cm, 4.9 cm and 5.0 cm. This is a reflection of the variety of axe and adze types used. One bowl (15264, *Pl. 17.50*) displayed evidence suggesting the use of three different blade profiles in its manufacture. Evidence for use of iron axes was confined to material from Trench I. These were seen on a variety of smaller objects [1017, 1217, 1433, 15597, *Pl. 14.35*; 1222, 14848, and an unnumbered piece). Well preserved facets were not readily apparent on the structural timbers associated with the square structure but the very flat character of most of the worked areas suggests the use of iron axes. A small number of items from Trench I (15576, 1058 and two unnumbered pieces) had slightly concave to almost flat facets. The blades that produced these marks were quite broad, at least 5.7–8.0 cm wide. They are unlikely to have been made by Bronze Age axes and more probably represent the use of large iron axes.

Small knives must have been used for a lot of the fine finishing work on many of the pieces such as the ladder (15580) and the hollow pegs associated with the large troughs. Nicely finished square, rectangular and round holes and notches are evident on many pieces including the troughs, the 'stretchers', 'hooks' and plank fragments. One of the stretchers (15267) showed the use of a chisel with a blade only 1 cm wide. Such a tool could have made all the holes.

The holes on the elder pegs from Trench I were made by pushing out the middle of the wood. This is possible with elder and is probably the reason why that species was chosen. The hollow pegs from Trench III were of a slightly different design with more pronounced shoulders. It was unclear what species they were, but it did not appear to be elder and the holes did not correspond to the tree ring pattern. This suggests that they were not pushed out but must have been made by boring or coring with a very fine tool. As the pegs in both trenches were performing the same task the differences in their manufacture probably reflect different makers and possibly different periods.

Many of the objects from both trenches had been made from split timbers. These were almost certainly made by using seasoned wedges and wooden mallets. Splitting wedges were found on the site and could have been used to split wood as well as, or instead of, blocks of salt. Half, quarter, radial and tangential splitting were all evident, mainly on the oak timbers.

Woodland character and management

The evidence of the timbers from Figa is also informative about the character and composition of the woodland in the area and how it was managed over time. There is some evidence of coppicing and the young age of many of the oak logs suggests intensive management of the local woodland. The young age of the trees will probably limit the potential of the site for dendrochronology but may be significant in understanding the local woodland and its exploitation.

Significance of the wooden assemblage from Figa

The collection of wooden structures and artefacts uncovered at Figa is of great importance for both our understanding of the late prehistoric salt industry and as evidence of prehistoric woodworking in a variety of forms.

The range of artefacts and woodworking techniques evident at Figa bears good comparison with the assemblages recovered from waterlogged prehistoric settlements such as the Alpine lake villages, the lake settlements of Poland, the Baltic states, and the UK (Glastonbury Lake Village). The objects and techniques are more varied and complex than the evidence from the more common waterlogged prehistoric site types such as trackways, bridges, ritual structures, burnt mounds and pits known in Europe. The tools used include axes and adzes, bronze chisels and knives. Some of the timbers must also have been split with wooden wedges and mallets.

The assemblage from Figa is very important because it provides testament to the wide range of tools and processes used in a form of prehistoric salt production not seen in other parts of Europe. Many of the objects uncovered are unique. These not only include the troughs and their nozzles but the 'hooks', pounders, 'stretchers' and the spiral log ladder. The incomplete ladder, the socketed axe and adze handles are also extremely rare finds. The props, wedges, bowls and small troughs are good examples of objects that do not often survive in the archaeological record. There is potential for understanding woodland management at the time, and for determining the range of axe types used. The preservation, extent, diversity and sophistication of the prehistoric wooden assemblage at Figa is highly unusual and very significant.

Site chronology

The chronology of the Băile Figa site is complex. Because of the nature of the site, in which minimal stratigraphical information is present, and the fact that few datable artefacts were found (e.g. pottery), it is the radiocarbon determinations that for the most part determine the chronology of the site. In addition, samples taken for dendro dating provide some supporting evidence.

66 radiocarbon dates have been obtained, the majority through grants from the UK's Natural Environment Research Council and carried out at the Oxford Radiocarbon Accelerator Unit (ORAU). A smaller number were obtained commercially from the Groningen laboratory; these were dated by the 'traditional' method.[4] Three dates were obtained from the Heidelberg laboratory as part of the dendrochronological analysis by Tomasz Ważny. *Table 4.2* lists all the dates.

Table 4.2. Radiocarbon dates from Băile Figa.

Lab no.	Year	Site no.	Fractionation	Tr.	Square	Object	Date BP
GrN-29955	2005	Trough				Trough 1 (Chintăuan)	2870 ± 20
GrN-29956	2005	Trough				Trough 1 (Chintăuan)	2840 ± 20
GrN-30475	2006	wood 215				Timber under Trough 1 (support)	2950 ± 50
OxA-19108	2007	1/1	d13C=-23.69	I	B1	Trough 2	3194 ± 27
OxA-19270	2007	1/14	d13C=-25.08	I	B1	Trough 2	3226 ± 28
OxA-19111	2007	1/8	d13C=-25.63	I	B1	Peg from Trough 2	3189 ± 27
OxA-19296	2007	1/5	d13C=-25.41	I	B1	Binding round Trough 2	3205 ± 28
OxA-19295	2007	1/3	d13C=-24.82	I	B1	Plant material from fill of Trough 2	3221 ± 29
OxA-19109	2007	1/4	d13C=-23.92	I	B2	Trough 3	3277 ± 27
OxA-19274	2007	1/18	d13C=-25.01	I	B2	Trough 3	3278 ± 29
OxA-19294	2007	1/2	d13C=-24.18	I	B2	Binding under Trough 3	3192 ± 28
OxA-19273	2007	1/17	d13C=-22.56	I	C2	Mallet	3837 ± 35
OxA-21451	2008	1/1157	d13C=-25.40	I	B1/C1	Bundle of sticks	3191 ± 29
OxA-21452	2008	1/1157	d13C=-26.03	I	B1/C1	Bundle of sticks	3158 ± 30
OxA-24837	2008	1/1509	d13C=-27.38	!		Wattle cut by shaft	3208 ± 27
GrN-40378	2007	1/1047		I	C3	Ladder	2175 ± 25
OxA-21450	2008		d13C=-23.36	I	C3/B3	Ladder	2289 ± 31
GrN-31521	2007	1/1051		I	C3	Timber over ladder shaft	2160 ± 50
OxA-21449	2008	1/1481	d13C=-24.86	I	CII	Box structure, eastern wall	2167 ± 28
OxA-24823	2008	1/1385	d13C=-24.30	I		Shaft support	2254 ± 27
OxA-24836	2008	1/1316	d13C=-24.81	I		Ladder (stepped timber)	2159 ± 25
OxA-24838	2009	1/1609	d13C=-24.21	I		Construction cut by shaft	2180 ± 26
OxA-19271	2007	1/16	d13C=-25.03	I	C4	Worked wood, end perforated, same as below	2206 ± 28
OxA-19272	2007	1/16	d13C=-25.05	I	C4	Same as above	2204 ± 26
OxA-19110	2007	1/7	d13C=-24.15	I	A1	E sector, Post A, 1182	1638 ± 28
OxA-19297	2007	1/6	d13C=-26.52	I	A1	Binding round post A. wood 1182	1493 ± 28
OxA-19298	2007	1/12	d13C=-25.92	I	A1	Binding round posts 1182 etc	1596 ± 27
OxA-24839	2009	1/1336/1538	d13C=-24.88	I		Construction cut by shaft	1538 ± 25
OxA-19280	2007	3/11	d13C=-27.22	III	C5	Elder branch near wood 130	4381 ± 28
OxA-19275	2007	3/1	d13C=-29.97	III	C2	Wattle	2732 ± 27
OxA-19276	2007	3/2	d13C=-26.21	III	A2	Wattle	2809 ± 29
OxA-19277	2007	3/3	d13C=-25.59	III	A5	Wattle	2858 ± 30
OxA-19278	2007	3/4	d13C=-27.05	III	A6	Wattle	2766 ± 28
OxA-19299	2007	3/13	d13C=-22.96	III	D4	Fragment of small tray	2852 ± 27
OxA-19279	2007	3/10	d13C=-23.44	III	F1	Wattle (modern)	123 ± 22

[4] The Groningen laboratory has since stopped using gas proportional counting and now only uses AMS.

Lab no.	Year	Site no.	Fractionation	Tr.	Square	Object	Date BP
OxA-19709	2007	3/15	d13C=-23.89	III	C2	Wood 299, perforated piece	30 ± 22
OxA-19747	2007	3/15	d13C=-25.25	III	C2	Wood 299, perforated piece	23 ± 24
GrN-30479	2006	wood 457		III		Pile, northern area	2940 ± 50
OxA-19386	2007	3/19	d13C=-23.03	III		Structure 2 (Timber 58)	2817 ± 31
OxA-19300	2007	3/20	d13C=-23.64	III		Structure 2 (Timber 61)	2842 ± 27
OxA-24820	2010	3/158	d13C=-23.22	III		Plank supporting wattle fence	2936 ± 27
OxA-24821	2010	3/161	d13C=-22.97	III		Plank supporting wattle fence	2921 ± 27
Hd-27309	2007	3/62 or 63	d13C=-24.5	III		Plank from Structure 2	2896 ± 30
OxA-24841	2010	3/154	d13C=-24.21	III		Massive timber on rock salt	2773 ± 27
OxA-24842	2010	3/155	d13C=-23.85	III		Massive timber on rock salt	2841 ± 25
OxA-24822	2010	3/784	d13C=-27.80	III		Timber from EBA pit above rock salt	4435 ± 30
OxA-24840	2009	wood 5040	d13C=-22.67	XV		Trough	2699 ± 29
OxA-19387	2007	wood 92	d13C=-23.95			Eastern side stream	3065 ± 30
OxA-19600	2007	wood 93	d13C=-24.02			Eastern side stream	3159 ± 26
OxA-19388	2007	wood 94	d13C=-23.74			Eastern side stream	3073 ± 29
OxA-19389	2007	wood 98	d13C=-24.05			Eastern side stream	3141 ± 33
Hd-27310	2007	wood 103/104	d13C=-25.8			Eastern side stream	3248 ± 23
GrN-30474	2006	wood 97				Eastern side stream	1560 ± 40
GrN-30476	2006	wood 265				Pile, central area	2800 ± 50
OxA-19308	2007	wood 301	d13C=-24.01			Central 'dam'	3058 ± 28
OxA-19309	2007	wood 301	d13C=-23.82			Central 'dam'	3067 ± 27
OxA-19307	2007	wood 302	d13C=-23.97			Central 'dam'	2993 ± 27
OxA-19306	2007	wood 304	d13C=-23.72			Central 'dam'	3005 ± 27
OxA-19305	2007	wood 305	d13C=-21.64			Central 'dam'	2901 ± 27
Hd-27335	2006	wood 307				Central 'dam'	2982 ± 22
GrN-30477	2006	wood 344				Pole at 'dam'	2990 ± 50
OxA-19304	2007	wood 378	d13C=-24.44			Southern 'dam'	2149 ± 27
OxA-19303	2007	wood 380	d13C=-24.53			Southern 'dam'	2227 ± 26
OxA-19302	2007	wood 383	d13C=-23.69			Southern 'dam'	2152 ± 27
OxA-19301	2007	wood 384	d13C=-25.08			Southern 'dam'	2174 ± 26
GrN-30478	2006	wood 385				Southern 'dam'	2230 ± 50

Of these 66 dates, the majority (43) fall in the Bronze Age. A smaller number (14) fall in the pre-Roman Iron Age. Of the remainder, one is much earlier and appears to be of final Eneolithic date; five are post-Roman / early medieval; and three are modern.

In *Fig. 4.57*, all the Bronze Age dates are tabulated, with the exception of one date that would seem to belong to the Early Bronze Age. They range from the 16th century cal BC (Trough 3) down to the 9th (Trough 4, and wattle from Trench III).

Trench I

The datable elements of Trench I fall into three distinct groups, with possible outliers at either end: Middle-Late Bronze Age, Iron Age, and post-Roman / early Medieval.

Early Bronze Age

A single radiocarbon date of 3837 ± 35 BP comes from a wooden mallet found among the other timbers of Trench I. There are no other indications that this trench falls so early, so the piece must be presumed to be left over from some early phase of activity such as is indicated by the early material from Trench III (below).

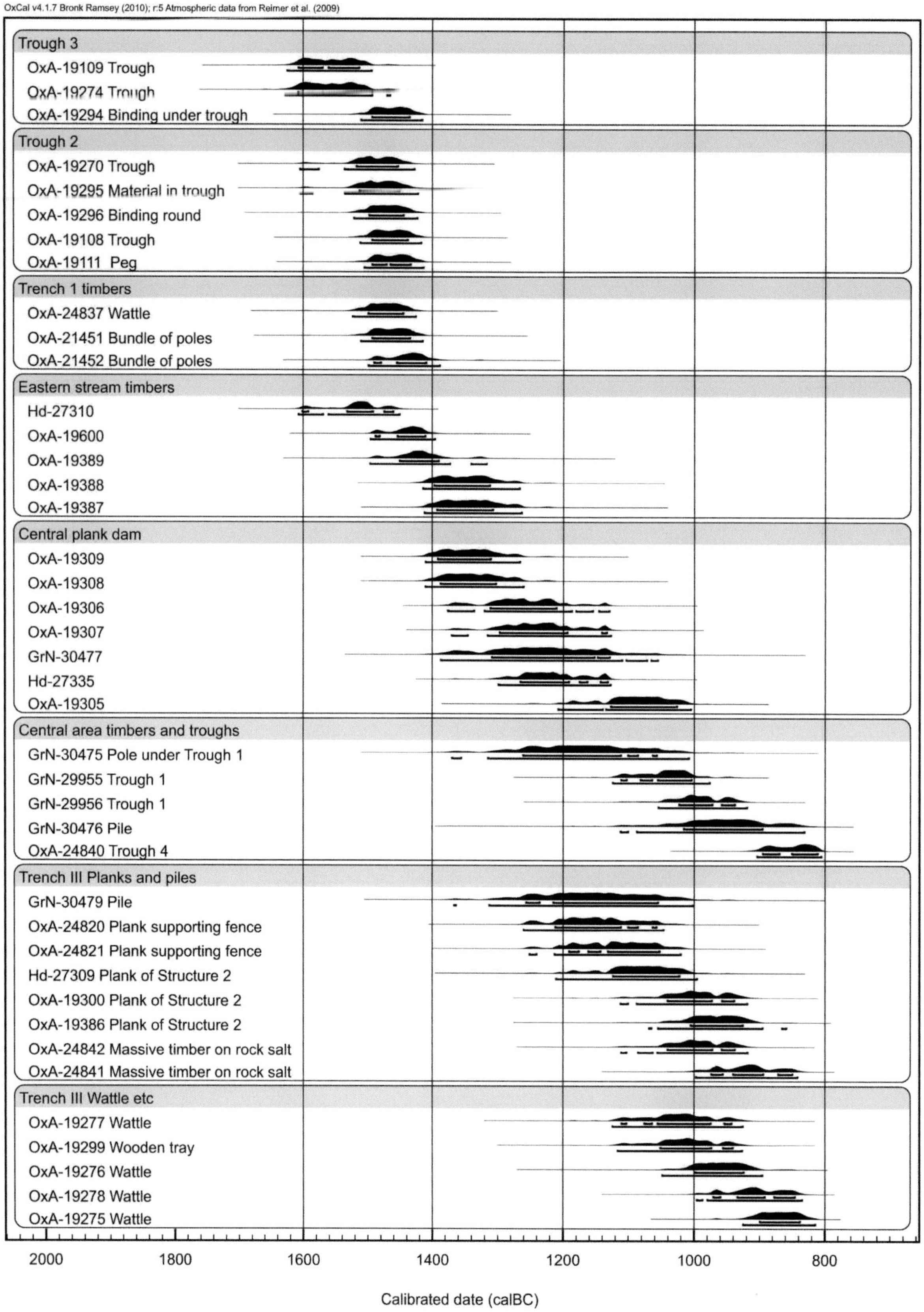

Fig. 4.57. Diagram showing all Bronze Age radiocarbon dates from Băile Figa, by structure or area.

Middle/Late Bronze Age

This was the main phase of activity in Trench I. *Fig. 4.58* shows the spread of radiocarbon dates from the trench. Trough 3, which belongs in the 16th century cal BC, is clearly earlier than the rest of the dated timbers, which all fall in the 15th century. Trough 2 falls in this period, as do the wattle elements that may have formed structures for storing the brine or rock salt.

The 359 sherds of Bronze Age pottery found in Trench I were all in secondary position, and are undiagnostic as to cultural affinity. If the radiocarbon dates are regarded as the relevant chronological indicator, then one would expect Wietenberg pottery to be present.

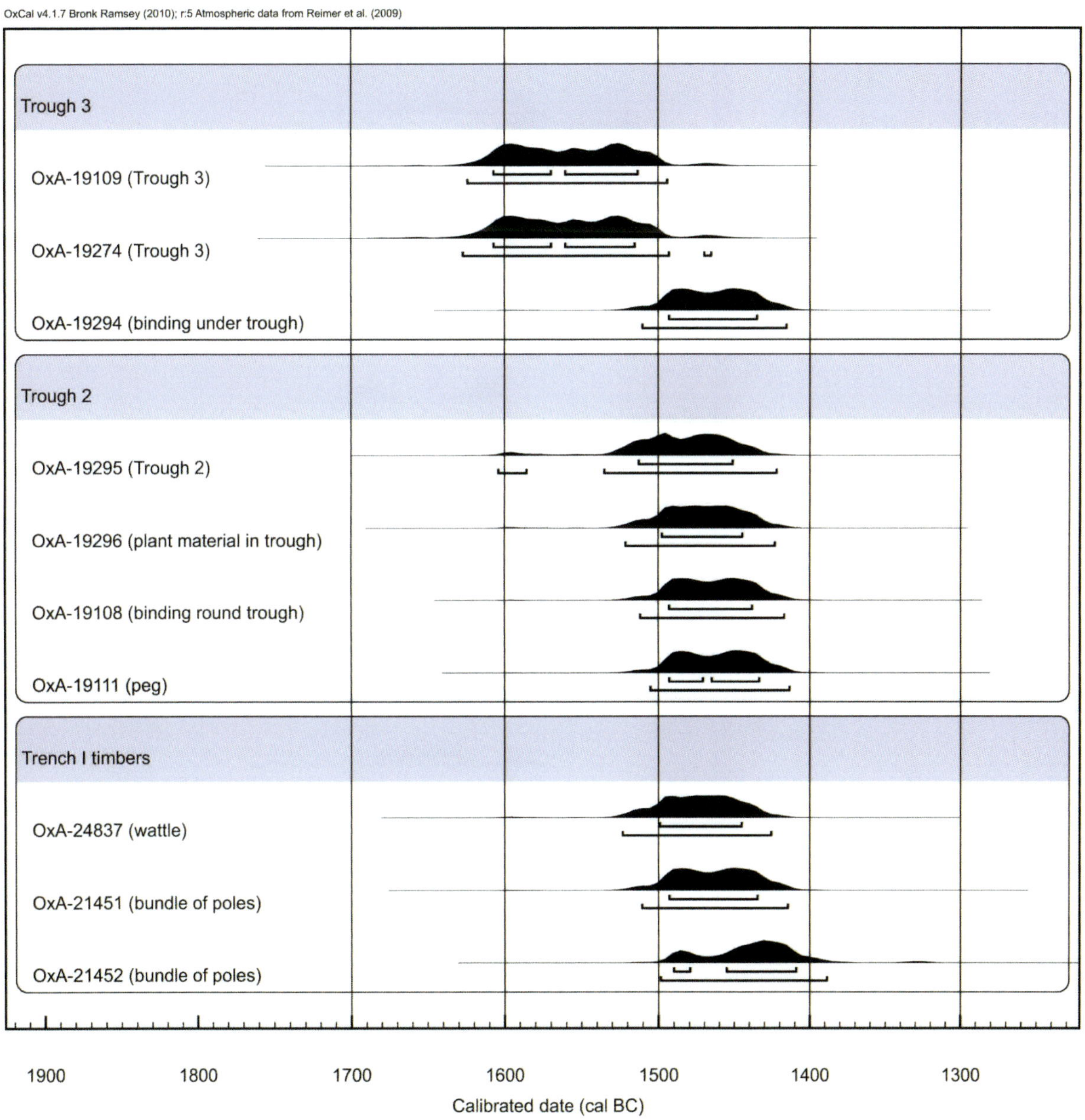

Fig. 4.58. Bronze Age dates from Trench I.

Iron Age

A significant number of features belong to the Iron Age in radiocarbon terms. *Fig. 4.59* shows that all of them fall in the fourth and third centuries cal BC, with a start date nearer 350 than 400. Because the calibration curve is flat in this period, no closer definition is possible. Thus the ladder and the wood-lined

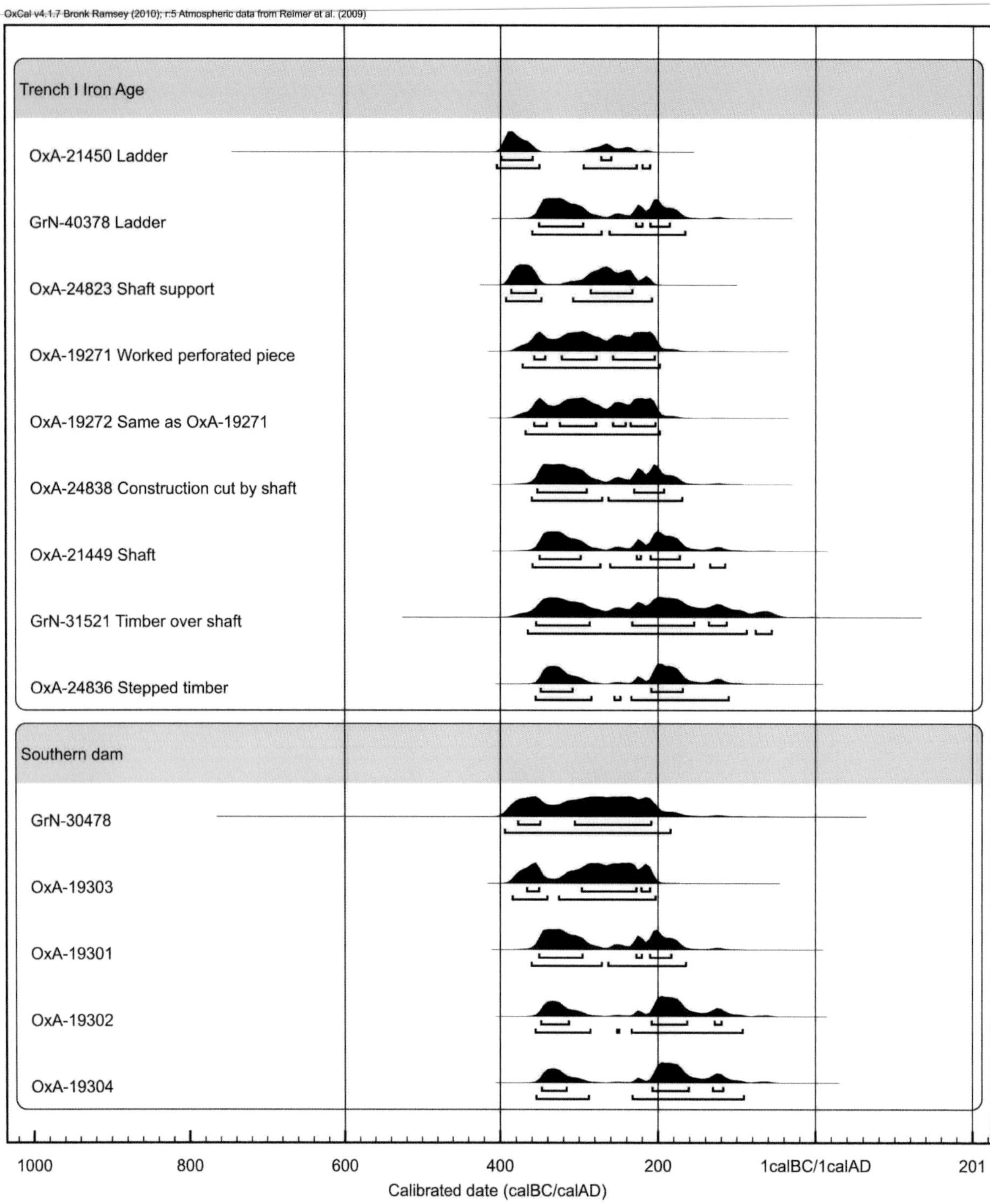

Fig. 4.59. Iron Age dates from Trenches I and XXI.

shaft are, in radiocarbon terms, identical, even though there may well have been more than a century of development and change in this period.

The dating of the 40 sherds of 'Dacian' pottery found close to the wood-lined shaft (box structure) agrees in broad terms with these radiocarbon dates.

Post-Roman / Early Medieval

Fig. 4.60 shows the five dates that belong to the post-Roman / early medieval period, four of them from Trench I. The spread is wide, but one can probably assume that the earliest date (1638 ± 28 BP) is the least relevant since it comes from a substantial post that may represent timber already old when utilised. On the other hand, two dates on bindings round these poles differ considerably in age; the later (1493 ± 28 BP) falling in the second half of the sixth century cal AD. The spread of dates suggests that while there was without doubt an early medieval phase of activity on the site, more samples would need to be dated in order to elucidate the sequence of events.

As explained above, the implication of these datings, on timbers at the same depth as the Bronze Age pieces, indicates that the area was constantly being reopened for continuing exploitation of the underlying rock salt.

Trench XXI (the 'southern dam')

Five radiocarbon dates on this construction (*Fig 4.59*) show that it falls in the same time period as the Iron Age phase in Trench I (fourth and third centuries cal BC). There is no pottery from the trench.

Trench XV

The single available radiocarbon date, on Trough 4, belongs to a much later date than the Trench I troughs, and falls in the 9th century cal BC. It is not yet known if other features in this trench belong to

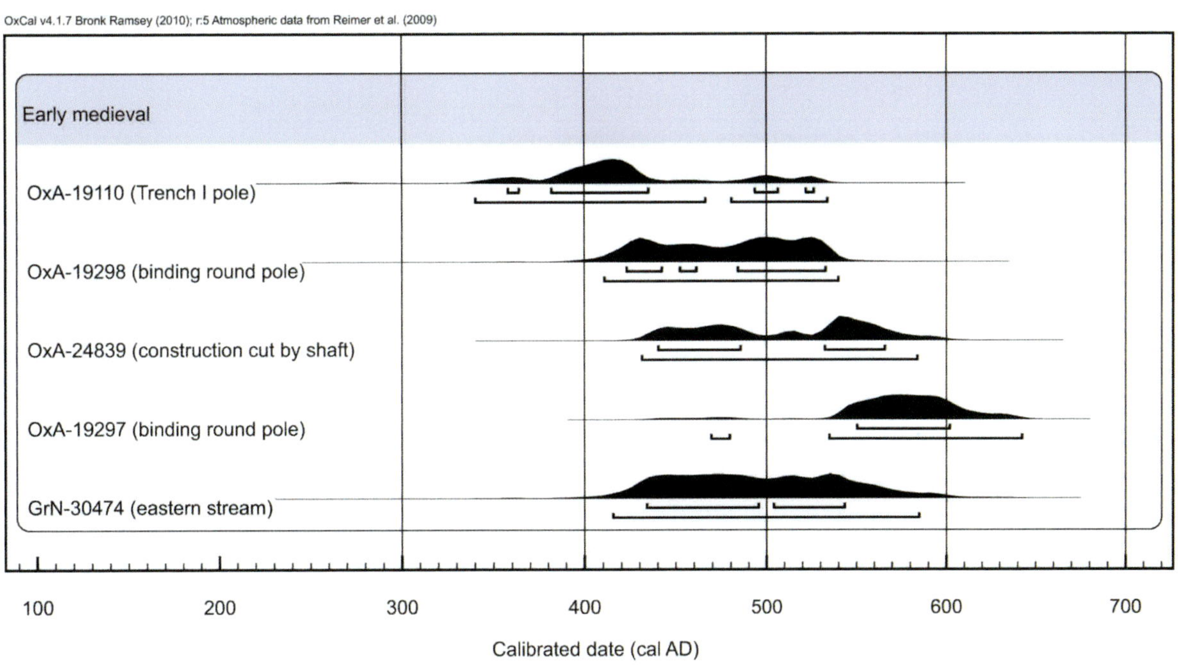

Fig. 4.60. Early medieval dates from Trench I.

a similarly late date. This trough lay in close proximity to that extracted by Ioan Chintăuan in 2005, and may have been used together with it, but its date is rather later. The development of troughs is considered below (Chapter 9).

Central and eastern areas

Fig. 4.61 shows the dates from the constructions dubbed the 'central dam' and the eastern side stream. Timbers in the central plank-built dam fall over a wide range from 1400 to 1100 cal BC, though one date (OxA-19305) is later and belongs to the 11th century. Troughs 1 and 4 appear to fall significantly later than this plank construction, especially Trough 4.

The range in the eastern stream is similarly wide. One timber from that area produced an early medieval date.

Trench III

The features found in Trench III belong to different periods: the final period of the Early Bronze Age, the Late Bronze Age, and the modern period.

Early Bronze Age

Two or three pits (1, 2 and 2a) belong to the final period of the Early Bronze Age, as shown by the many sherds of this date. These suggest a short-lived settlement phase. The pit dug into the rock salt containing Early Bronze Age pottery (including brushed and textile-imprinted sherds) might suggest the exploitation of rock salt on a seasonal basis at this early date.

In terms of absolute chronology, the evidence is contradictory. Two pieces of wood from Trench III have produced surprisingly early radiocarbon dates: an elder branch with no certain context falls in the range 3090–2914 cal BC (95% probability, while the twigs from Pit 2a, associated with Early Bronze Age pottery, falls between 3300 and 2900 cal BC (*Fig 4.62*). Both these dates are much earlier than one would expect for this pottery style in the Carpathian Basin, where a date in the last two-three centuries of the 3rd millennium BC would be considered normal. The elder branch might be seen as fossil timber that was lying around in the stream, incorporated into the salt works at a later date.

The detailed description of this pottery is presented above.

In Transylvania this kind of pottery is assigned to the Iernut group (Ciugudean 1996; 2003) and/or the Zoltan group (Cavruc 1997); in Banat it is widespread within the Gornea-Vodneac (Lazarovici 1998) and Foeni groups (Gogâltan 1996); in Oltenia in the Orleşti-Silea group (Petre-Govora 1988; 1995). Recently Cristian Popa has uggested uniting all these similar groups with different names into a single archaeological unit under the name of Gornea-Foeni group (Popa – Totoianu 2010, 54–5). In Hungary and Slovakia, pottery of this kind is widespread mainly within the early phase of the Hatvan culture (Furmánek – Marková 2008, 45 Obr. 13; Görsdorf *et al.* 2004, 81 Fig. 2).

No radiocarbon dates are yet known in Romania for the Gornea-Foeni group, but the stratigraphical contexts and the morphological features suggest that in Transylvania this group falls between the end of Schneckenberg (regionally Jigodin) culture at one end, and the beginning of the Wietenberg and Costişa-Ciomortan cultures at the other. For the age of Costişa-Ciomortan group in Transylvania – which might be considered as a *terminus ante quem* for the end of the Gornea-Foeni group – there is only one

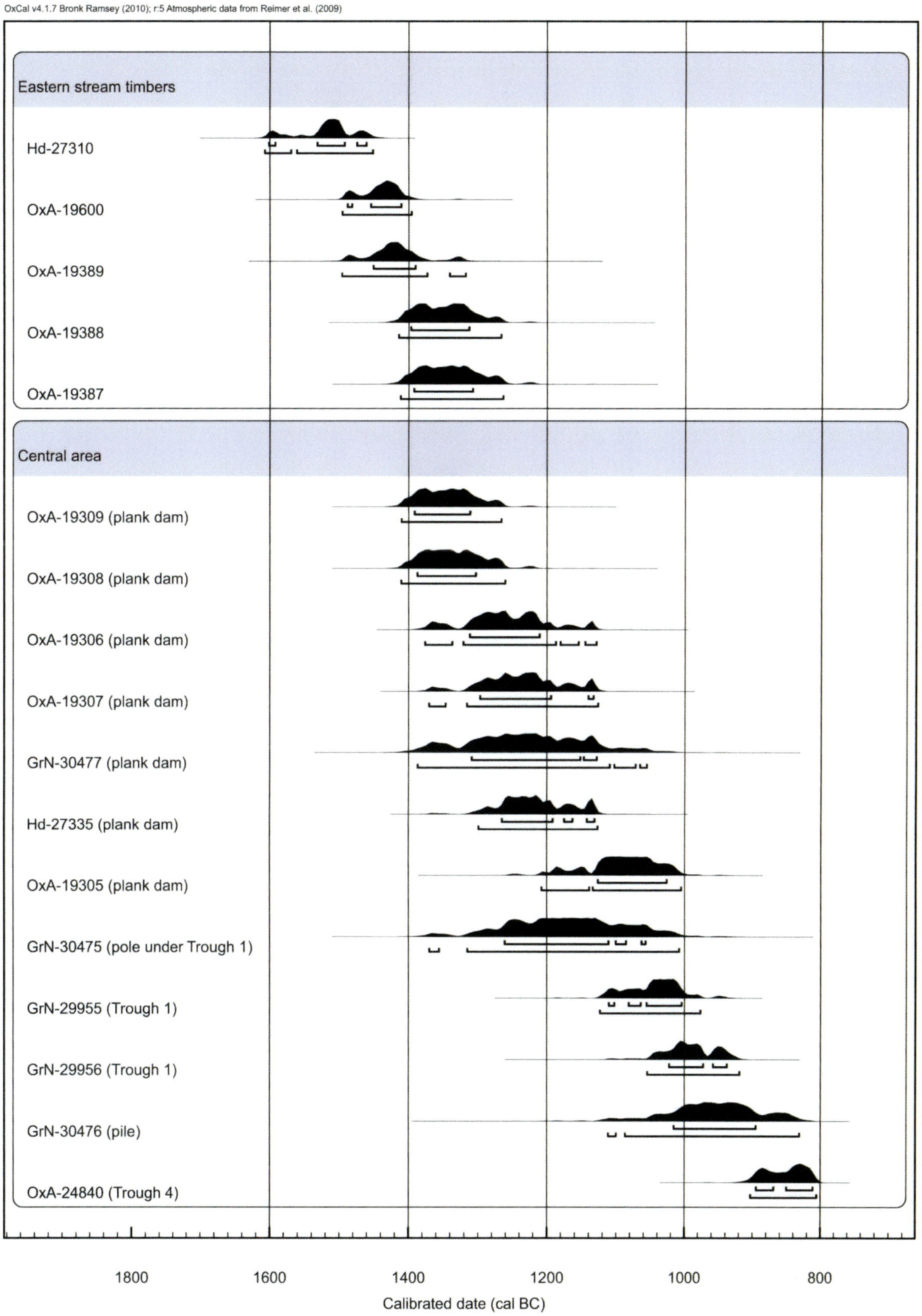

Fig. 4.61. Dates from central and eastern areas at Băile Figa.

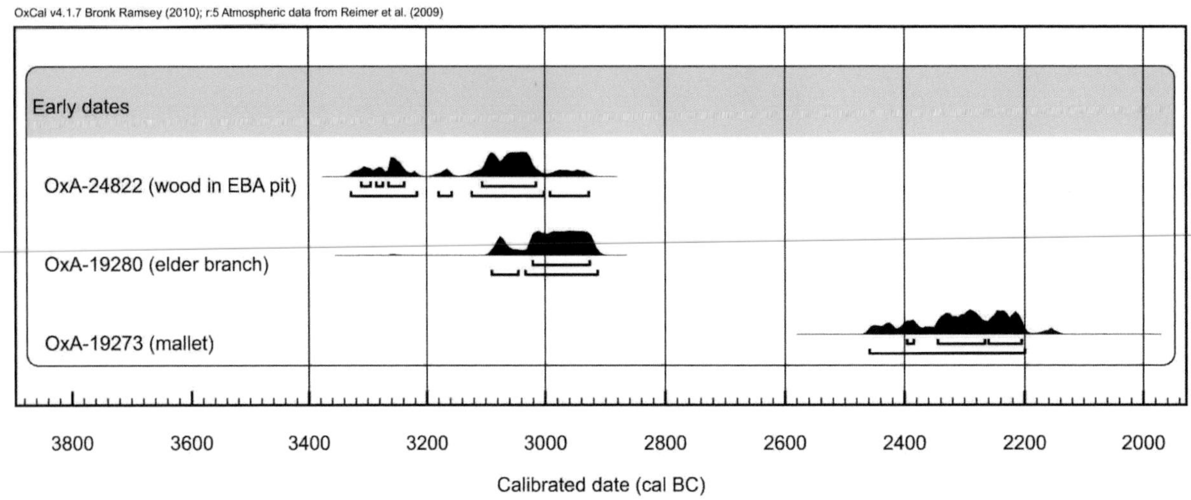

Fig. 4.62. Early dates from Figa Trench III.

available radiocarbon date from Transylvania, from the Păuleni-Ciomortan settlement: 3620 ± 16 BP, or 2031–1925 cal BC at 95% probability.[5]

Radiocarbon dates suggest that the earliest phase of the Hatvan cuture falls within the last third or last quarter of the 3rd millennium cal BC (GÖRSDORF *et al.* 2004, 89). Recent work indicates that the final stage of the Early Bronze Age in Transylvania (BT III) – through its contemporaneity with the Nagyrév, Mureş, Nitra and Hatvan cultures – would date between 2300/2200 and 2000/1900 BC (POPA – TOTOIANU 2010, 106). If that is the case, the Early Bronze Age pottery from Băile Figa should date to the last two or three centuries of the 3rd millennium BC.

The gap between this dating and radiocarbon dates from Trench III is thus a little short of a millennium! It remains to be seen whether the date is relevant or not for the Early Bronze Age finds in Băile Figa. The date 3837 ± 35 BP for the mallet found in Trench I looks much more appropriate to the date of the EBA pottery found in Trench III.

In view of this, it appears that there are two phases within the EBA material from Trench III: an early one, with two radiocarbon dates, that falls around 3000 cal BC; and a later one, based on a single date from Trench I and the conventional dating of the Gornea-Foeni type pottery, falling in the late 3rd millennium BC.

Late Bronze Age

The most consistent dates in Trench III come from Late Bronze Age structures. As discussed above, there are stratigraphical indications that Structures 5 and 2 were the earliest in the trench; the plank-built construction of Structure 2 differs from the wattle structures that fall later. Structures 1 and 3 cut Structure 2, so must fall later than it.

Fig. 4.63 shows the spread of radiocarbon dates for piles and wattle constructions in Trench III. The overall range extends from before 1200 to 800 cal BC, with both groups potentially extending over that entire timespan.

The radiocarbon evidence is modelled in *Fig. 4.64*. It shows that in general all the dated structures fall within a relatively short timespan: Structure 5 in the 11th century cal BC, Structure 2 around

[5] This date was obtained within the Project PN II – program Idei, cod ID_664 – "Atlasul arheologic al României din paleolitic până în anul 1000 p. Chr.", financed by UEFISCSU according to the contract 137 / 01.10.2007).

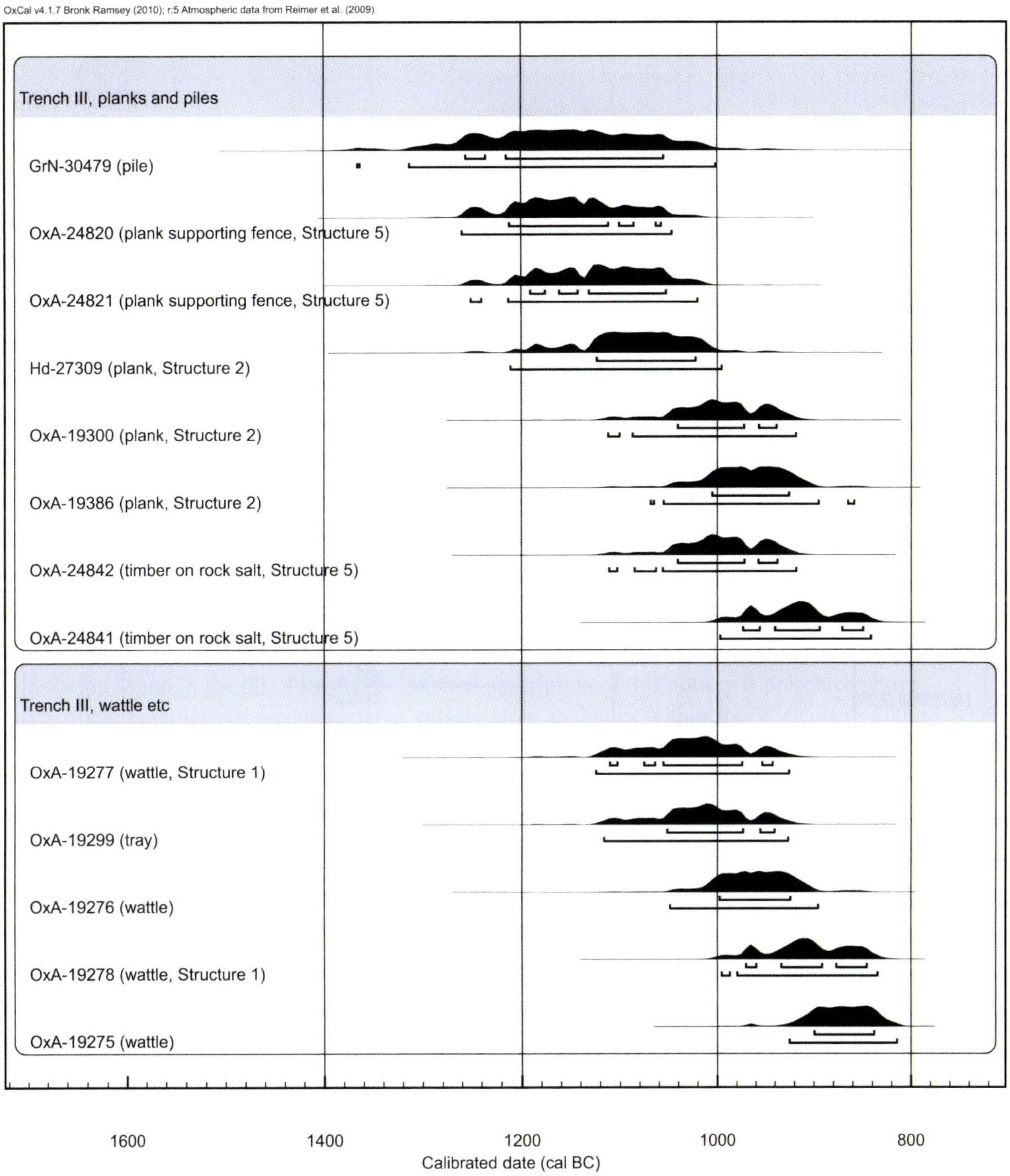

Fig. 4.63. Middle and Late Bronze Age dates from Figa Trench III.

1000 cal BC, and Structure 1 in the 10th century cal BC. It must be stressed, however, that more dates are needed to confirm this model, especially since timbers from each individual structure show significant divergence from one another.

Unfortunately the pottery found in Trench III that might be assigned to the Late Bronze Age is all undiagnostic, so that no comparison with the radiocabon dating can be made.

from various parts of the site where timbers with an adequate number of annual rings were present; not all subsequently turned out to be usable. A smaller amount of sampling was done by Dr Istvan Botár of the Csíki Székély Múzeum in Miercurea Ciuc (Csíkszereda).

Wazny's report appears at Appendix 3. He shows that four groups of timbers were identified, of which the fourth consists of pieces that do not match with any others. The remaining three can be assigned to different use phases of the site.

Group I consisted of 20 cross-matching samples, with a total length of 281 years. Most of the samples came from the radially cut oak planks forming parts of Structure 2 in Trench III, and from another group of timbers in the central stretch of the stream. Sapwood is only present on five samples so the felling date of most timbers is unknown, but it seems clear that these plank fences or dams are of the same period in terms of tree-rings. From the radiocarbon dates above it is known that two timbers in Structure 2 (though not, unfortunately the dendro-dated ones) gave dates in the 29th century BP (1050–920 cal BC at 2σ).

Group II consisted of seven cross-matching samples, with a total length of 95 years. They came from the central dam, which is largely composed of quartered timbers, similar to those in Structure 2 in Trench III. The radiocarbon dates for this feature lie in the 31st and 30th centuries BP (1380–1250 cal BC at 2σ). Thus although the technology of the squared timber walls is similar, there is a significant difference in time, such that timbers of one do not fall in the same group of year-rings as the other.

For Group III, consisting of only three samples covering 102 years, and situated in the eastern tributary, we have the benefit of a radiocarbon date that was obtained for wiggle-matching (Hd-27310): sample 63 gave a date of 3248 ± 23 (1607–1451 cal BC, 2σ). This is 80 or so years earlier than the other radiocarbon dates from this area (31st and 30th centuries BP), but since the tree was sizeable this is not problematic, especially as it is not known from which part of the growth cycle the sample in question was taken.

Although further analysis of these results is awaited, in general it is clear that the chronological order of the dates is this: Group III, Group II, Group I. These groups relate to the site organisation as follows:

Group 1: Trench III, Structure 2
Group 2: Central dam
Group 3: Eastern stream

In due course it should be possible to assign exact years to some of these timbers, though the absence of bark and sap rings on most is a problem. However, the creation of an unbroken tree-ring sequence back into the Bronze Age for the Carpathian Basin will be a task for some years to come.

Figa chronology: conclusions

The diagram *Fig. 4.57* gives a clear picture of the general trend of the development of the Figa site. The earliest constructions on the site are those at the southern end, where the brine spring rises, and investigated in Trench I. These dates all fall in the 16th–15th centuries cal BC. Apparently as early, though continuing later, are timbers from the eastern tributary stream (the discovery of elements of a trough in this area in 2012 would support the idea that this technology began early on the site).

In the 14th–13th centuries, activity shifted northwards to the central area of the site (the plank-built "dam") and perhaps other features. This area of the site has not been excavated, so that no details are available on the nature of the exploitation.

The third main phase of activity was that represented by the structures in Trench III (the northern part of the site). These show quite a wide span, but seem to start in the 13th century, but fall mainly between 1200 and 800 cal BC. The plank-built structures are earlier than the wattle ones by a century or more.

One of the latest dates on the site is represented by Trough 4 from Trench XV, which falls in the 9th century cal BC. This is the last Bronze Age phase of activity; after that there is a gap of several hundred years until activity resumed in the 4th–3rd centuries cal BC, with a much later, poorly defined, phase of use between 400 and 800 cal AD.

The dating thus shows a dynamic development of the site, probably starting at the spot where the main spring emerges from the hills to the south, and moving northwards – as the need for salt on an industrial scale grew, perhaps. The scale of activity we return to in Chapter 10.

Săsarm, district Chiuza, "Valea Slatina"

This locality was also brought to our attention by Ioan Chintăuan in 2005. He had previously listed it in his catalogue of mineral sources in Bistriţa-Năsăud county (CHINTĂUAN 2002,153–6; CHINTĂUAN – RUSU 1988, 273–4). It consists of a salt stream and a well-head (*Figs 4.65–4.66*). The well-head lies 1.8 km north-north-west of the village of Săsarm, 7 km in a straight line due north of Figa, in a valley enclosed on three sides by significant hills rising to over 400 m to east and west. The stream rises near the summit to the south-west at around 350 m asl and runs north-north-east to join a larger stream, the two together flowing east to join the Someşul Mare near Săsarm village.

Large numbers of timber posts are present at the confluence of the two streams, and in most of the course of the stream bed.[7] The well-head is situated some 130 m from the confluence; north-west of it a block of uneroded material separates the main stream from a small side stream. A visit in October 2005 noted an uneroded piece of land near the well containing three areas of blackish soil mixed with ashes and containing Bronze Age, probably Early Bronze Age, sherds (*Fig. 4.67*) (ANDRONIC *et al.* 2006, 70–72, plan and photos on pp. 75–76). Five metres north of these spots, in the vertical bank on the west side of the stream and under a deposit of yellow clay about 3 m thick, there is a strip of black soil mixed with ashes. This strip is 25 cm thick and about 4 m long, and it contains many sherds, again probably Early Bronze Age. We also observed that some isolated Bronze Age sherds were visible in the western bank of the main stream bed about 40 m south of the well-head.

In the western bank of the hollow around the well-head, among a group of timbers which seem to be part of a mining gallery, two isolated Iron Age sherds were found by Dr Vayda (Năsăud Museum), a fragment of an amphora and one sherd of pottery of the type known as 'Celtic'. Isolated fragments of medieval wheel-made pots are also present in the hollow. About 30 m north-west of the hollow a group of sherds from medieval wheel-made pots was identified.

Subsequent visits to the site showed that the area covered by the Early Bronze Age sherds is much larger than previously noted. The black soil strip containing Early Bronze Age sherds continued in the western bank of the main stream bed about 2 m south of the well-head. About 100 m south of the well-head, in the eastern bank of the stream, Early Bronze Age sherds are also present. All the spots which contain Early Bronze Age pottery seem to belong to the same cultural layer, which was disturbed and sometimes covered by soil caused by human activity in subsequent periods. In most cases the soil appeared to include black ash, which suggests intense burning activity on the site. Could this have been related to the evaporation of brine?

In 2006 a total station survey was conducted on the northern part of the site, and the visible timbers catalogued (*Fig. 4.68*).[8] Because of the extensive nature of the site, and the fact that some constructions

7 Regrettably, some of this wood has now disappeared as there has been activity to improve the track leading from Săsarm village to the farms north of the confluence; the area is not protected as an archaeological site (noted 2010).

8 The catalogue is kept in the Museum of the Eastern Carpathians, Sf. Gheorghe. A copy will be deposited in Bistriţa museum.

Fig. 4.65. General view of Săsarm, showing well and surrounding area.

Fig. 4.66. Săsarm, well with local peasant collecting brine.

Fig. 4.67. Săsarm, sherd scatter near the well.

were obviously modern (reinforced concrete is visible at one point), only the northernmost part was surveyed in detail, though general notes were made of the main groups (complexes) of wood further to the south. In general the impression gained was that the stream at Săsarm is much more active than any other we have come across, the slope being much steeper (a drop of some 110 m vertically over 650 m horizontally) with the result that the sides of the valley are eroded and the stream bed deeply scoured. Except at and near the bottom of the valley, near the well, we found no artifacts at Săsarm – which we attribute to this erosion process.

Fig. 4.68. Săsarm, northern area, topographical map.

A hillock lies between the well and the stream junction. In the cut that separates this hillock from the more southerly valley side large timbers are visible, and at least one area where a shaft, revetted with wattle, goes into the hillside (*Fig. 4.69*). A little further north, on the hillside to the west, lie a number of fashioned timbers, including a channelled piece (*Fig. 4.70*). These elements have not yet been dated, and may well emanate from modern workings.

The wood pieces upstream of the wellhead are many and complex, including lines of timber piles running across the stream (*Fig. 4.71*), shafts going in to the valley side, piles and other timbers. We identified some seven separate 'complexes' of timber (not all the elements in any one complex belong to the

Fig. 4.69. Săsarm, wattle fence supporting gallery cut into the stream side.

Fig. 4.70. Săsarm, channeled wood to the north of the well area.

Fig. 4.71. Săsarm, wooden construction remains north of the well area.

Fig. 4.72. Săsarm, wooden trough or channel observed in 2012.

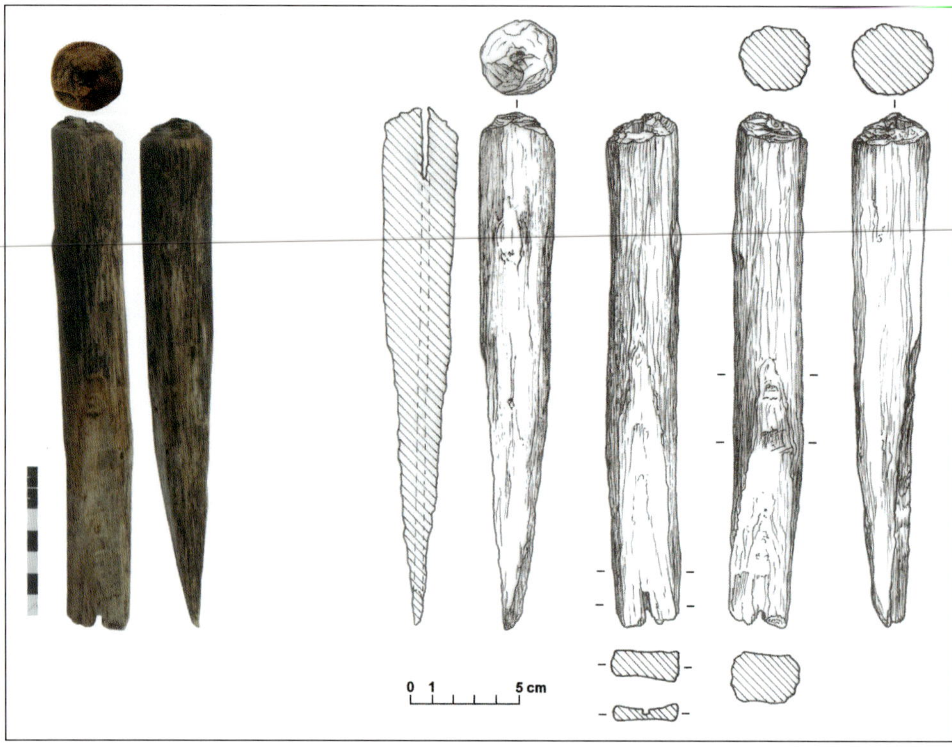

Fig. 4.73. Săsarm, wooden peg found in the stream bed in 2012.

same construction) in this part of the stream. These features deserve full examination in the future, as they probably come from a range of dates – some certainly modern, others much older.

On a visit in 2012 to sample brine from the well, and to collect salt-impregnated sherds, close to the spot with medieval pottery about 30 m north-east of the well-head, the end of a large curved wooden object was noted, possibly a trough (*Fig. 4.72*). The projecting end has been cut off straight, and it was not possible on a brief investigation to see any holes – but it appears that the central part of what would be the base is missing, so this is not conclusive. The body is made from a massive tree-trunk, hollow inside, and the outer surfaces appear to be dressed smooth. The thickness of the wood is, however, only about 4–5 cm, which would be slight for a trough. In addition, a peg was recovered from the stream bed (*Fig. 4.73*). This provides unambiguous evidence that the trough technique was used at Săsarm as at the other sites considered here, even though no certain trough has so far been recovered.

In 2009 samples were taken for dating and funding for three was provided by the Natural Environment Research Council (via the ORADS facility). The samples came from posts sunk in the stream bed, in the vicinity of the well-head. The dates are as follows:

Table 4.3. Radiocarbon dates on piles from Săsarm (calibration OxCal v.4.1.7).

Site no.	Lab no.	$\delta^{13}C$	Date BP	Date cal BC 1σ	Date cal BC 2σ
Wood 5	OxA-24824	-23.46	3101 ± 28	1419–1320	1435–1305
Wood 98	OxA-24825	-25.00	3014 ± 27	1370–1214	1385–1132
Wood 50*	OxA-24843	-25.27	3148 ± 28	1451–1401	1496–1326
Wood 50*	OxA-24844	-25.31	3119 ± 27	1431–1326	1449–1314

* Wood 50 was dated twice for laboratory reasons. The combined value of the two dates is 1452–1379 at 92% probability (OxCal R_Combine procedure).

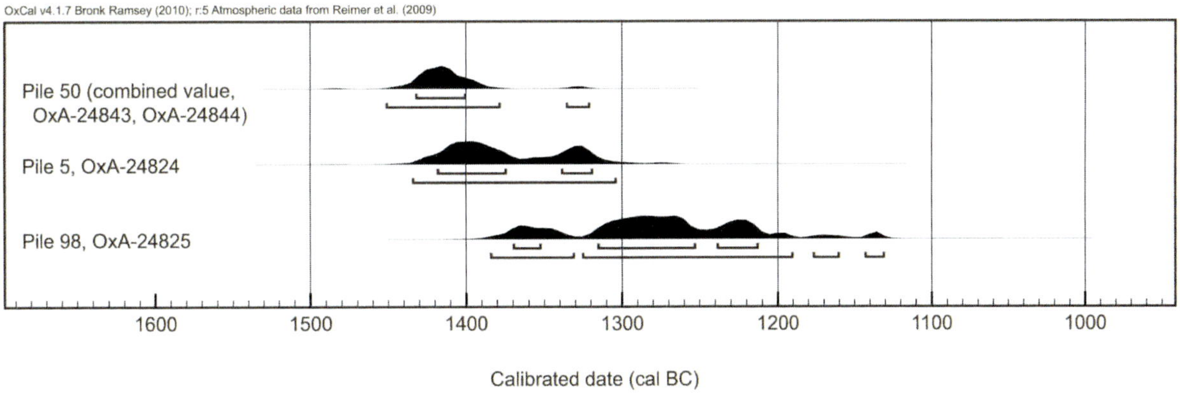

Fig. 4.74. Probability distribution of Săsarm radiocarbon dates.

As can be see from *Table 4.3* and *Fig. 4.74*, the three timbers all date to the Bronze Age and fall in the same period as those from Trench I at Figa. While Wood 50 is potentially earlier than Wood 98, not enough dates are yet available for any conclusions to be drawn about the chronology of the site, other than that salt-working appears to have taken place during the Middle Bronze Age.

Caila, commune Şintereag, "La Sărătura"

The site at Caila had been listed as a salt source by Chintăuan (CHINTĂUAN 2002, 129 f.; CHINTĂUAN – RUSU 1988, 256). The archaeological elements were noted in 2007 by Andrea Chiricescu, ethnographer at the Museum of the Eastern Carpathians, Sfîntu Gheorghe. The site lies 1.5 km west of the village of Caila and 10.75 km in a straight line east of Figa, in a salty stream that drains a marshy area lying in a bowl-like depression; the stream runs northwards, joining the Roşua, a small river that flows westward to join the Şiu at Şintereag. A well-head in a hut of the usual form lies on the stream bank (*Fig. 4.75*).

Fig. 4.75. Caila, general view of the stream and well.

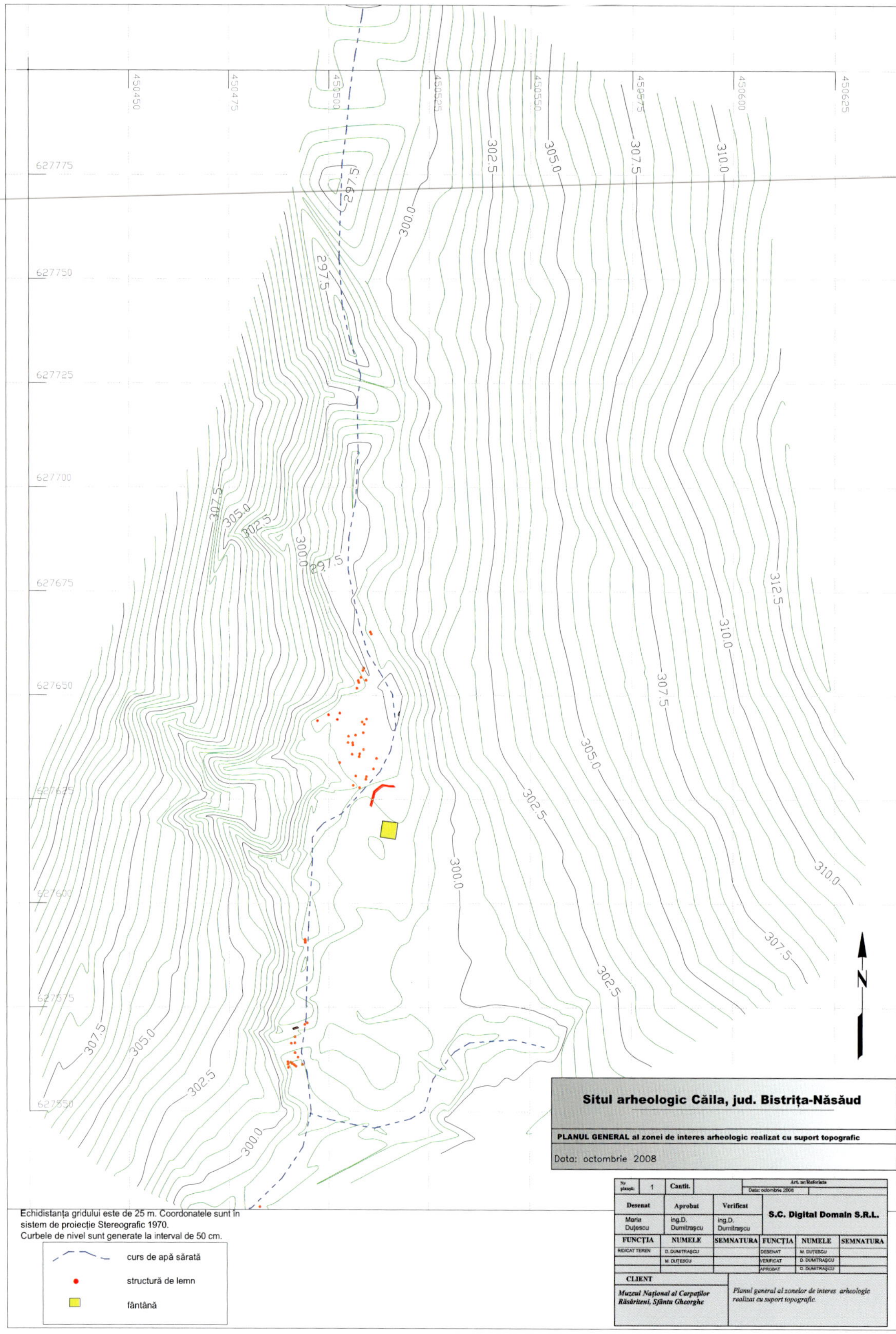

Fig. 4.76. Caila. Topographical plan showing the well and major timbers.

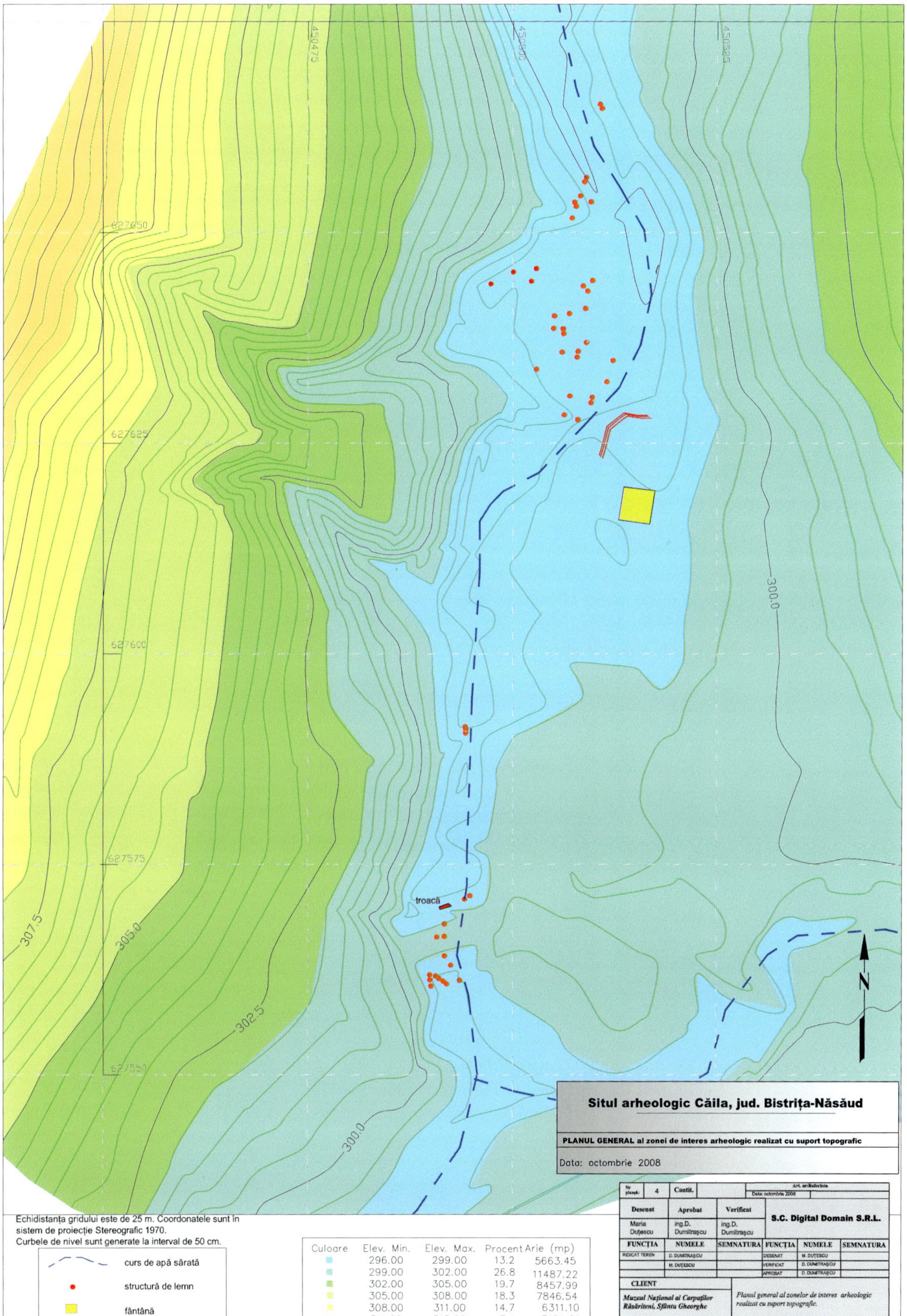

Fig. 4.77. Caila, detailed plan of the area with timbers.

The stream contains many timbers over a stretch some 200 m long (*Fig. 4.76*). There is a concentration of timbers, large and small, immediately to the north of the well, and another at the southern end of the site (*Fig. 4.77*). These do not form any visible structure, except that in a few places posts are apparently placed in a line (*Fig. 4.78*). In at least one place there appear to be the remains of wattles (*Fig. 4.79*), and in 2008 bindings were observed in the eroding stream bank (*Fig. 4.80*; barely visible in 2009).

Fig. 4.78. Caila, timbers in the stream bed.

Fig. 4.79. Caila, wattle remains in the stream bed.

Most remarkable was the discovery of a trough protruding from the low western stream bank (*Figs 4.81–4.82*). At the time it was first noticed, just one square base hole was visible; as time has gone on more of the bank has been uncovered and more holes are now visible. In addition, on our first visit to the site, a large piece of timber with square holes cut into it, evidently part of a trough base, was recovered (*Fig. 4.83*).

Fig. 4.80. Caila, twisted bindings visible in the stream side (2008).

Fig. 4.81. Caila, view of the stream with trough end as first discovered (2008).

Fig. 4.82. Caila, trough as visible in 2009.

Fig. 4.83. Caila, trough base found in the stream bed, 2008.

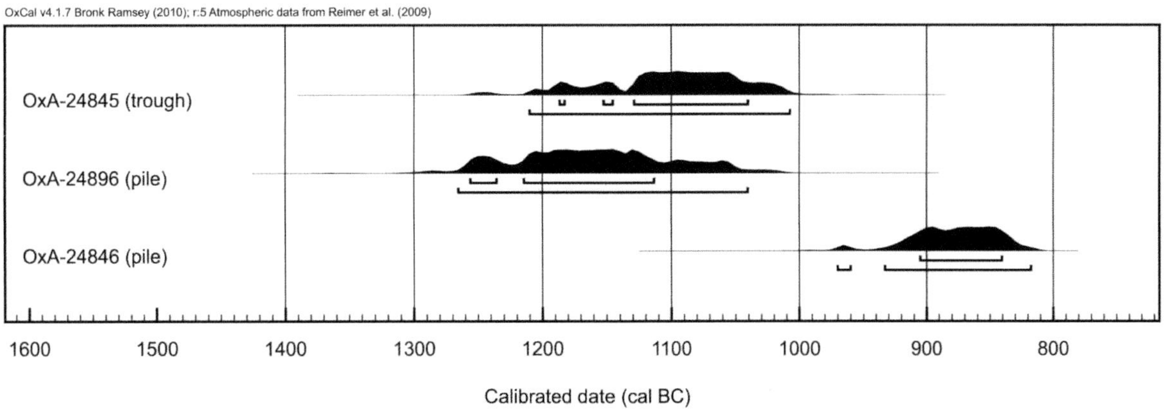

Fig. 4.84. Caila, probability distribution of radiocarbon dates.

Samples taken from the trough and two of the piles were submitted for dating in 2010; the results can be seen in *Table 4.4* and *Fig. 4.84*.

Table 4.4. Radiocarbon dates on timbers from Caila (calibration OxCal v.4.1.7).

Site no.	Lab no.	$\delta^{13}C$	Date BP	Date cal BC 1σ	Date cal BC 2σ
Trough	OxA-24845	-23.78	2907 ± 27	1187–1041	1211–1008
Pile	OxA-24846	-25.52	2740 ± 28	906–841	971–818
Pile	OxA-24896	-25.60	2943 ± 32	1257–1114	1266–1041

This trough lies very close in date to those from Figa. It is also identical in age to one of the piles; the other pile dates later by at least two centuries, probably indicating an extended period of use at the site.

Blăjenii de Jos, commune Şintereag

2.3 km north of the village of Blăjenii de Jos, and 5 km north of Caila, lies the botanical reservation "La Sărătura", a 20 ha area created a few years ago to preserve the unusual halophile flora, including the Baltic-central European species *Armeria maritima* Wild (http://transilvania-ro.romaniaexplorer.com/page_11109.html, accessed 10 January 2011). The immediate area is rich in salt springs, and includes a salt well-head (*Fig. 4.85*). This is of the usual square form, made of horizontal adzed oak beams, with a well inside; outside, the tops of two rows of piles project a short way out of the earth. In drainage channels nearby extensive remains of wattle may be seen, and at isolated places individual poles project from the stream bed (*Figs 4.86–4.87*).

The area was noted by Chintăuan as a source, but he makes no mention of ancient timber (Chintăuan 2002, 127 ff.; Chintăuan – Rusu 1988, 256).

Too little remains of this complex to say anything more than that its surface remains are similar to those from nearby Caila and Figa. Satellite photos from 2003 available on Google Earth show that at that time the area was extensively dug over, presumably by local people for salt extraction. From the wood remains visible, it seems likely that this was originally also the site of historic production.

1 km south of the village at the locality "La Şes" a Late Bronze Age hoard of five axes and two fragments was found, while two similar axes were thrown away by the finder (http://www.cimec.ro/scripts/arh/rar-index/sel.asp?Lang=EN&nr=8&NrSel=0&IDRap=1608) (accessed 24 July 2012).

Fig. 4.85. Blăjenii de Jos, timbers in the ground beside the well.

Fig. 4.86. Blăjenii de Jos, wattle in the stream side.

Sărățel, locality Iocul (47° 02′ 54″ N, 24° 26′ 37″ E)

The area was noted by Chintăuan as well as earlier authors (CHINTĂUAN 2002, 137 ff.; CHINTĂUAN – RUSU 1988, 260 ff.), and a Roman votive altar comes from nearby, recording a slave by the name of Atticus who was the agent of one P. Aelius Marius, a *conductor pascui et salinarum* (RUSSU 1956);

a similar situation to that at Sânpaul (above). Chintăuan noted the salt lake at the localities Făget or Bobeica, and other spots with brine springs and halophyte plants. In the leisure park there are a number of former salt pools, some containing wood (*Fig. 4.88*). A short way to the east, low-lying ground near the river contains a number of wooden structures, including what appears to be a roughly square construction (*Fig. 4.89*). According to those who have inspected the area in the past, there used to

Fig. 4.87. Blăjenii de Jos, wattle scatter.

Fig. 4.88. Sărățel, pond (former salt mine), with wood preserved.

Fig. 4.89. Sărățel, wooden box construction.

be many more wooden structures visible; recent earth-moving is apparent, and this has presumably destroyed or at least covered the wood.

No samples have been taken from this spot for dating so that the age as well as the character of the structures is completely unknown.

Other sites

Chintauan has noted many other places in Bistrița-Năsăud county with salt springs or wells of various kinds, many covered in the usual way by wooden buildings, but only one where wooden poles were in recent times preserved in the ground: Pinticul Tecii (district Teaca, 14 km south-east of Sărățel) (CHINTĂUAN 2002, 144 ff.; CHINTĂUAN – RUSU 1988, 265–6). We have visited this spot but do not know if the piles depicted might be the remains of ancient workings as at Figa, or the foundations of a well-house; appearances suggest the former.

At Dumitra, 10 km north of Bistrița between Bistrița and Năsăud, there is an area of exposed rock salt in the centre of the village, as well as a well-house (CHINTĂUAN – RUSU 1988, 252–3) (*Figs 4.90–4.91*). There is no indication of ancient exploitation of this source, but on a visit by Andrea Chiricescu in 2007 a local man was filmed working the rock with a 15 kg iron axe, and interviewed to establish how and why he was working (DEÁK 2008). This illustrated graphically how hard it is to break up the rock even with modern tools. The same has been noted at Figa at the time the rock salt was discovered and a mechanical digger could do no more than make scratches in its surface (see Chapter 8).

Beyond Bistrița-Năsăud county to the east, a large salt source at Orşova in Mureş county has a brine well with many timbers in the nearby mud, including *in situ* wooden piles and channeled pieces (CHIRICESCU 2006, 164–5; 168 photos 6–8) (*Figs 4.92–4.94*). We have inspected the site but not collected any samples for dating. Nothing is currently known about the date of the timbers, and there is no evidence at present that they are ancient.

Transylvania – discussion

The evidence obtained in our own recent fieldwork, combined with that from previous work, indicates a remarkable situation in terms of prehistoric salt exploitation. While the finds from Valea Florilor have long been known, their chronological context has only recently become clear with the radiocarbon

Fig. 4.90. Dumitra, brine well house.

Fig. 4.91. Dumitra, rock salt exposure in the centre of the village, still worked by locals.

Fig. 4.92. Orşova, brine well house.

Fig. 4.93. Orşova, channeled timber.

Fig. 4.94. Orşova, timbers visible in stream bed.

dating of material from that site, the sites in Bistriţa-Năsăud county, and the material from Királyvölgy. As far as the method of production is concerned, much remains to be elucidated, particularly in terms of how the troughs were used and what the purpose of the wattle and plank constructions was. These matters may become clearer if more extensive excavation becomes possible at Băile Figa and elsewhere.

It is a matter of interest and debate to what extent the Transylvanian material is different from that in adjacent areas. We assume it is unlikely to be unique, especially as at least one trough has been found north of the Tisza in present-day Ukraine. No troughs have so far been found in Moldavia, but there are plenty of indications that wattle constructions were used, probably for revetting adits, for instance at Cucuieţi, jud. Bacău, 'Slatina Veche' (Roxana Munteanu in ANDRONIC *et al.* 2006, 62–4, 86; MUNTEANU *et al.* 2007). The extensive preservation of wood at Figa and Caila is unusual on a European scale, but it seems unlikely that it would not be repeated in Moldavia or other nearby provinces. We cannot say at present whether this is merely the result of chance, but it is worth noting that several teams have been working in Moldavia for some years without coming across the wood remains present in Transylvania.

By the same token, it is not currently possible to say whether the discoveries in Bistriţa-Năsăud county (evidently also present at Valea Florilor) might be repeated in Mureş or Harghita counties, where we have shown that wooden remains from ancient and early modern mining activity survive at Sânpaul.

There is no indication at present that the trough technique outlived the Bronze Age, and certainly none that the Dacian and Roman periods used it – indeed, the evidence from Figa Trench I suggests that by that time, the technique used was that of digging shafts down onto the rock salt, presumably to extract lumps of rock for manipulation elsewhere. But the numerous salt sources in those counties, and the as yet undeveloped state of research into ancient salt, makes it likely that something of the sort will turn up. There are of course other major salt sources in Transylvania, notable at Ocna Sibiu and Ocna Mureş, in the vicinity of which archaeological work has begun in recent years; we await the results of these researches with interest.

Salt sources in relation to archaeological sites in Bistriţa-Năsăud county

Marion Uckelmann

As part of the analysis of the situation at Băile Figa and the other prehistoric salt production sites in the Beclean area of Bistriţa-Năsăud county, a set of GIS maps was created. The programme used was gvSIG OA Digital edition 2010, and the information on site and salt sources came from a variety of sources, including a series of articles by Gheorghe Marinescu (Dănilă – Marinescu 1974; Marinescu 1980; 1984; 1988; 1993; 1994; 1995; Marinescu – Dănilă 1974; 1976), records in the *Repertoriul Arheologic Naţional*, and information derived from Bogdan Sandric, then of CiMEC (Institute for Cultural Memory, Bucharest)[9].

Fig. 4.95 shows all archaeological findspots from different periods in Bistriţa-Năsăud county. The majority of finds came from the lower river valley areas, in the western and southern parts of the county. The few finds in the north-eastern parts are mainly single finds. The Roman sites, on the upper line of finds in the north-west, are all towers and belong to a larger protected area, against threats from the higher regions.

The totals by period for Bistriţa-Năsăud county (up till 2011) are as follows:

- Neolithic: 76 (16 find spots; 21 settlements; 39 single finds)
- Neolithic/Eneolithic: 14 (all single finds)
- Neolithic/Bronze Age: 29 (in map listed under Neolithic: 4 settlements; 25 single finds)
- Eneolithic: 52 (17 settlements; 33 single finds; 1 find spot; 1 unknown)
- Eneolithic/Bronze Age: 25 (in map listed under Eneolithic: 3 find spots; 2 settlements; 20 single finds)
- Bronze Age: 391 (91 find spots; 6 fortified settlements; 9 cemeteries; 22 hoards; 3 salt production sites; 118 settlements; 5 settlements with graves; 110 single finds; 25 unknown find circumstances)
- Bronze Age/Hallstatt: 4 (in map listed under Bronze Age; 3 single finds; 1 hoard)
- Hallstatt: 62 (6 fortified settlements; 4 cemeteries; 1 salt extraction site; 47 settlements; 3 single finds; 1 unknown find circumstances)
- La Tène: 52 (12 fortified settlements; 8 cemeteries; 1 salt extraction site; 2 settlements with graves; 27 settlements; 1 find spot; 1 single find)
- Roman: 155 (14 find spots; 7 fortifications; 3 cemeteries; 4 hoards; 1 salt production site; 74 settlements; 5 settlements with graves; 20 single finds; 27 towers)
- Migration/Early Medieval: 42 (1 fortified settlement; 4 cemeteries; 30 settlements; 6 settlements and graves; 1 find spot)
- Medieval: 13 (8 fortifications; 5 settlements)

[9] Now no longer in existence.

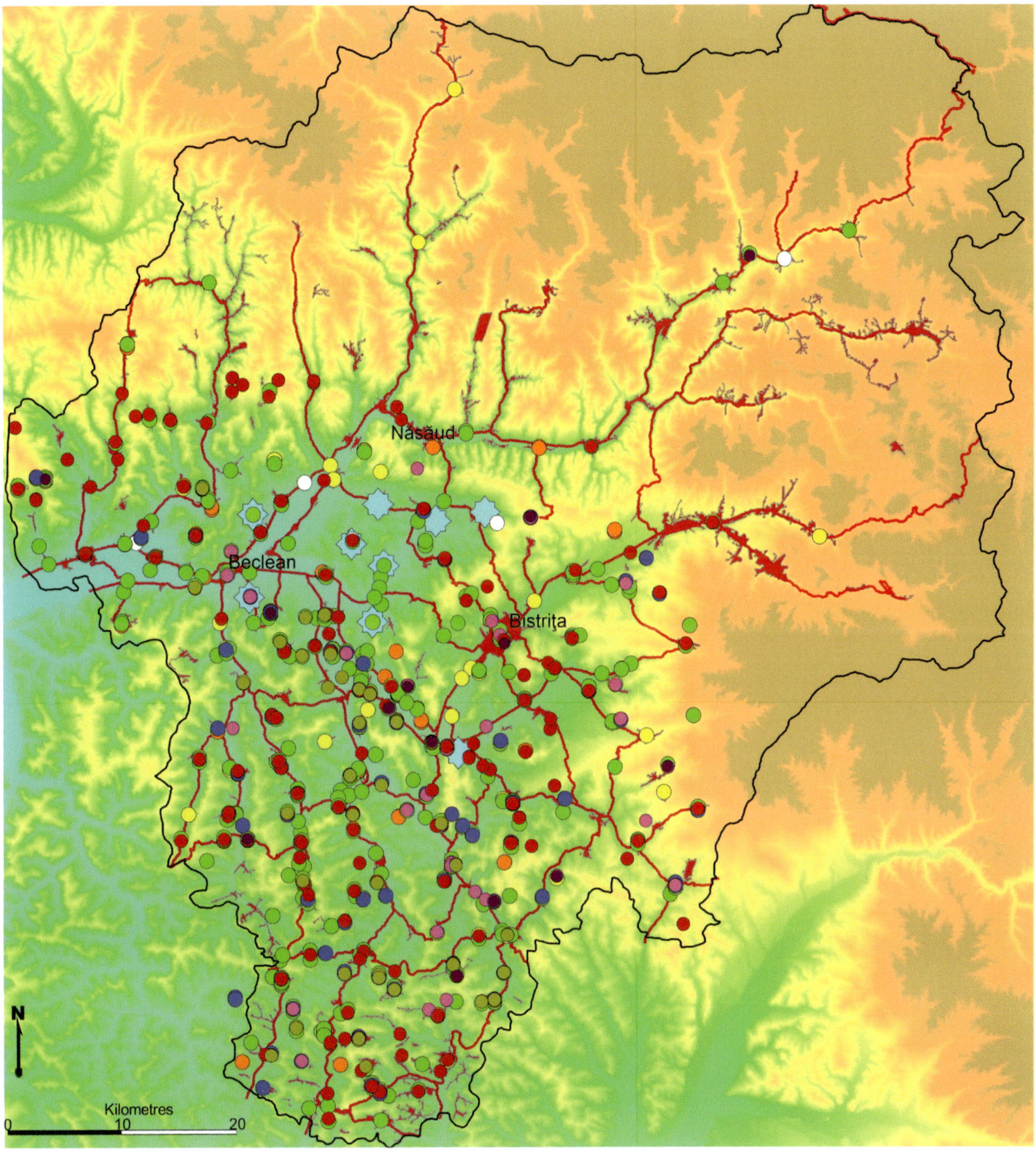

Archaeological finds and salt springs in Bistriţa-Năsăud county, Romania

- Medieval
- Migration and Early Medieval
- Roman
- La Téne
- Hallstatt
- Bronze Age
- Eneolithic
- Neolithic
- Prehistoric
- Salt springs

Fig. 4.95. Findspots in Bistriţa-Năsăud county, all periods.

9 sites can only be assigned a prehistoric date (7 settlements, 2 cemeteries).

Bronze Age sites are thus by far the most strongly represented in the county. In terms of site numbers, especially settlements, this period saw denser settlement than either before or after.

Fig. 4.96 shows Bronze Age finds and salt springs in Bistriţa-Năsăud county. The dominant Bronze Age finds are shown here according to find context. Settlements are quite evenly distributed over the lower areas, as are single finds and findspots. Cemeteries and settlements with graves are less common, and loosely distributed, with more in the southern part. There are only a few fortified settlements. Hoards are mostly found in the western parts, north of the Someş, with a few in the south. There are three salt production sites: Băile Figa, Săsarm and Caila; settlements are found a few kilometres away. One fortified settlement near Beclean (Coldău) is only a few kilometres away from Băile Figa and Săsarm.

Fig. 4.97 shows finds and salt springs between Beclean and Bistriţa. The finds in this area, as in the rest of the county, are found in the valleys. Only occasional finds and sites are located on higher areas. Most finds turned up near roads and in settled areas, and many were found during construction works. Bronze Age finds are dominant. Most salt springs show prehistoric activity nearby, mostly from the Bronze Age (Băile Figa, Săsarm, Taure, Blăjenii de Jos and Caila).

Fig. 4.98 shows Bronze Age finds and salt springs between Beclean and Bistriţa. Settlements, single finds and findspots are distributed over the whole area, with concentrations on the modern settled areas. Settlements are found a few kilometres away from the three salt production sites. One fortified settlement near Beclean (Coldău) is only a few kilometres away from Băile Figa; another lies south of Caila, near Şieu-Măgheruş, while two settlements lie close to salt springs (Taure and Blăjenii de Jos).

Fig. 4.99 shows finds and salt springs between Beclean and Caila. Bronze Age finds are most common, and Migration/Early Medieval period finds are often found on the same site.

Fig. 4.100 shows Bronze Age finds in the micro-region around Băile Figa. Settlements are not far away from the known salt production sites, and the number of findspots indicates a larger amount of Bronze Age activity in the area. Two settlements are close to salt springs (Taure and Blăjenii de Jos). The single finds are mainly axes. The hoard from Blăjenii de Jos consists of five socketed axes (cf above).

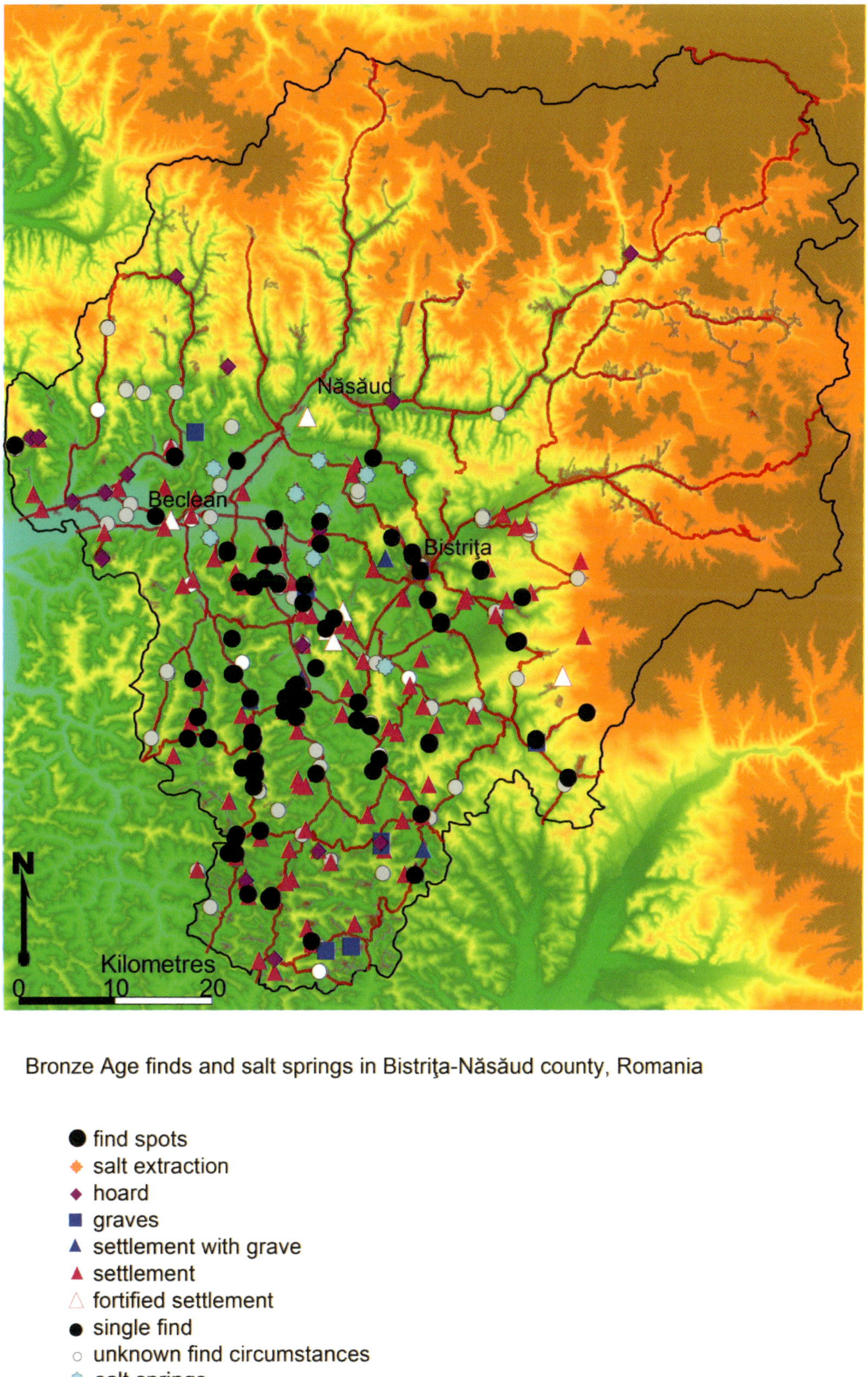

Fig. 4.96. Bronze finds and salt springs in Bistrița-Năsăud county.

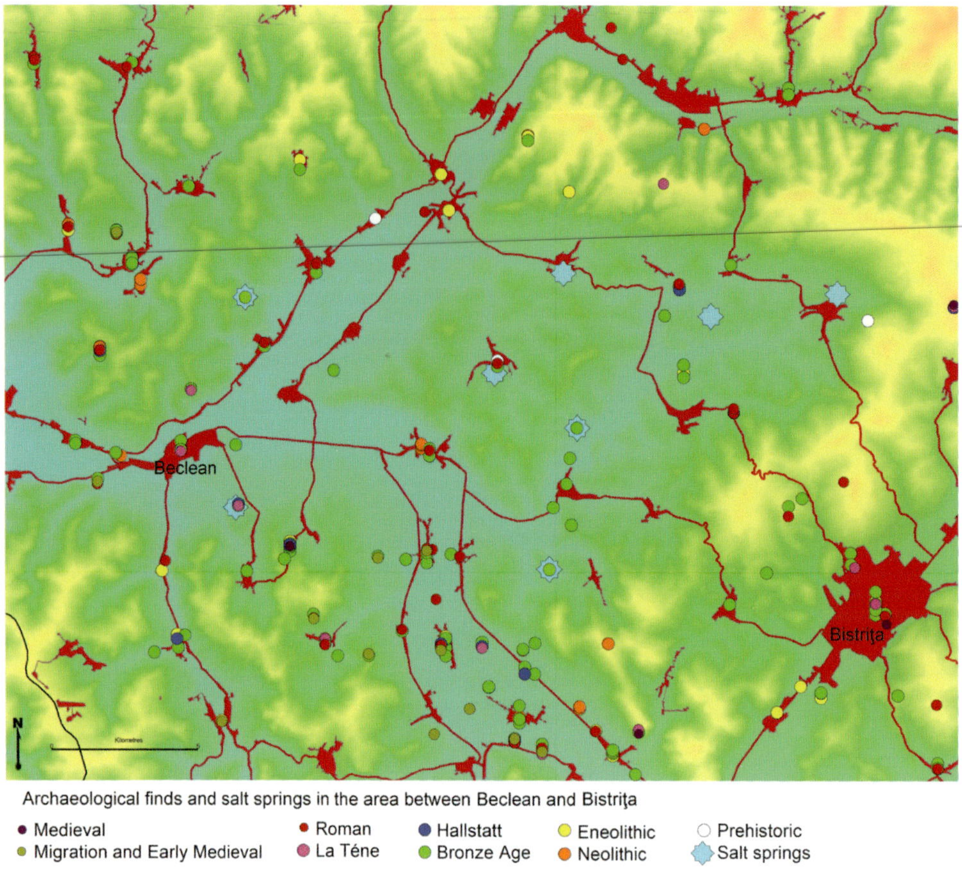

Fig. 4.97. Findspots in the area between Beclean and Bistriţa, all periods.

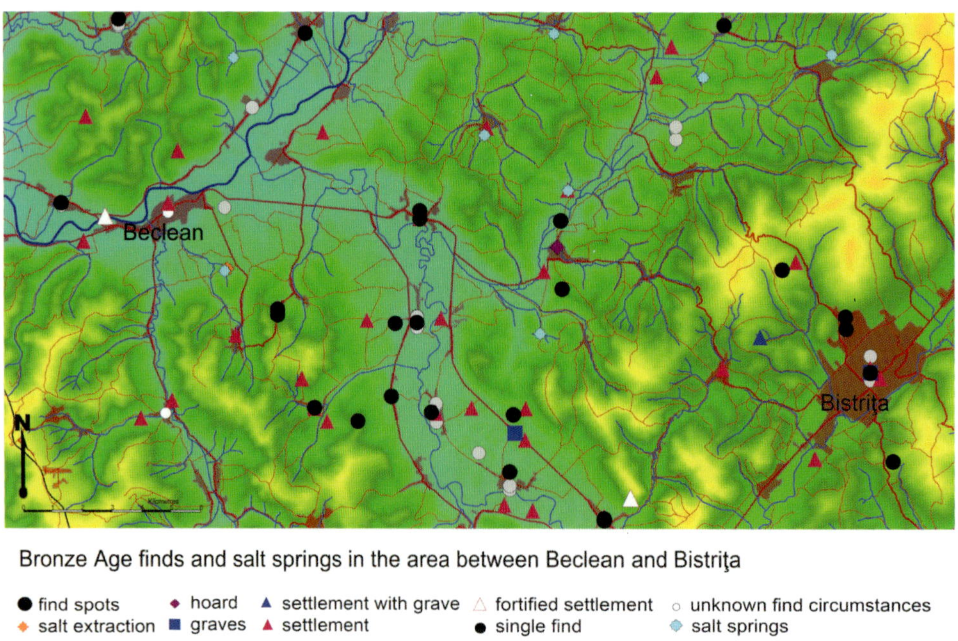

Fig. 4.98. Bronze Age finds between Beclean and Bistriţa.

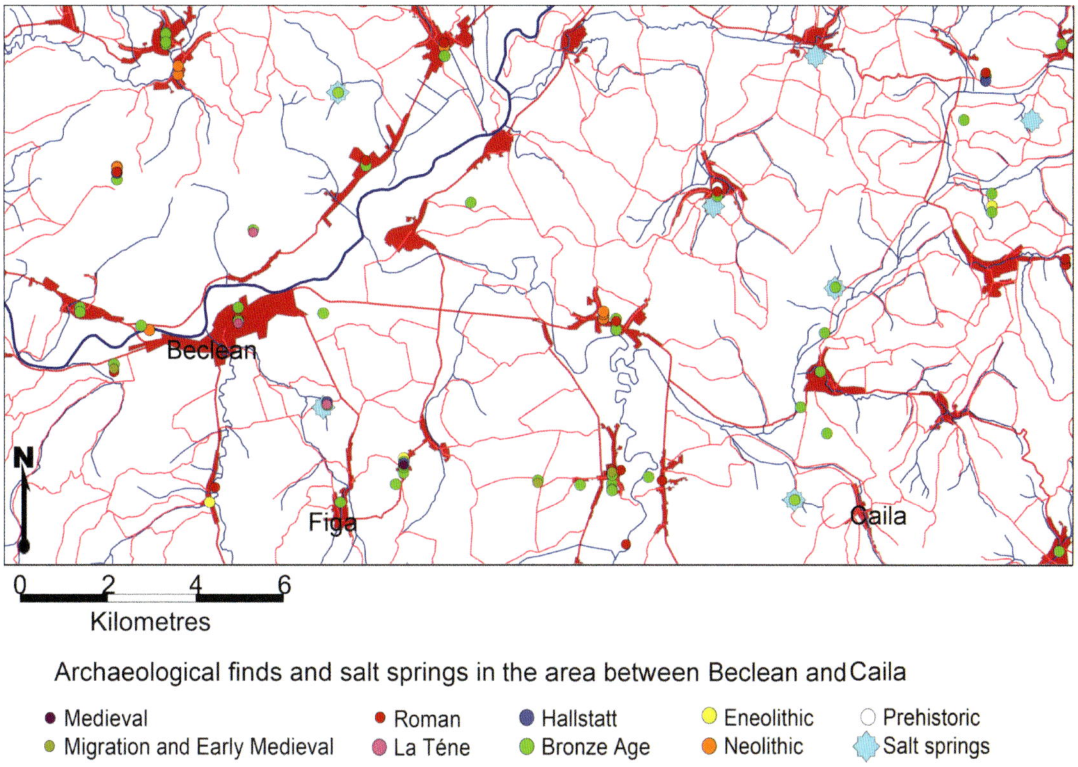

Fig. 4.99. Findspots between Beclean and Caila.

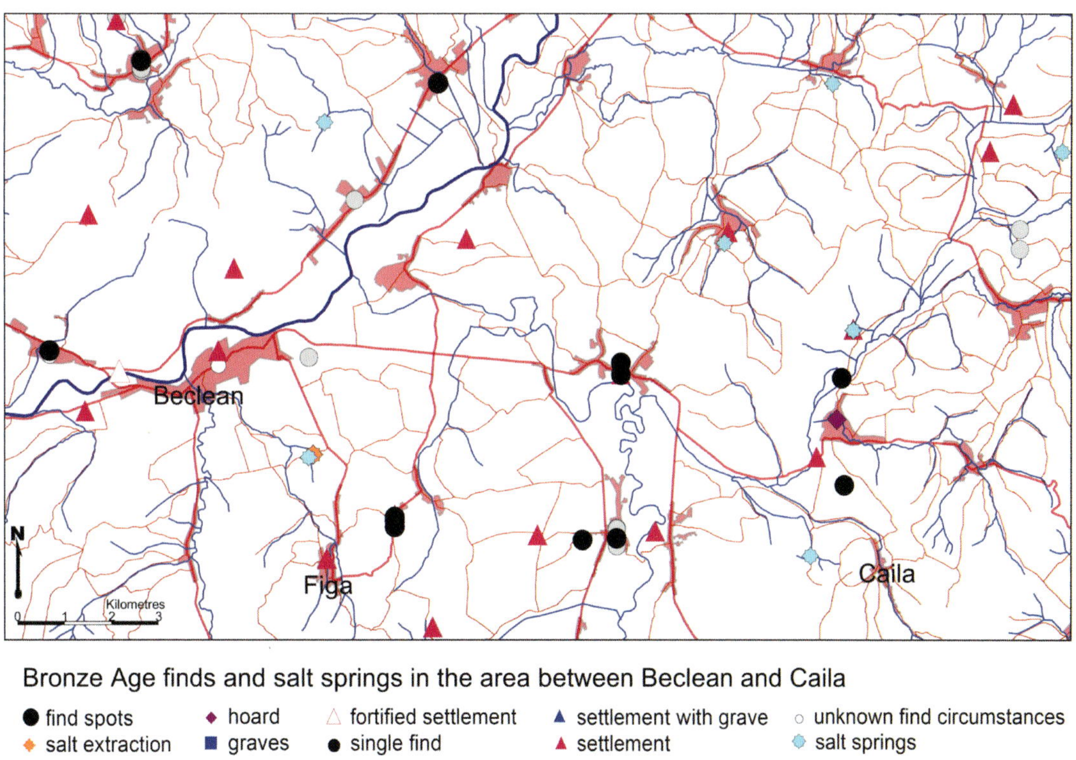

Fig. 4.100. Bronze Age finds around Băile Figa.

Chapter 5
Ukraine

ANTHONY HARDING – VALERII KAVRUK

Present-day Ukraine is rich in salt, but all the deposits being worked nowadays are in the east of the country, in Donetsk oblast. The Artemivsk deposits are the largest in the country and were one of the biggest in the former Soviet Union; in 1979 the salt produced there accounted for 90% of all production in the Ukrainian SSR (*Encyclopaedia of Ukraine* I, 1984, cited in http://www.encyclopediaofukraine.com/pages/A/R/Artemivskrocksaltdeposits.htm). According to this source, salt has been produced by evaporation there since the 16th century, being mined industrially from 1881. We have not investigated these reports, nor have we visited the Donetsk salt sources, but it would appear that it could be worthwhile to do so.

Historically, attention focuses on salt production in the west of present-day Ukraine, partly in Transcarpathia and partly in Precarpathia (Galicia; Ivano-Frankivsk and Lviv oblasts). In Transcarpathia the most important deposits worked in recent times are those at Solotvino on the river Tisza, bordering Romania; there was significant production here until a few years ago. According to the Encyclopaedia of Ukraine two underground mines were still active in 1982, and in 1960 production accounted for 10% of the total produced in the whole Soviet Union (http://www.encyclopediaofukraine.com/display.asp?linkPath=pages\S\O\Solotvynarocksaltdeposits.htm). It is known that industrial production started at the end of the 18th century under Austrian rule and continued on a large scale throughout the 19th century. Today production has stopped and the mining area is abandoned.

In the Austrian period production also took place at other locations in this area; some of these are described below.

In the Precarpathian zone north of the Carpathians (historically Galicia) there are many salt sources. The origin of the name 'Galicia' (the latinized version of Slav 'Halych') is disputed; although it appears similar to other place names stemming from the *hal-* root, indicating 'salt', most sources seem to prefer an etymology deriving it from a name akin to 'Gallia' (Gaul, i.e. Celtic), or Khwalis or Kaliz, the name of the people who occupied the area in the Migration period. The historical province of Galicia was largely within Poland; what is in present-day Ukraine is historically eastern Galicia, with western Galicia covering the south-eastern Polish district of Przemyśl. There is potassium salt in the Lviv and Ivano-Frankivsk oblasts (Stebnykiv, Dobrogostivsk, Boryslav, Kalush-Golyn, Trostyanets, Turo-Velytsk), and table salt in Precarpathia and Transcarpathia (Drogobych, Kalush, Dolyna, Solotvino) (DOBÁK 2010).

Salt production in the Lviv oblast, and especially in the area around Drogobych, was very important in the 19th and earlier 20th centuries under Austrian and Polish administration; nowadays it has all but ceased. Place-names like Stara Sil' ('Old Salt') or Stara Ropa ('Old Brine') indicate the one-time presence of production that has long disappeared. The regional museum in Drogobych today has a section devoted to salt production and a number of relevant historic artefacts, but nothing relating to ancient production. It is hard to create a map of salt sources now, since most of them no longer survive. Pre-Second World War Polish sources provide some information, with brine springs being recorded at Chyriv, Stara Sil', Sambir, Vykoly, Yasenytsya Sil'na, Rudavka, Blazhiv, Holovs'ke, Jordanivka and Zubrytsya (DICKER 1924),[1] while names like Solyanuvatka clearly indicate an association with salt.

[1] These are the modern Ukrainian names; the Polish sources of course give the Polish versions, which are mostly very similar. Locations are as follows: Chyriv (district Sambir), Stara Sil' (district Sambir), Sambir, Vykoly [Vykoty? near Sambir], Yasenytsya Sil'na (district Drohobych), Rudavka – locality between Terlo (Терло) and

Similarly the coat of arms of the town of Gorodok (historically Horodok Solyanyi, "Salt town") shows a salt furnace, and the name of the town is derived from its importance in the salt trade. There is clear potential for a full investigation of the survival of historical and ancient salt production in this area.

Sources in Ivano-Frankivsk oblast are more dispersed, and in many the knowledge of former salt production has vanished. Apart from the places mentioned above, there are a few where archaeological investigations have taken place, as discussed below. Internet sources also give information about other places, such as Utoropy, between Kolomiya and Kosiv and thus not far from Tekucha (see below). Here the salt works, which had been run under the Austrian imperial monopoly, were closed in 1786, but brine springs are evidently still used by the local people.[2] We observed the same at Tekucha, where local people provided hospitality entirely from local sources, including the use of brine from the local springs for pickling.

The most important archaeological sites in Ivano-Frankivsk oblast connected with salt production are Tekucha and Loyeva (see below).

The debate over salt in prehistoric Ukraine

In recent years a debate has been engendered concerning the need for salt in prehistoric communities – mainly Neolithic – in Ukraine (CHAPMAN – GAYDARSKA 2003). This applies particularly to the vast sites of the Tripolye culture, where large numbers of houses can be present – up to hundreds in some instances. This implies not only a large human population but also a large animal population to feed it. Since both humans and animals need a certain intake of salt, and there are no salt deposits in the western steppe zone in Ukraine, it has been supposed that salt might have been moved from Moldavia to supply these sites. On the other hand, MIRCEA – ALEXIANU (2007) have pointed to the abundant evidence for salt production by means of the lagoons (*limany*) along the northern Black Sea coast, especially at the mouths of the great rivers (Dniester, Bug, Dnieper).

The debate is hard to resolve in the absence of conclusive information about the Ukrainian sources themselves, in particular the potential contribution of sea salt from the Black Sea to the diet of Tripolye (and later) communities. Although the Black Sea has low salinity, there are salt lagoons in various parts of its shoreline – particularly in Bulgaria but also in Ukraine, especially the Sivash Sea (lagoon) on the north-east coast of Crimea, and various websites today advertise the sale of Black Sea sea salt (for instance at Krasnoperekopsk, south of the Sivash Sea). It would be highly unlikely that the Greek colonists of the Black Sea would not have utilised this source of salt, though information on this aspect appears to be lacking. Although the waters of Sivash contain many impurities that would be undesirable for dietary salt, with repeated washing and heating acceptable table salt can be produced – and no doubt was in antiquity.

These facts cannot be taken as conclusive in any way, but the presence of such potentially rich salt sources may be taken as at least suggestive that long-distance transport of Moldavian salt to the east

Voloshinovo (Волошиново), district Sambir; Blazhiv (Блажів) district Sambir; Holovs'ke (Головське), district Turka; Jordanivka (Iordanivka) (Йорданівка, Iорданівка), district Mostys'ko; Zubrytsya (Зубриця), district Turka; all in oblast Lviv.

2 'We drove to a part of the village the locals call the 'Bania.' This is where natural brine bubbles up in Utoropy. For centuries it was boiled in the 'banias' till the water evaporated and the salt could be gathered. The salt industry was one of the oldest industries in Ukraine and for a long time it was one of the most important one. It first emerged in this area, which for centuries was the main salt-producing region of medieval Ukraine. Today there are few traces of the famous Utoropy salt industry. The salt works had brought great riches to the various magnates and officials involved in the trade, but the Austrian government had a monopoly on salt. It closed the works in Utoropy in 1786 and they never reopened. However, to this day village people boil the local brine to get salt for home use. We each were given a cup of salt as a memento from Utoropy' (http://www.brama.com/yara/tree-ut.html).

does not need to be seen as necessary, either in prehistory or in historic times. These matters have been considered for the Usatovo and related cultural groups by Svetlana Ivanova, who also provides a list of salt lagoons in the north-western Pontic area (IVANOVA 2010, 73).

The sites

A. Zakarpats'ka oblast' (Transcarpathian region)

Detailed maps and lists of sites where salt is known to have been extracted in historical times in this region prepared by Dr Kobal' served as the basis for the visits in 2008 (*Fig. 5.1*).

1. Tyachiv district

Solotvino (Slatina, Aknaszlatina) on the Tisza was a major centre of salt mining in recent times. Today the mining area is all but abandoned though there is still a spa; the town mainly serves the border crossing to Romania at Sighetu Marmației (there is still a Romanian minority in this area north of the Tisza). The museum of salt mining was opened for us and we were able to study historical pictures and artefacts from the salt mining.

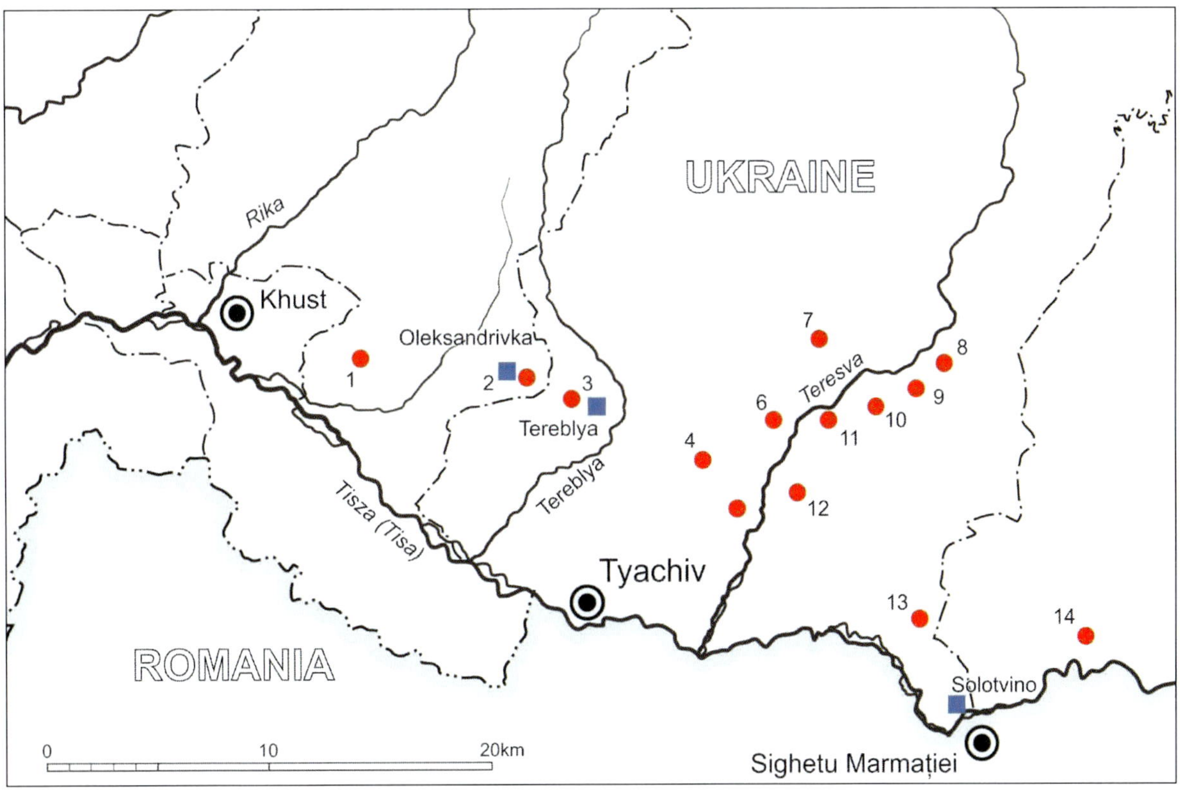

Fig. 5.1. Rock salt sources and brine springs in Hust and Tyachiv districts, Transcarpathian Oblast, Ukraine (information kindly provided by Dr Josip Kobal', Uzhgorod). Green squares: Rock salt; Red circles: brine springs. Names in brackets (below) are the historical Hungarian names.

1. Boronyavka (Baranya); 2. Oleksandrivka (Ósándorfalva); 3. Tereblya (Taláborfalu); 4. Okruglya (Kerekhegy); 5. Dobryans'ke/Nyagovo (Nyágova); 6. Vilkhivka (Irhócz); 7. Novoselitsya (Taraczújfalu); 8. Kalini (Alsókalinfalva); 9. Solone (?); 10. Ganichi (Ganya); 11. Neresnitsya (Nyíresháza); 12. Ternovo (Kökényes); 13. Apsha (Ápsa); 14. Veliki Bichkiv (Nagybocskó).

Although no ancient finds survive from Solotvino, a newspaper account from 1867 refers to the finding of a wooden trough, ladder and hammer (mallet) from the mine (SZATHMÁRY 1867).[3] A sketch (*Fig. 5.2*) shows a complete trough with two closed ends and ten pegs in the base, as well as a block of rock salt with three cylindrical depressions in it. It is not completely sure that these are different from those from Királyvölgy (see below) since there are a number of points of similarity in the two accounts (trough, ladder, hammer and rock salt with depressions; also the fact that the Királyvölgy finds were brought to and stored at Solotvino, according to Preisig's testimony) (PREISIG 1877). It may be objected

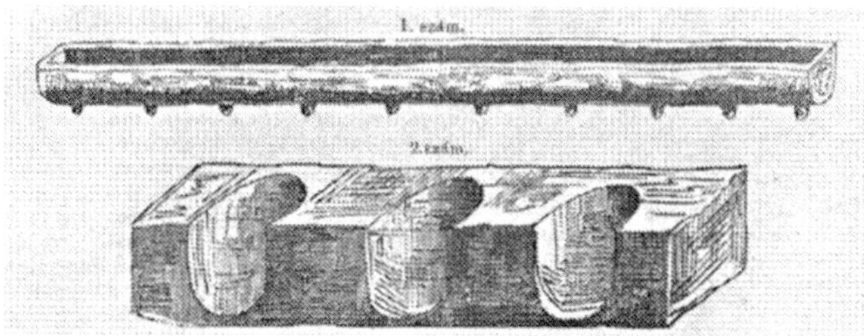

Fig. 5.2. Trough and rock salt found at Aknaszlatina (Solotvino) (SZATHMÁRY 1867).

Fig. 5.3. Photograph of the Királyvölgy objects in the Museum of the State Geological Institute, Budapest, 1909.

[3] We thank Dr Carol Kacsó for sending us a copy of this remarkable and hitherto unknown report, the result of his determined detective work in the archives.

that the rock salt shown by Szathmáry is different from that illustrated by Preisig, having only three depressions whereas Preisig's has four; but exactitude in these illustrations was demonstrably not a prime consideration, since Preisig's drawing is at variance with the surviving piece in the number of pegs present (five, whereas eight are present). Whatever the truth of the matter, it certainly appears that the Szathmáry trough was closed at both ends and probably had around ten pegs. Nothing is known about the other finds described by Szathmáry, but it is possible that they are the ones now in Sopron.

Novoselitsya/Tisolovo: Kiralyvölgy or Valea Regilor (Königstal, King's Valley) is the spot where wooden objects (trough, ladder, mallet, ropes etc), were removed from the site in 1817 during the extension of a salt extraction shaft. These were taken to Budapest (the area was part of Austria at the time) and were only known from a photograph in the 1909 museum catalogue of the Hungarian State Geological Institute in Budapest, from where they have long disappeared (HARDING 2011; PREISIG 1877; VEZETŐ 1909). (*Fig. 5.3*).

Old Hungarian maps show the place-name Királyvölgy as being situated on a side-valley of the Luzhanka stream, itself a tributary of the Teresva, a major river flowing south-west to the Tisza (*Figs 5.4–5.5*). The nearest hamlet is Tisolovo on the Luzhanka, the nearest larger village Novoselitsya on the

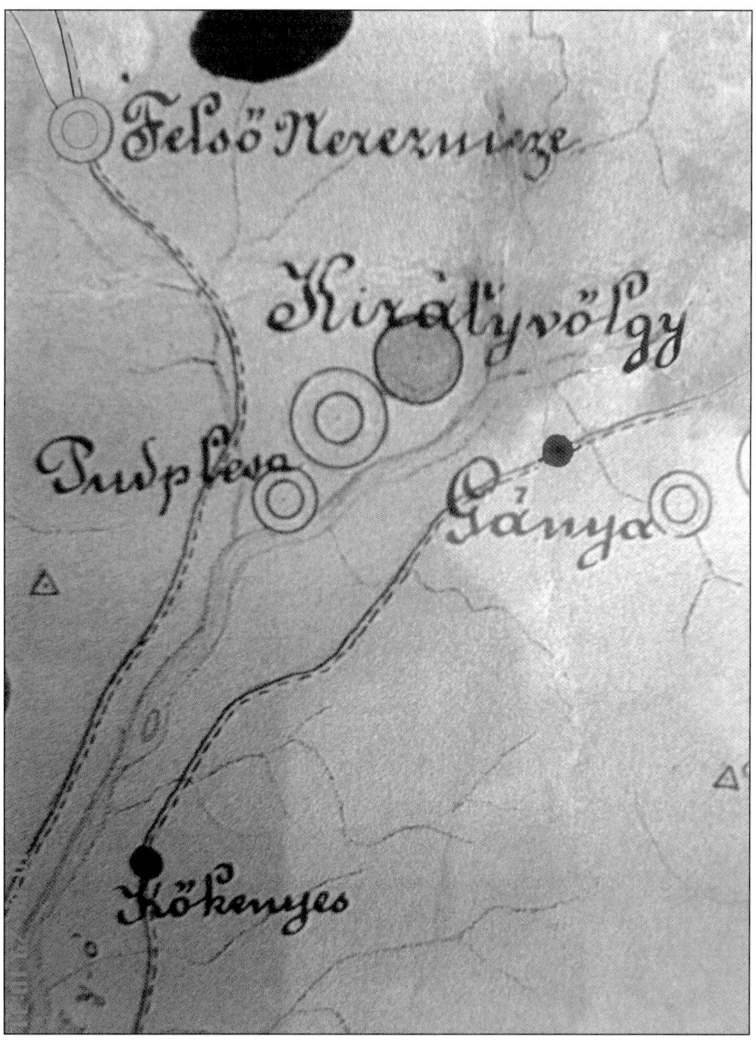

Fig. 5.4. 19th century map held in the Hungarian Central Mining Museum, Sopron, showing the position of Királyvölgy.

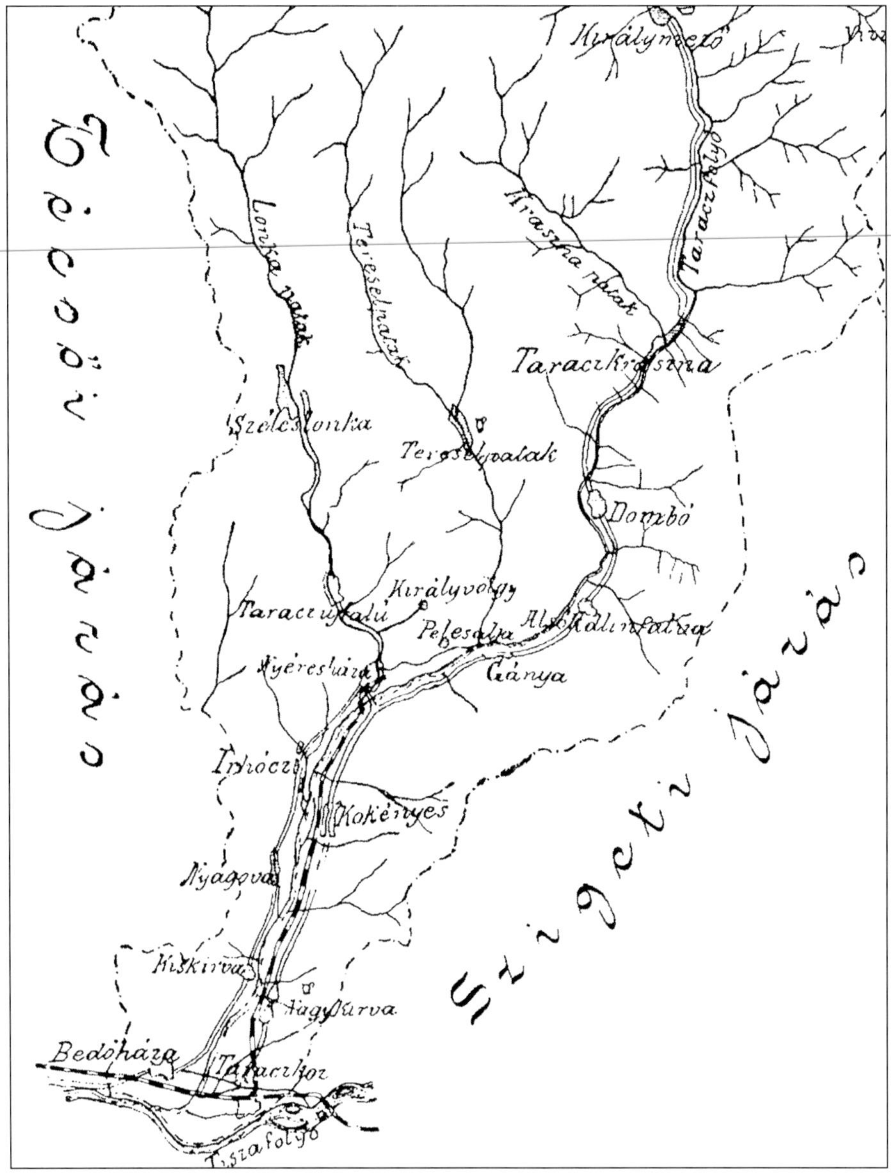

Fig. 5.5. 1901 map showing the position of Királyvölgy (VÁRADY 1901).

Teresva. The area was visited, when it was apparent that the remains of the former extraction site are visible as a large water-filled shaft (currently being developed as a spa-hotel) (*Fig. 5.6*), several smaller shafts, and canalised springheads (48° 08' N, 23° 47' E). Timber was noted in a side stream and in the main stream beside the spa development, as well as in at least one smaller nearby pit (*Fig. 5.7*). Sherds of Gava type and Roman period pottery were found nearby. No other signs of ancient exploitation are currently visible.

Subsequently it emerged that the wooden objects extracted in 1817 are now in the Központi Bányászati Múzeum (Central Mining Museum) in Sopron, Hungary. The objects were studied on a visit in 2009 and samples taken for radiocarbon dating. The objects and resulting dates have been published separately (HARDING 2011), but the main conclusions are repeated here.

Fig. 5.6. The main water-filled shaft at the site of Királyvölgy (2008).

Fig. 5.7. One of the small pits containing wood at Királyvölgy (2008).

The surviving objects are as follows (all measurements in cm):

Trough, incomplete, with 8 square holes in the base (*Fig. 5.8* upper). Inv. No. 2000.106.1 (Preisig Fig. 17–18). L. 136, Diam. 34, Ht 29, top opening 18.5 where split, 16.5 where not split; hole nearest complete end 2.8 x 2.4; depth of holes (thickness of base) 5.5–6.0.

Radiocarbon date: GrN-32399 2890 ± 50 BP

The trough is incomplete, having been sawn off straight down one side and obliquely across the base and the other side, presumably to facilitate extraction from the mine (it was evidently in this state when Preisig saw it, though curiously his drawing shows a V-shaped notch in the upper edge of one side – this is no longer present, the whole side having been removed). The wood is completely dried out, and the trough has a large crack running along the length of the base. One peg is wedged into one of the holes, its base is apparently round, diam 2.7. The surviving loose peg presumably belonged with the trough, though it might equally have come from the trough base fragment (below).

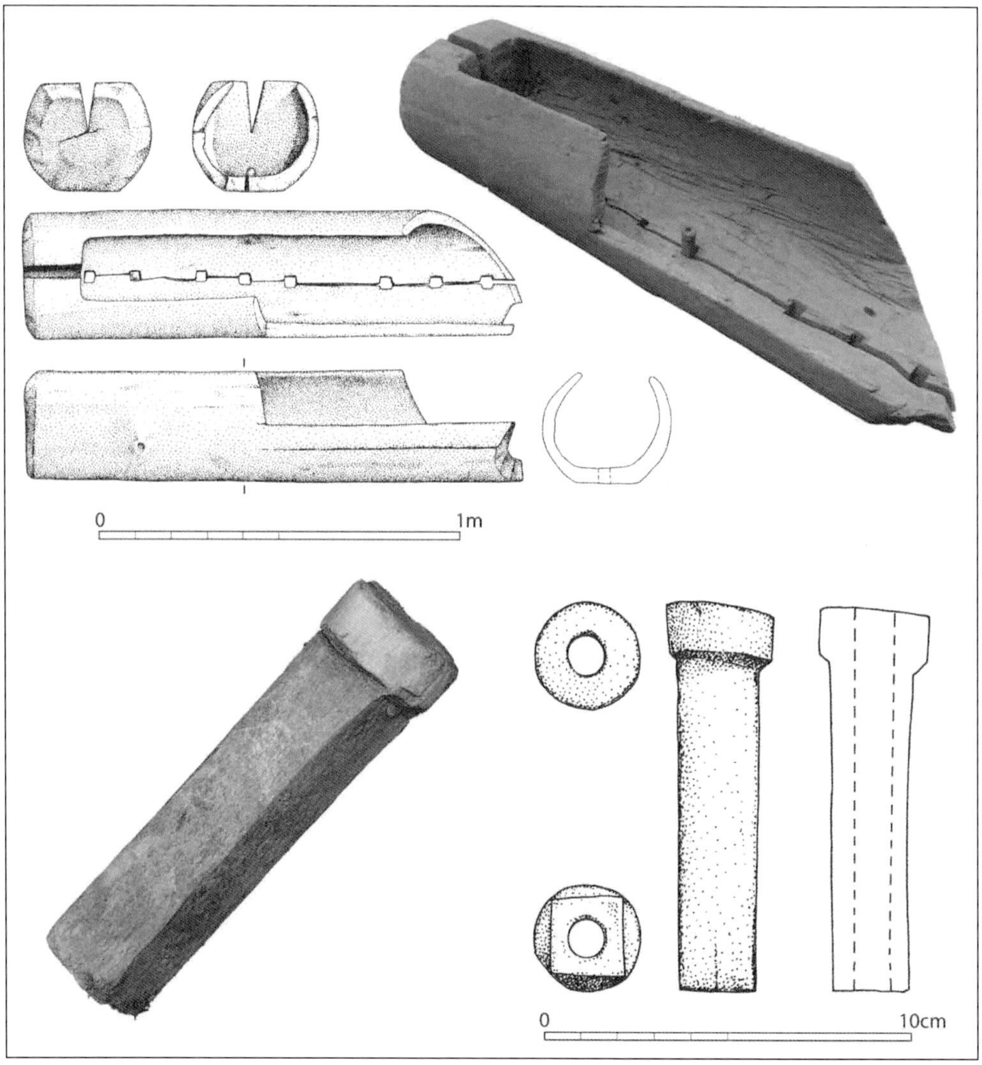

Fig. 5.8. Trough and surviving peg from Királyvölgy, in the Hungarian Central Mining Museum, Sopron.

Peg (*Fig. 5.8* lower). L 10.3, square section, 1.7–2.0 wide, diam of collar 2.8, depth of collar 1.3–1.5, diam of perforation 1.0.

Preisig shows five pegs in the holes of the 'Wasserverteilungsrinne' (see below)

Mallet (*Fig. 5.9* lower). Inv. No. 2000.106.4 (Preisig Fig. 22–23). L. 33.0. Di 14.0 at one end, 12.5 at other end.
Radiocarbon date: GrN-32400 3060 ± 65 BP

Ladder consisting of 2 side pieces and 4 rungs (*Fig. 5.9* upper left; *Fig. 5.10*). The side pieces swell laterally at the points where the perforations for inserting the rungs were made. The rungs have tenons to fit into these perforations. The rungs are shaped (or worn) in a concave fashion. In the end of one of them an iron nail is present, probably to assist the reconstruction of the ladder when it was displayed in Budapest.
Side pieces: Inv. Nos. 2000.106.5 and 6 (Preisig Fig. 26–27)

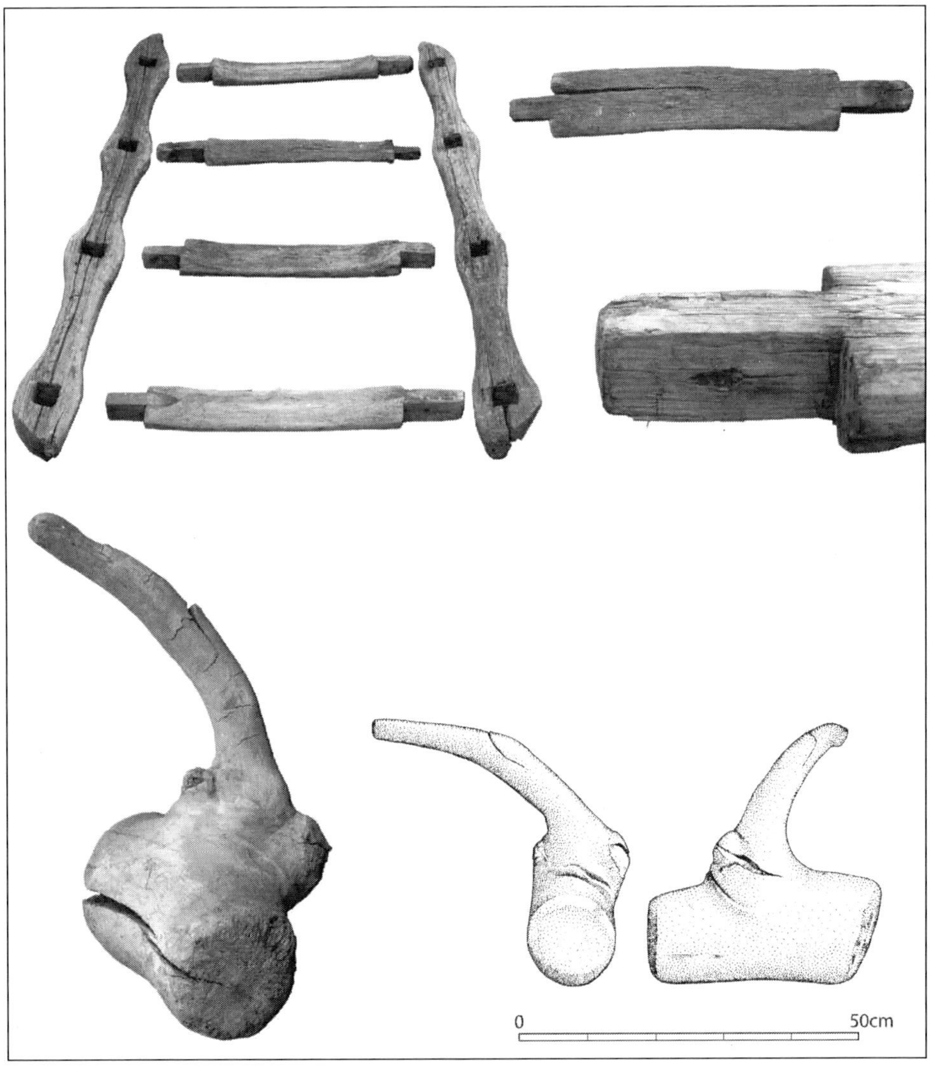

Fig. 5.9. Ladder and mallet in Sopron.

L. 236.5 and 233.5, W. 18.5 at widest points, Th. 9.0; holes for rungs 8.0 x 7.5, 8.0 x 7.0 etc
Radiocarbon date: GrN-32398 2845 ± 35 BP

Rungs: Inv. Nos. 2000.106.7, 8, 9 and 10. Two were measured (none had museum numbers or could otherwise be distinguished).

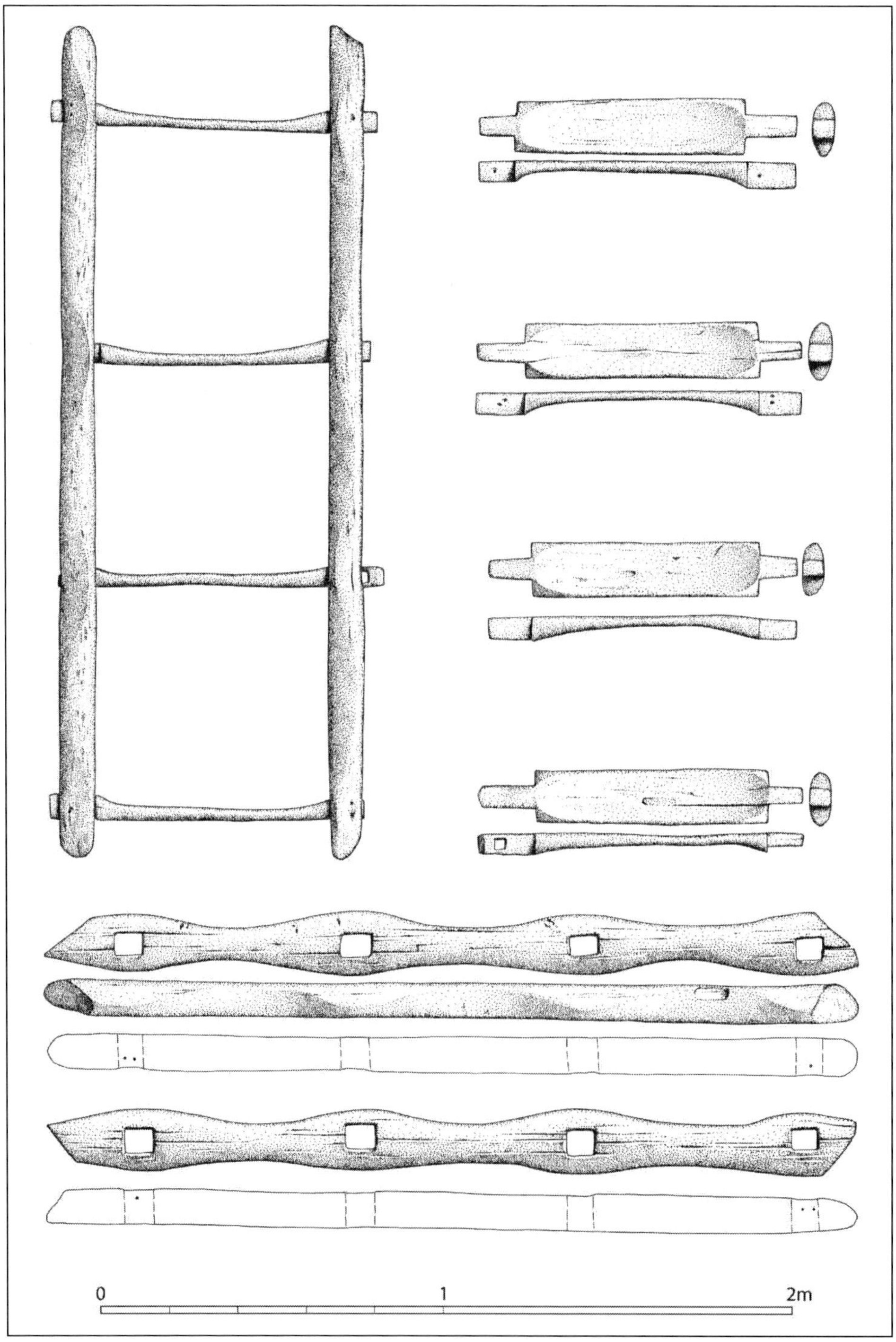

Fig. 5.10. Ladder, rungs and side-pieces in Sopron.

L. 92.5, W. 14.5, Th. at one end 6.0
L. 87.5, W. 15.5, Th. at sides 7.0, Th. on step 3.0
Radiocarbon date: GrN-32397 2850 ± 45 BP

Ladder side piece (*Fig. 5.11* lower). Inv. No. 2005.136.1 (not illustrated by Preisig). Has three square or rectangular holes and a possible fourth, cut off at the end. L. 161, W. 9.5, Th 6.5, split, Holes 4 x 2.5, 3.5 x 2.5, 6 x 3.5, mid-hole to mid-hole 49
Radiocarbon date: GrN-32396 1170 ± 45 BP

Trough base fragment (*Fig. 5.11* upper). Inv. No. 2000.106.2 (Preisig Fig. 15–16, described as 'Wasserverteilungsrinnen', and shown with pegs in the holes). The identification as a trough base is based on the examination of several other troughs in Romania. Six holes survive and part of a seventh. Signs of shaping on the underside. L. 107, W (surviving) 10, Th. 4.0–5.5, distance between holes 12, 15, 12.5, 14, 15, 12.5, holes 2.7 x 2.2, 2.2 x 2.0, 2.2 x 2.3, 2.7 x 2.2, 2.5 x 2.3, 2.5 x ?
Radiocarbon date: GrN-32394 2860 ± 35 BP

Channel (piece curved in cross-section) (*Fig. 5.11* middle). Inv. No. 2000.106.3 (not illustrated by Preisig). L. 117, W. (max) 15, curving, Th. of wood 2.5
Radiocarbon date: GrN-32395 2820 ± 50 BP

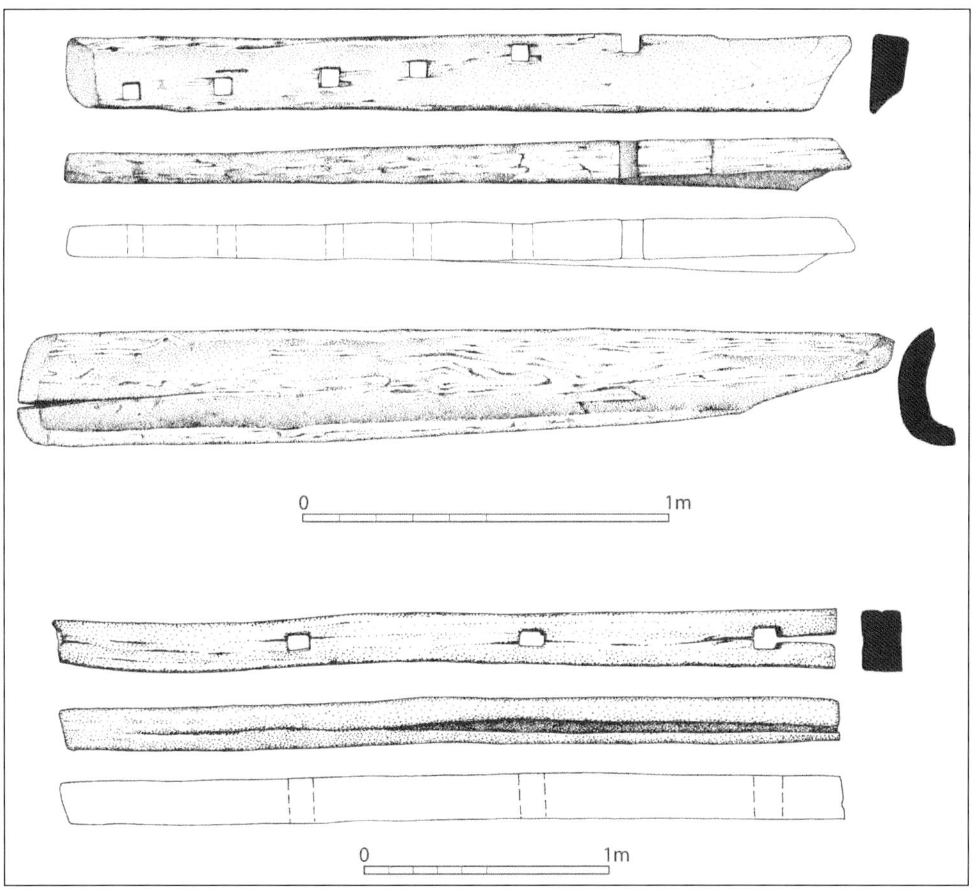

Fig. 5.11. Trough base and channeled piece in Sopron.

Radiocarbon dating

Samples were taken for radiocarbon dating and submitted to the Centre for Isotope Research of Groningen University. The wood of the objects had completely dried out, even to the extent that no salt was evident; there did not appear to have been any conservation treatment.

Table 5.1. Radiocarbon dates from Királyvölgy with calibrated date ranges. Calibration has been carried out using OxCal v. 4.1.6 (2010).

Object	Lab no	Date BP	Date cal BC 68.2% prob.	Date cal BC 95.4% prob
Mallet	GrN-32400	3060 ± 65	1414–1224	1453–1126
Trough base	GrN-32394	2860 ± 35	1112–948	1130–919
Channel	GrN-32395	2820 ± 50	1041–910	1123–843
Rung	GrN-32397	2850 ± 45	1111–931	1192–902
Large ladder side	GrN-32398	2845 ± 35	1051–934	1121–916
Trough	GrN-32399	2890 ± 50	1191–1000	1259–927
Ladder side piece	GrN-32396	1170 ± 45	AD 779–939	AD 720–980

These results are shown diagramatically in *Fig. 5.12*.

Remarkably, all the samples except the ladder side piece produced very similar radiocarbon ages, in spite of the chequered history of the objects, which gives one confidence in the integrity of the assemblage. Only the mallet is significantly earlier than the other finds. The combined value of the samples other than the mallet and ladder side piece is 2853 BP, giving a range 1050–975 cal BC at 68.2% probability and 1120–930 cal BC at 95.4% probability (1120–970 cal BC at 85.2%).

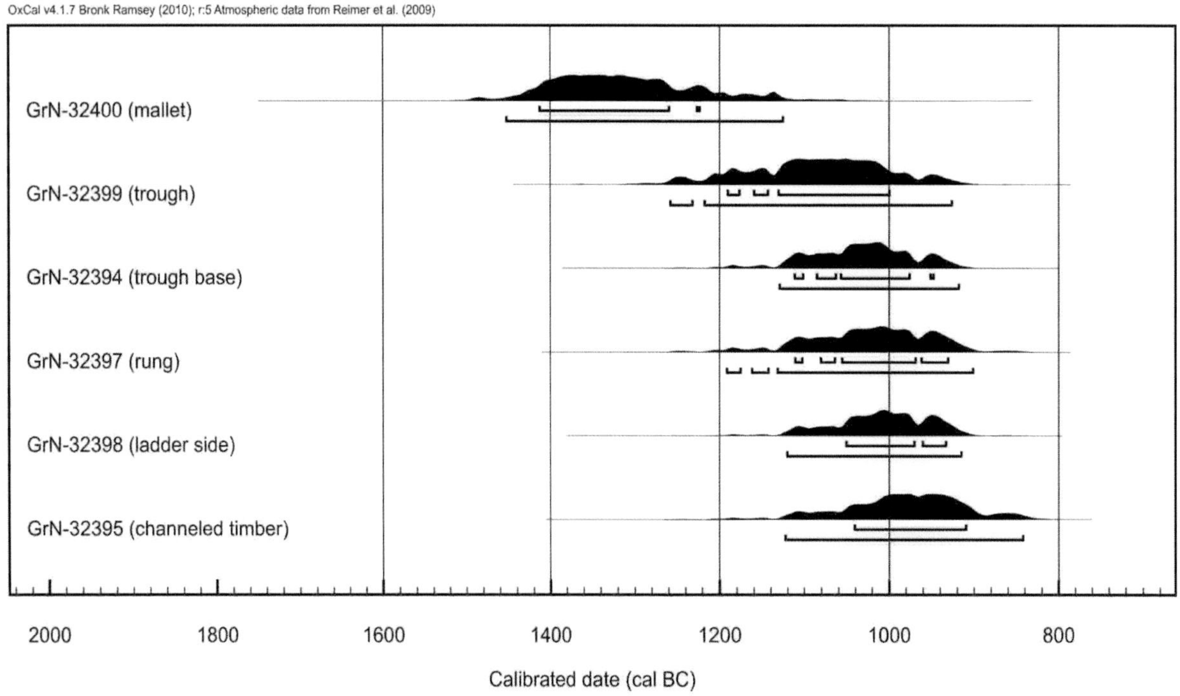

Fig. 5.12. Probability distribution of the radiocarbon dates from the wooden objects from Királyvölgy held in Sopron.

The dates fall in precisely the same time period as the bulk of those obtained at Băile Figa in Transylvania, where the main period of usage for the wooden installations fell between 3200 and 2800 BP. As discussed above, this is the period of exploitation of most of the Transylvanian sites.

The one date that is more recent comes from the fragmentary side of a putative ladder, and at 1170 ± 45 BP falls in the medieval period, at 720–980 cal AD (92.4% probability). It is thus somewhat later than the early medieval dates from Figa, which fall between 400 and 600 cal AD, and earlier than a date from wood at the nearby salt mine of Tereblya at 670 ± 15 BP (1280–1385 cal AD) (see below). Little can be deduced from this other than that salt mining very probably continued at these sites throughout the centuries, from prehistory to modern times.

Hanici (Ganja) (48° 07' 13' N, 23° 51' 55' E): a brine springhead lies in a private garden, with a disused shaft nearby.

Dobryans'ke (formerly *Nyagovo*) (48° 04' 44' N, 23° 42' 03' E): a former mining shaft, now a reed-bed, contains a spring-head. Nearby is an abandoned shaft. A salt spring lies a few hundred metres above this (48° 04' 28' N, 23° 42' 08' E).

Tereblya (48° 07′ N, 23° 34′ E): behind the village is an area of ponds where the records indicate salt extraction took place in the Austrian period (*Fig. 5.13*); a wellhead lies nearby (*Fig. 5.14*). Numerous pieces of timber are visible in the water and the soft mud around it. At the time of the visit it was impossible to sample the wood because of the problem of access, but subsequently Dr Kobal' visited during the winter when the ponds were frozen and took a number of samples. Two were submitted for radiocarbon dating. The position of these is known approximately, though their true context is of course unknown (both were surface timbers) (*Fig. 5.15*). The results indicate usage of the site in the medieval and early modern periods (*Table 5.2* and *Fig. 5.16*), rather than the Austrian period; while not suggesting a prehistoric date, it certainly shows that such sites have potential for further investigation.

Fig. 5.13. Salt ponds (former mine shafts) at Tereblya, district Khust.

Fig. 5.14. Brine well at Tereblya.

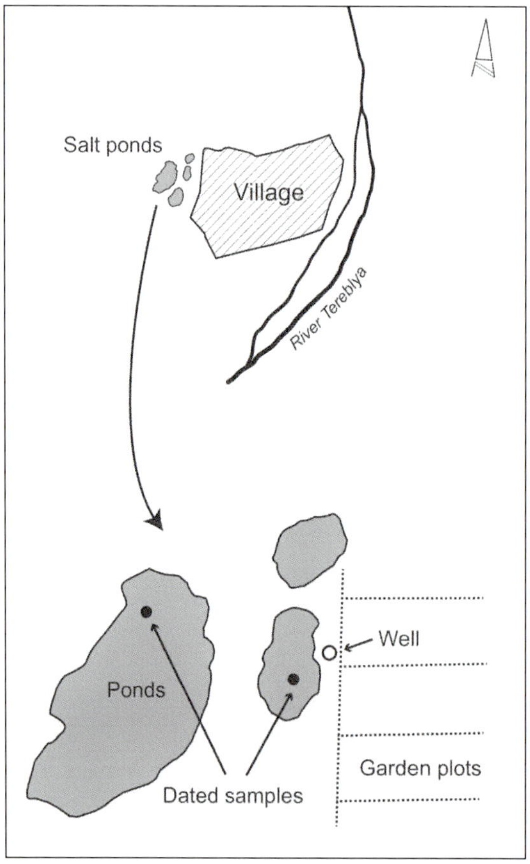

Fig. 5.15. Sketch map by Dr J. Kobal' showing the location of the ponds from which samples were taken for radiocarbon dating. Upper: general situation; lower: detail of the sample and well location (not to scale).

Table 5.2. Radiocarbon dates from Tereblya (calibration using OxCal v.4.1.7).

Lab no.	Date BP	Cal AD 68%	Cal AD 95%
GrN-32181	355 ± 15	1477–1619	1465–1632
GrN-32182	670 ± 15	1284–1380	1280–1385

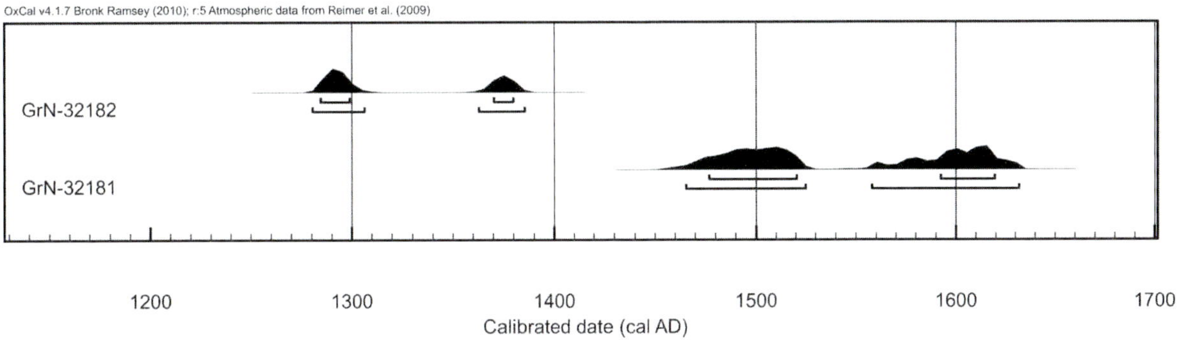

Fig. 5.16. Probability distribution of the radiocarbon dates from Tereblya.

Fig. 5.17. Well-head at Nizhniy Rozsul'nik.

Nizhniy Rozsul'nik (48° 05' 57' N, 23° 36' 18' E): across the river Tereblya from the previous site is a wooden wellhead and a salt stream (*Fig. 5.17*)

Okrugla (48° 05' 27' N, 23° 39' 46' E): a small water-filled shaft is present at this village.

2. Hust district

Oleksandrivka: there is a brine wellhead in the village (48° 08' 01' N, 23° 29' 12' E), still in use, and nearby a large hollow, all that remains of an extraction shaft (48° 08' 04' N, 23° 30' 18' E). The village hall has a small museum on its upper floor, including ethnographic items, a number of wooden salt containers, and a dish of local salt.

The nearby wooden church possesses a metal salt-boiling cauldron.

Boronyava (48° 09' 55' N, 23° 24' 56' E): there is a wet salty area here and a small shaft in the wood nearby.

B. Ivano-Frankivsk oblast'

1. Nadvirna district

Loyeva: site II, excavated in 1981, 1983 and 1988 by Larisa Krushelnitska, lies on the west side of the village on the flank of a small hill, above a stream (KRUSHELNITSKA 1993, 58–68; plans repeated in KRUSHELNITSKA 2002a; 2002b) (*Fig. 5.18*). Although there is nothing to see at the present day, the plans and finds from the excavation provide a clear picture of the site (*Figs 5.19–5.20*). Numerous burnt clay surfaces surrounded a wooden box-like construction which was carefully constructed and seems to have been used in the salt production process, perhaps for storing brine. According to the excavator, most pot was used for brine boiling, but we have not seen the pottery in question.

2. Kosiv district

Tekucha: the sites excavated by Professor Krushelnitska and Dr Dmitro Pavliv lie on the north side of a tributary of the river Prut (48° 22' 40' N, 24° 46' 57' E) (KRUSHELNITSKA 1993, 68–93) (*Fig. 5.21*). Below it, adjacent to the river, there is a series of brine springs. Excavation of site I showed an extensive area of burnt clay, with suggestions that brine boiling took place in the construction; large quantities of pottery were found, much of it secondarily burnt. Site III is said to be one of the most significant sites for brine boiling (KRUSHELNITSKA 1993, 79). The pottery illustrated in the report appears to be of early Gava type. A little further away, in the village, is a brine well-head (48° 23' 28' N, 24° 48' 03' E)

Fig. 5.18. Loyeva, Nadvirna district: view of the site (2008).

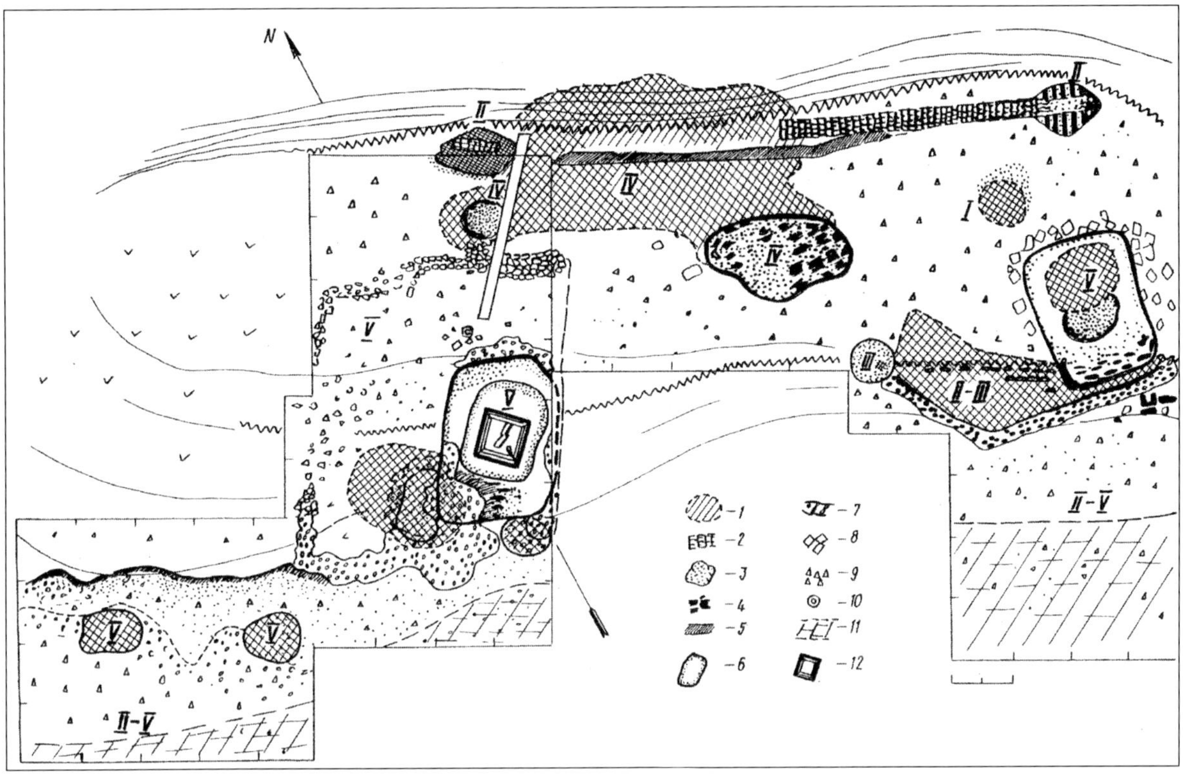

Fig. 5.19. Plan of the excavated area at Loyeva, showing timber-lined shaft (KRUSHELNITSKA *2002b).*

Fig. 5.20. The timber-lined shaft at Loyeva (KRUSHELNITSKA *2002b).*

(*Fig. 5.22*). This is clearly a rich salt production area, though we were unable to find any other surviving traces of prehistoric working.

According to Krushelnitska (op. cit.), other sites are connected with the evaporation of brine for salt production, Gnilitsa and Girske among others.

Fig. 5.21. Tekucha, Kosiv district: view of the site.

Fig. 5.22. Brine well at Tekucha.

C. Lviv oblast'

1. Stariy Sambir district

Stara Ropa/Stara Sil': both place-names indicate the one-time presence of brine springs and salt, but local informants at Stara Ropa told us that these were in the stadium area and are now covered over.

Ukraine: Assessment

The only part of Ukraine where there is specific evidence for prehistoric salt extraction is Transcarpathia, and specifically the Solotvino area. This is probably a matter of chance, and the fact that the technology used in the mines in this area involved large quantities of wood, some of which has survived across the centuries. The evidence for the Precarpathian area of Galicia is still very uncertain, particularly the Lviv oblast, where the Drogobych district was once home to numerous salt mines.

The wide distribution of resources in modern Ukraine and the large area involved mean that the task of elucidating ancient salt production is potentially huge. Svetlana Ivanova (Archaeological Institute of Ukraine National Academy of Sciences) has made suggestions about the significance of salt resources for the Eneolithic cultures of the south-west of the country; Dr Josip Kobal' is one of the few archaeologists active in the Transcarpathian region, and has few possibilities for conducting his own monitoring of sites; there are no HERs or similar resources in this part of Ukraine, remote as it is from Kiev. North of the Carpathians the situation is, if anything, worse: the huge area, the tiny number of archaeologists active in fieldwork, and the general lack of resources mean that it will be many years before one can speak of a situation comparable to that in neighbouring regions to the west and south. The study of prehistoric salt exploitation in Ukraine is certainly possible but still has obstacles to overcome.

While we did not find wooden structures in 2008 that are certainly comparable to those now turning up in some numbers in Transylvania (and known or suspected to be present in Moldavia), we were able to show that there is considerable promise in salt research in parts of the sub-Carpathian region. The finds from Királyvölgy have long been known and indicated that a rather specific technology was used in ancient salt production in that region; the recent study of the objects confirms that they belong to the Bronze Age and are very similar to what is present in Romania, at Figa in particular. Several sites produced wooden objects that may well have been part of structures connected with salt extraction in the ancient rather than the modern past. This may be true for Tereblya and Novoselitsya.

In Transylvania, our efforts to find certain traces of ancient working took some years, and we believe that it will be with future fieldwork in Ukraine and Slovakia that more progress will be made. In particular, it is necessary that ethnographers join with archaeologists to investigate salt production in recent times, as it has been through this means that the most interesting sites in Romania have turned up. As it is, there appears at present to be very little interest in Ukraine in historical salt research, and none at all in archaeological salt research. Generating this interest will be a task for the coming years; our visit in 2008 will have been a significant step in this direction. Dr Kobal' is now aware of what may be possible, and is planning fieldwork to investigate the situation in the Solotvino area.

There are rock salt deposits in eastern Slovakia which, according to local informants, because of their depth could not have been accessible to prehistoric technology. In 2008 it was not possible to check the reliability of these assertions, which would have involved a considerable time in the field, but it seems likely that as in Poland any trace of ancient activity would have been lost during the period of industrial production, especially in the 19th century and first half of the 20th.

1. Prešov district

Prešov: within the city limits are several spots where salt has been extracted in historic times, and where some working still continues at the present day. At *Solivar* (formerly Sóvár) the extensive former salt works are preserved as a museum and World Heritage Site (DUCHOŇ 2007). The sources indicate a history of salt production going back into the late medieval period, when documents mention a well (*puteus*), mine pit (*fodina*) and salt boiling (*coctura salis*) (BUTKOVIČ 1978). The spots inspected include the open area north of Solivar and Solná street (the former *Sol'ná Baňa*) in Šváby. No signs of ancient activity are visible; north of Solivar there is an extensive area of disturbed ground, with modern or clearly recent activity still in progress. The area was walked for some distance to the north, to include the stream running down into the salt production area, but nothing ancient was visible.

Išla: three brine wells are present in the grounds of an activity centre.

2. Vranov nad Topl'ou district

Hlinné: a simple brine well lies beside a stream at 48° 56' 22' N, 21° 34' 22' E. The stream was walked over a distance of more than 1 km but no ancient remains were visible. According to Ing. J. Kotul'ák the stream is too 'dynamic', i.e. runs down the hill too rapidly and with too much water, for brine production to be possible.

Sol': The area between the villages of Sol' and Rudlov, along the river Topl'a, lies above a large deposit of salt, stated to be 270 m in depth and 405 m across (KÓNYA – MICHNOVIČ 1997, 10), the rock having a salt content of around 80%.

Various archaeological finds have been made in the area over the years (KÓNYA – MICHNOVIČ 1997, 14 ff.). The most important of these took place in 1989, when a small excavation was conducted by the late Stanislav Šiška (Institute of Archaeology, Slovak Academy of Sciences) near two brine springs at the locality Surovice in the parish of Sol', and Neolithic finds were recovered (ŠIŠKA 1997). Šiška interpreted this as the first clear evidence for a prehistoric interest in the Slovak salt sources. While it was not possible to define Šiška's excavation spot exactly, visits were made to several possible spots; in the fields west of the village there are today several brine well-heads. A typical brine spring, not canalised, lies near the farm at 48° 55' 03' N, 21° 35' 31' E; at least one wooden post is embedded in the ground, but otherwise no signs of ancient activity are present (*Figs 6.2–6.3*).

At *Dlhé* and *Poša* there are deposits of rock salt at some considerable depth, without any indication of exploitation in prehistory (information from Dr Eva Horvathová; not visited).

3. Bardějov district

Cigel'ka: several brine well-heads and briny springs lie in the fields north-west of this village close to the Polish border (49° 24' 47' N, 21° 08' 18' E; 49° 24' 49' N, 21° 08' 21' E), another in the village itself (49° 24' 03' N, 21° 08' 21' E), and still others in the woods north of the village. The most interesting find,

director Dr Július Béreš. Institute staff, notably Dr Eva Horvathová and Dr Klára Marková, assisted us in every way. They had prepared maps and other information for us and guided us to the most important locations in eastern Slovakia where salt is found. We were also accompanied by Ing. Jaroslav Kotul'ák, who through his intimate knowledge of the geology and hydrology of the area was able to lend invaluable assistance.

however, was a well-head at 49° 24’ 07’ N, 21° 08’ 31’ E (*Fig. 6.4*). This lies beside a stream, and in the stream were large numbers of timbers that had formed a series of constructions (*Figs 6.5–6.7*). While some of these may be part of a bridge, or connected with a nearby factory, the resemblance to wooden constructions we have seen in northern Romania was striking. Samples were taken for ^{14}C dating, and one that was submitted gave the following result:

Fig. 6.2. Sol’, district Vranov nad Topl’ou: salt spring.

Fig. 6.3. Sol’, wood at the salt spring.

Table 6.1. Radiocarbon date from Cigel'ka.

Lab no.	Date BP	Calib range & probability		
GrN-32180	225 ± 20	AD1645–1675 (45.6%)	AD1777–1800 (39.9%)	AD1941–1954 (9.9%)

Fig. 6.4. Cigel'ka, district Bardějov: brine well.

Fig. 6.5. Cigel'ka: wooden posts in the stream bed.

While this is only one date, it suggests that the post in question (and potentially the whole installation) is most likely to fall before AD1800 – with the highest probability that it dates to the 17th century, and after that the 18th century. There is a small chance that the timber is modern. It is not known what the installation was, but it is hoped that a full survey of the spot might be conducted by the Archaeological Institute when conditions permit.

Fig. 6.6. Cigel'ka: wooden constructions in the stream bed.

Fig. 6.7. Cigel'ka: wooden constructions in the stream bed.

4. Trebišov district (Zemplín area)

Kuzmice: a brine well-head in the fields east of the village was inspected.

Michal'any: there are several brine well-heads in this area, notably at 48° 30' 20' N, 21° 37' 50' E.

At none of the sites in the Zemplín area was there any sign of ancient workings.

5. Michalov'ce district

Rock salt deposits are present at *Zbudza* (information from Dr Eva Horvathová; not visited).

6. Sobrance district

Rock salt deposits are present at *Zalužice* (information from Dr Eva Horvathová; not visited).

7. Snina district

Brine springs are known at *Zboj*, very close to the Polish and Ukrainian borders (information from Dr Eva Horvathová; not visited). Many springs are known across the border in Poland a short distance to the north (see below). These were not visited during our fieldwork in 2008.

Central Slovakia

Námestovo district

Brine springs are present at *Oravská Polhora* near the Polish border, in the Beskydy mountains (information from Dr Eva Horvathová, not visited). This area lies south-west of the salt massif in Małopolska, as seen most vividly at Wieliczka.

Slovakia: summary

The visits to salt production sites or known salt deposit areas in 2008 did not produce any direct evidence for salt working in the prehistoric period; in most cases where there are records working was under way in the late medieval period. The presence of Neolithic material near salt springs at Sol' is undoubtedly suggestive, but not conclusive. No Bronze Age material is known to be associated with any of the salt sources, nor are there indications of Iron Age working. It must be stressed, however, that the inspections were cursory. Slovak archaeologists are well aware of the possibilities, but have not yet discovered specific sites with evidence of pre-medieval salt working.

Poland

The salt deposits of Poland are widespread and prolific (CZAPOWSKI – BUKOWSKI 2010; cf Chapter 2). Although those of Litle Poland (Małopolska) are much the best known in the wider world, today's production is based in Great Poland (Wielkopolska), where the mine at Kłodawa provides for all of the country's needs; and in the North of the country there are extensive deposits around the Puck Bay, and lesser ones near Kołobrzeg. No signs of historical production are known at Kłodawa, but elsewhere in Wielkopolska and Kujavia (Kujawy) there are abundant records of production since the medieval period. The mine at Inowrocław, for instance, was a major producer into modern times, until it was closed in the 1990s; although salt is no longer mined there, there is still an evaporation facility operated by Solino s.a. (Inowrocławskie Kopalnie Soli), and the park in the town surrounding the spa has a newly constructed 'graduation tower' (*Salzgradierwerk*, *tężnia solankowa*) for the evaporation of brine. Excavation in this area in the early 1970s produced installations for salt production of the 10th to 13th centuries AD.

Little Poland (Małopolska)

Historically this is the part of Poland where most evidence for ancient exploitation has been believed to be situated. The famous mines at Wieliczka, Bochnia and elsewhere are well-known tourist attractions; at Wieliczka the Salt Museum (Muzeum Żup Krakowskich) contains exhibitions on the historical development of the saltworks, which include a small element of Lengyel culture material believed to be briquetage (small conical pots in a coarse fabric). This material has been presented in the literature many times (JODŁOWSKI 1968; 1971), but little relating to prehistoric exploitation has emerged from this institution in recent times.

By contrast, rescue excavations in advance of motorway construction south and east of Kraków discovered a series of features containing briquetage of the Lausitz culture (KADROW 2003, 212–218, Figs 14–18), which through the kind offices of Professor Sławomir Kadrow Harding has examined in the Archaeological Museum of Kraków (*Fig. 6.8*). Interestingly, these findspots lie off the main salt

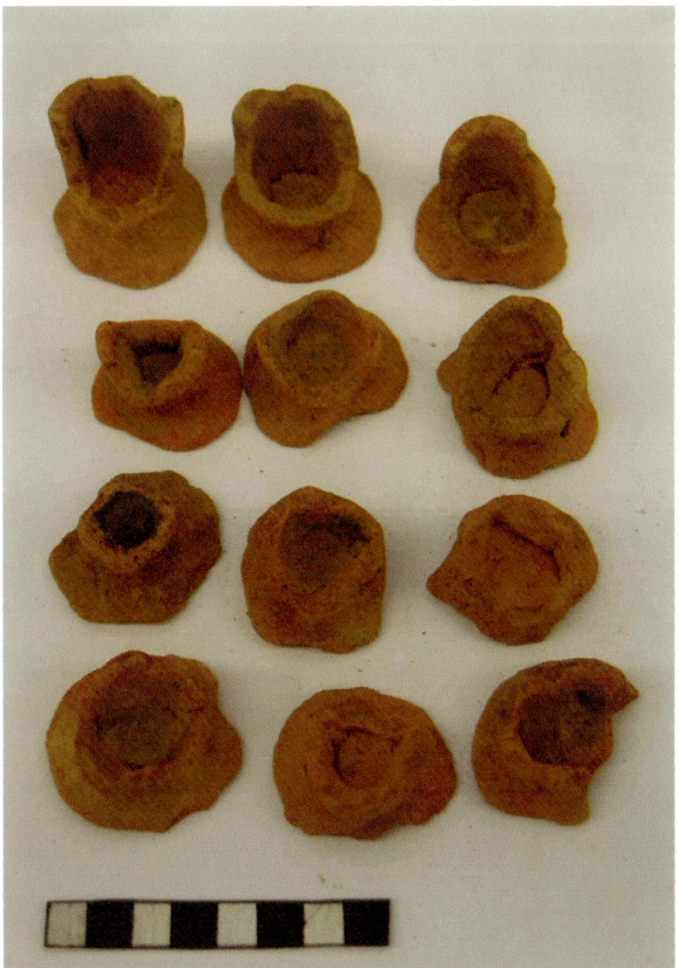

Fig. 6.8. Bronze Age briquetage from Zakrzów near Kraków (photo: Marcin Przybyła).

deposit, indicating some kind of local transport of salty materials to the places where evaporation was to take place.

There are numerous salt springs in the south-east of Poland, actual or known historically, and many more are supposed from toponym evidence (JODŁOWSKI 1971, Map 1). We have not yet investigated these. According to Jarosław Fraś from the Wieliczka museum, many brine wells in the Kraków area (and presumably elsewhere) were capped in concrete in the Austrian period to prevent unauthorised exploitation.

Literature from pre-war Poland, when what is now western Ukraine was part of Poland, shows how a range of salt sources were exploited in the border area, from Sanok eastwards to Sambir and Drohobych (DICKER 1924; KUNYSZ 1964–65 [1968]). As explained in Chapter 5, little or nothing remains in Ukraine of the exploitation of these sources. The same is true of Poland, but we are indebted to Marek Skubisz and Jarosław Fraś of Lunula s.c.for images of salt springs and salt lakes near Sanok, and in particular near Tyrowa Solna, that are evidently active at the present day. We are not aware that any of them contain wood that might be ancient, but these matters await detailed field investigation.

Gedl's account of the Late Bronze Age occupation of this area makes no reference to the possibility of salt exploitation (GEDL 1996). It thus remains quite uncertain how important these sources might have been in ancient times, though our experience of Ukraine and Romania would suggest that it is highly unlikely the sources were unknown and unexploited.

Fig. 6.9. Salt spring at Zgłowiączka, district Lubraniec.

Great Poland and Kujavia

At the present day, this is where most salt is obtained, and principally from Kłodawa. The discussion by CZAPOWSKI – BUKOWSKI (2010) shows where the salt lies and where it has been exploited, but in none of these cases is there known evidence for exploitation earlier than the medieval period. Inowrocław is one of the historical centres of salt production, though the mine there closed some years ago; the town museum features the history of salt. Z. Bukowski has argued (BUKOWSKI 1963; 1985), on the basis of toponyms, geographical proximity and artefactual evidence, that salt production in Great Poland was the source of the great flourishing of sites in the Hallstatt period, suggesting that spool-shaped objects at sites such as Biernatki and Sobiejuchy represented briquetage. This seems unlikely in the absence of more data to substantiate the notion.

Anna Majchrzycka, formerly a Masters student of the Adam Mickiewicz University, Poznań, has studied the salt sources of this area where there are indications of medieval exploitation (MAJCHRZYCKA 2009). Thanks to her, with the kind assistance of Professor Janusz Czebreszuk, Harding was able to visit a number of historically known salt sources in Kujavia and Great Poland. These included several spots near Zgłowiączka (district Lubraniec), Wrąbczyn (district Zagórow), and Łęczyca. While salt springs are still visible in some cases, there are no signs of ancient working of any kind (*Fig. 6.9*).

Poland: summary

At present it is not possible to demonstrate the existence of prehistoric salt exploitation in the centre and north of the country, though it remains likely that this took place. In Little Poland, there is some evidence for Late Neolithic exploitation at Barycz near Wieliczka, and briquetage from sites near Kraków. The south-east of the country, in particular the area around Sanok in Przemyśl vojvodship, would appear to merit field examination on a larger scale than has occurred so far, but the apparent absence of any wooden installations in such salt streams and springs as are known there makes it appear unlikely that any major prehistoric workings will turn up.

Chapter 7
The exploitation of rock salt using wooden troughs. Experiments conducted at Băile Figa in 2008–2010

DAN BUZEA

Introduction

From ancient times salt has been considered a particularly valuable natural resource. Today we take advantage of its special properties, regardless of where we find it: on the surface or deep underground, or in brine (salt water springs). We use salt to cook and preserve food, for therapeutic purposes, in industry and research, etc. Even though we now know how to exploit salt and to benefit from its properties, we cannot help wondering what the situation thousands of years ago was like. How was salt exploited and used in prehistoric times?

The National Museum of the Eastern Carpathians, Sfîntu Gheorghe (Romania), tried to find an answer to this question by turning to experimental archaeology, and taking into consideration one of the most important archaeological discoveries of the Sub-Carpathian area of Moldavia (Romania), dated to the 6th–5th millennia BC, at Lunca-Poiana Slatinei, jud. Neamț (MONAH 2008; WELLER – DUMITROAIA 2005), and from Beclean–Băile Figa, Bistrița-Năsăud County, where one of Europe's oldest salt mines, dated to the 3rd–2nd millennia BC, was discovered (this volume).

The most important results revealed so far by the archaeological investigations are four wooden troughs discovered in the stream at Beclean–Băile Figa (see chapter 4). One of these troughs was excavated in 2005 by the geologist Ioan Chintăuan. The other three were discovered in 2007 and 2008, during excavations conducted by the National Museum of the Eastern Carpathians and Exeter University.

The salt massif found here is particularly hard. During the excavations, workers attempting to dislodge pieces of salt often failed, as the tools they were working with (iron pick-axes), broke when they struck the rock (CAVRUC 2010, 22). During rescue excavations in 2008, several attempts to collect salt using the scoop of a mechanical excavator were made. Despite the machine's power, only superficial scratches were made in the surface of the rock salt.

Until recently, few plausible and convincing arguments had been adduced for the various hypotheses on how the troughs functioned in producing salt. In articles published in recent years on salt exploitation during the Bronze Age in Transylvania and Maramureș, only archaeological considerations have been taken into account (CAVRUC 2010, 17). Preisig's interpretation of the troughs as appliances for directing water jets onto the rock salt in order to create holes or depressions in it – upheld by V. Wollmann, rejected by I. Chintăuan, and treated with scepticism by A. Harding – has recently received support from V. Kavruk (CHINTĂUAN 2005; HARDING 2009; PREISIG 1877; WOLLMANN 1996).

Optimal conditions for conducting an experiment were achieved only in 2010, when the Experimental Archaeology Camp headed by Dan Buzea ran alongside the excavation.

The first experiments

The first experimental trough was made in 2007 on the pattern of those discovered during the excavations in Trench I. The trough had these dimensions: length 120 cm, width 26 cm, base width 17 cm, internal length 100 cm, internal width 13 cm, internal depth 18 cm; the piece was cut from a single spruce log

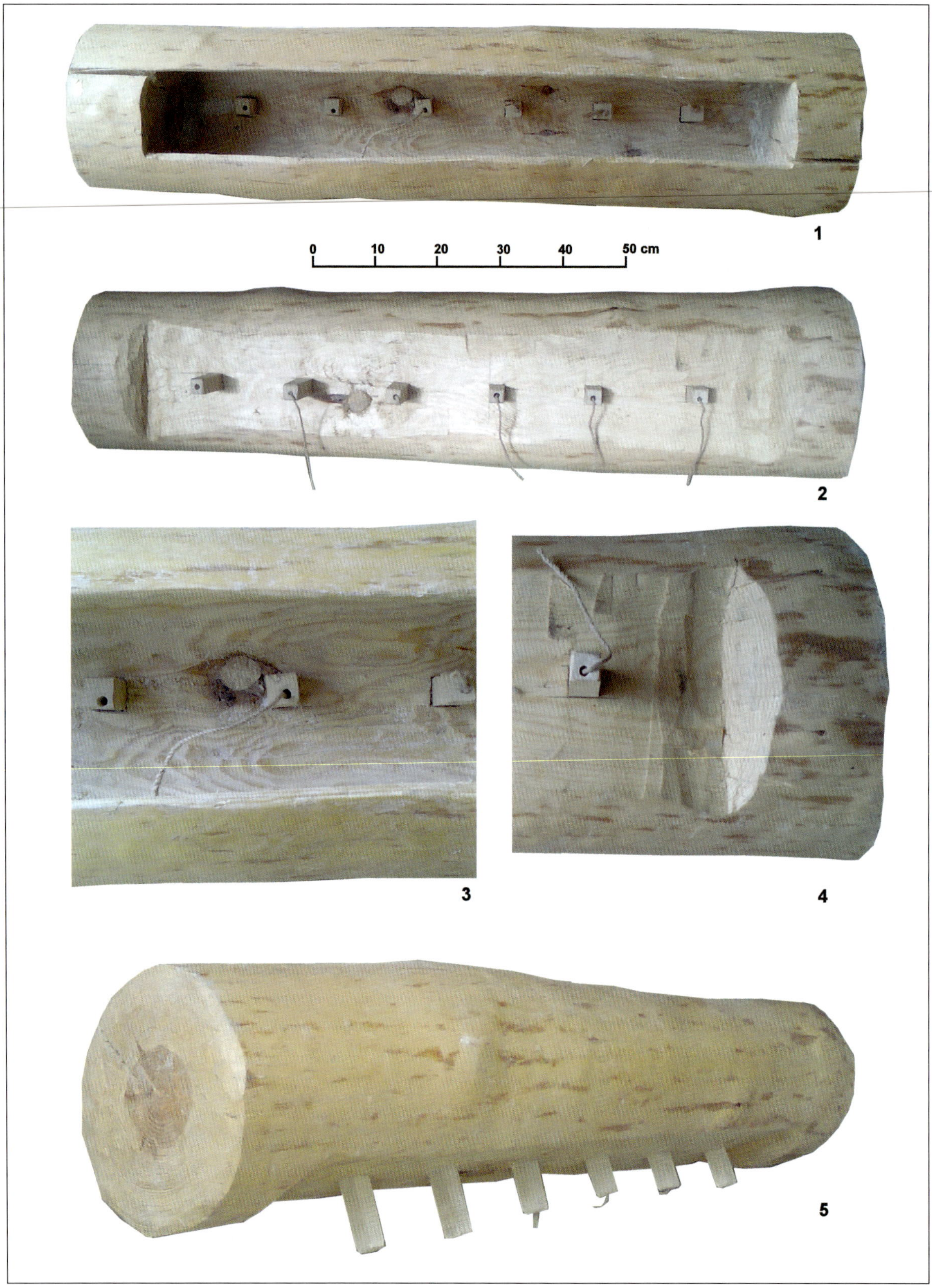

Fig. 7.1. Trough made by Gheorghe Puşcărău in 2007. 1. From above; 2. From below; 3. Interior detail; 4. Detail of the base; 5. General view. Photos: M. Domboşi.

that had a diameter of ca. 30 cm (*Fig. 7.1*). The shaping was done with an iron adze, iron hammer and a set of iron chisels. The trough was made in around 16 hours by Gheorghe Puşcărău, an amateur carpenter and wood carver.

In the bottom of the trough, along a median longitudinal line, six rectangular perforations (sides measuring *ca* 2.5 cm) were made at equal distances (*ca* 10–15 cm) from each other. A wooden plug was pushed into each of the perforations; prior to their insertion, the plugs were drilled along their longitudinal axis and a hemp cord inserted.

The trough was meant to be used in experiments scheduled for the 2008 season. Because that year no salt deposits were available, the experiments were postponed until the next season. The trough was stored in unsuitable conditions and during summer 2009, its base cracked.

The experiments of 2009

In 2009, we were able to commission a new trough from Marian Marius, a wood carver and keen amateur archaeologist.

For making the trough, the following materials were used:

- Trough: spruce log with a diameter of 40 cm;
- Plugs: oak;
- Cord: twisted hemp.

The resulting trough measured 180 cm long, 55 cm wide, and internally 160 cm long, 20 cm wide and 20 cm deep (*Fig. 7.2*). The hollowed out part had a rectangular cross-section. Seven rectangular holes (sides *ca* 4 cm) were made in the bottom of the trough, along the median axis, at equal distances from each other (*ca* 10 cm). A wooden plug was inserted into each of the perforations.

The plugs were made of oak, and were of similar size and shape; they were rectangular in shape and their lengths varied between 18 and 20 cm. At the top they were tapered, while in their upper part they were rectangular in section (4 x 4 cm) and had a collar that was considerably wider than the perforation in the bottom of the trough. They were drilled along their longitudinal axis (0.5–1 cm in diameter) using a gimlet (borer).

A three-twist hemp cord was introduced into each plug. The cords were of *circa* 30–40 cm in length, and at the upper end of the plug they were knotted.

The trough was then filled with water, and the manner in which the holes in the plugs could be opened or closed by means of the hemp cord was tested. When the knot was detached from the upper opening of the plug, the water from the trough trickled along the hemp cord, forming thin jets of water.

In 2009 the experiment was halted because a proper surface containing sufficient exposed rock salt where the trough could be set up was not available. Time constraints also meant that there was insufficient time for experimenting with the installation. The trough was stored in a cellar near the site, in a slightly damp environment.

The experiments of 2010

Between 10 and 20 August 2010, The National Museum of the Eastern Carpathians, with the support of its partners (the 'Cucuteni for the 3rd Millennium' Foundation from Bucharest, the 'Carpații Răsăriteni' Cultural and Scientific Association, and Exeter University, UK) and the financial support of the National Cultural Fund Administration of the Romanian Ministry of Culture, organised, as part of the 'People

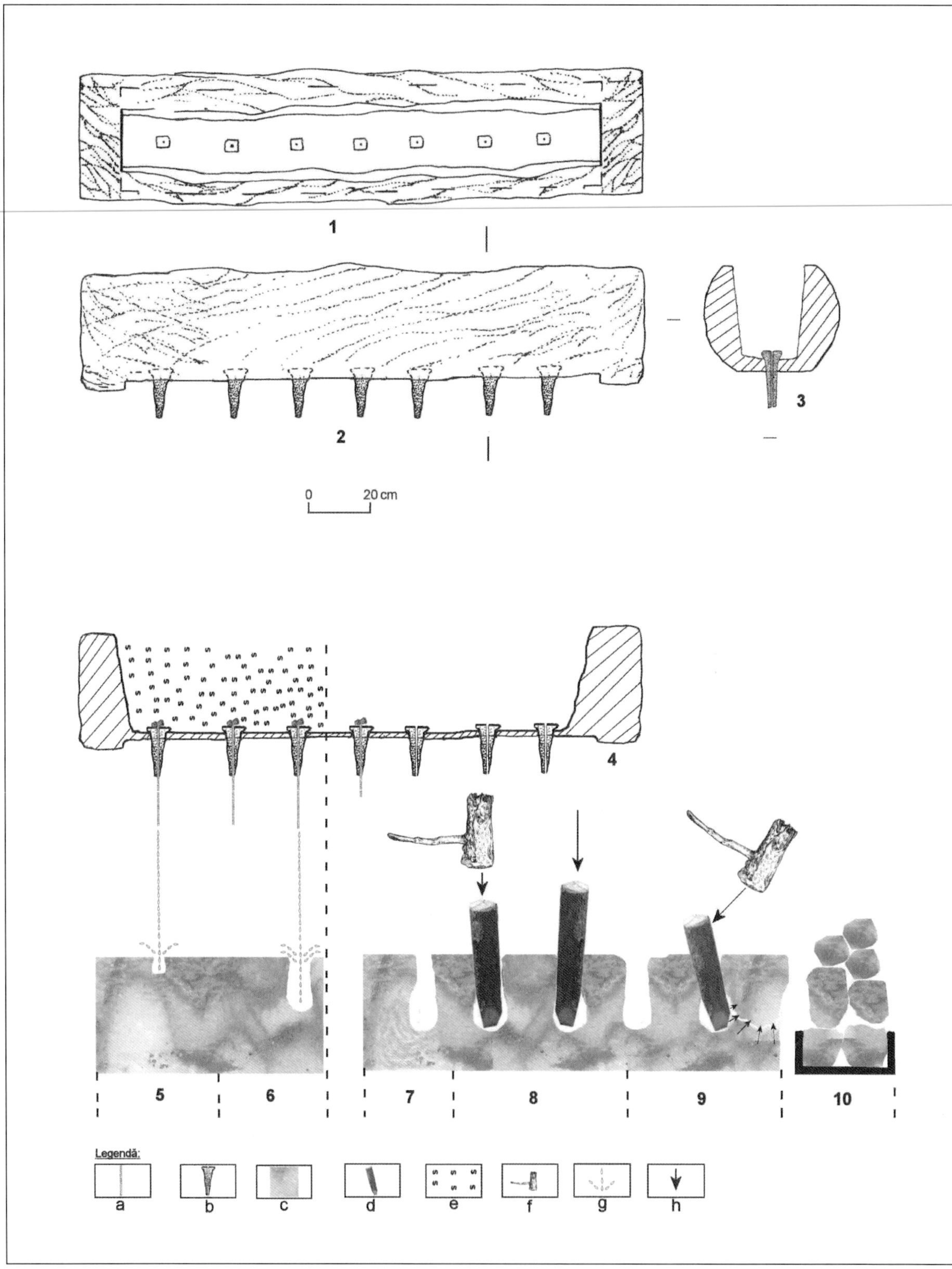

Fig. 7.2. Trough made by Marian Marius in 2009. 1–3. Plan, side view and cross-section; 4–10. The experiment in action; 4. Trough with water passing through the plugs; 5–6. Formation of depressions in the rock salt by the jets of water; 7–8. Insertion of wooden wedges into the holes and hammering; 9. Hammering the wedges sideways to break off chunks of rock; 10. Dislodged pieces of rock salt.

of the Salt' Project, the Experimental Archaeology Camp from Beclean–Băile Figa, Bistrița-Năsăud County.

The camp was designed to provide answers about the methods of prehistoric salt exploitation, by means of a number of experiments. Two methods were investigated:

- The production of salt by *briquetage*, that is, by boiling brine in coarse ceramic vessels until the water evaporates and salt crystals are left behind;
- The 'trough' method, using the replica trough described above.

The present paper is concerned with the second of these (BUZEA 2010).

The archaeological situation at Băile Figa, as it existed on 10 August, 2010, provided optimal conditions for setting up and performing the experiment, as this was the first time since the excavation began in 2007 that a large enough surface of the salt bedrock was exposed. This surface, fully explored from an archaeological point of view in 2010, was located in Trench III (systematically investigated between 2007 and 2010) in the northern part of the site, on both banks of the Sărat stream.

Mr Gheorghe Pușcărău was again co-opted into the project team to make the wooden installation on which the trough would sit. During the course of one day, he fashioned the legs from oak posts. An adjustable system for fixing the trough was set up at both ends; at each end, the legs of the trough were inserted, each consisting of three wooden posts secured with hornbeam splints (*Fig. 7.3*). To feed the trough with water, a 7 m long and 10 cm wide channel was carved from oak (*Fig. 7.4*).

During the days preceding the deployment of the trough, other wooden tools necessary for the experiment were produced: a wooden sledgehammer (maul) and several wooden wedges (chocks) sharpened at one end. The experiment itself lasted two days.

Day 1: 18.08.2010 – During the first day, the trough was suspended on the adjustable legs, at a height of 80 cm, over the rock salt. In this area, the salt surface was flat. After the trough was set in place, the salt rock was washed thoroughly to remove mud and dirt.

Fig. 7.3. Gheorghe Pușcărău setting up the stand for the trough in 2010.

Fig. 7.4. Water running along the 7 m oak channel into the trough.

The trough was aligned along an east-west direction, and a flow of water from the Sărat stream was started, using 10 litre plastic containers. The trough received a constant supply of water for eight work hours, and the total water fed to it amounted to *circa* 300 litres.

During this first day, all seven plugs were opened, and the water ran along the cords, forming thin streams that trickled constantly onto the surface of the salt rock. Even though the installation functioned properly, seldom requiring intervention to unclog the plugs from inclusions in the water, the expected results failed to materialise. Thus, after eight hours of continuous operation, only very shallow circular perforations were noticeable where the water hit the salt rock. Six depressions (from the seven plugs) were produced, of which the largest had a diameter of *circa* 2 cm and a depth of 1 cm. Given this unsatisfactory result, it was clear that such a method was grossly inefficient.

The same day, a second attempt was made using salt water, and the hemp cords were removed from the plugs so as to produce a more powerful stream of water. This attempt was also a failure, because the water flowed in a disordered fashion, splashing over the surface and not hitting the rock in one single place. At the same time, the trough required a much larger supply of water.

After the first day of the experiment, the results were far from encouraging, and the suggested method was starting to find less favour.

Day 2: 19.08.2010 – This time the trough was installed where the salt bedrock sloped slightly, from south-east to north-west in the trench. The trough was once again suspended *circa* 80 cm above the rock. To easily feed it with water, a channel 7 m long, was fitted to the trough; it carried fresh water (rainwater) carried in plastic containers by members of the team, from a pool near Trench III (*Fig. 7.4*).

By 11 a.m. the trough was in place, and from noon a constant supply of fresh water (approximately 300 litres) was poured into the channel, which was positioned so as to ensure a steady flow of the water straight into the trough (*Fig. 7.5*).

This time, we decided to open only five (nos 1, 2, 4, 5 and 6) of the trough's seven plugs (numbered 1–7 from left to right). After approximately two hours from the start of the experiment, the results were beginning to be visible to the naked eye. The water flowed smoothly along the inside of the plugs, guided by the hemp cords, forming thin water jets which quickly started to perforate the rock salt. After two hours of operation, the dimensions of the depressions were impressive: a diameter of 4–8 cm and

Fig. 7.5. The trough in action.

a depth of 5–8 cm (*Fig. 7.6*). The fresh water bored easily into the salt rock, while at the same time the running water created narrow cracks and channels as it dribbled over the surface of the rock.

One problem that occurred during this experiment, somewhat disrupting the operation of the trough, was the clogging of the upper ends of the plugs with inclusions (vegetation, loose mineral particles, insects, etc.) from the fresh water. This was rectified by simply moving the hemp cord about, dislodging the blockage and restoring the flow of water.

After around eight hours from the start of the second day, five circular depressions were created on the surface of the salt, with these dimensions: plug 1: diameter 8 cm; depth 13 cm; plug 2: diameter 6 cm; depth 15 cm; plug 4: diameter 8 cm; depth 15 cm; plug 5: diameter 6 cm; depth 12 cm; plug 6: diameter 8 cm; depth 17 cm. The 3rd and 7th plugs were kept stopped.

Without any action on our part, the water running out of the holes created by the water jets created two cracks (channels) in the salt rock. The first, from the 1st and 2nd plugs, was *circa* 1.5 m long and

Fig. 7.6. Depressions formed in the rock salt by the water jets.

Fig. 7.7. Wedges hammered into the depressions to break up the rock.

15–20 cm deep. The second formed between the 4th, 5th and 6th plugs, and had a length of *circa* 1.5 m and a depth of *circa* 10–25 cm.

For the next stage of the experiment, the trough was dismantled. Several wooden wedges (chocks), *circa* 10 cm in width and sharpened at the top, were inserted into the depressions and channels on the surface of the rock salt (*Fig. 7.7*). The wedges were driven into place using the wooden sledgehammer (maul). Where the distance between the cracks and perforations was *circa* 5–10 cm, the sledgehammer broke off large chunks of salt rock, approximately 30 x 15 x 5 cm in size.

We also inserted wooden wedges into the two channels (cracks) in the salt rock, with several wedges in each channel, hitting them hard with the sledgehammer in an attempt to dislodge large blocks of salt rock (60 x 30 x 30 cm); the attempt was however unsuccessful, despite the best efforts of the team to dislodge such large pieces. After one hour of repeated attempts to extract the blocks, the experiment was halted.

September 2010: During September, staff of the National Museum of the Eastern Carpathians took part in work to secure the excavation site at Băile Figa.

During the four weeks after the August experiments, the archaeological area was covered by water from the Sărat stream, and the salt rock was submerged throughout this period. After the water was pumped out and the salt rock cleaned, the experiments were resumed by Mr Marius Domboşi, who attempted to obtain massive lumps of rock salt by using the wedges driven into the channels, which were well preserved under water. On this occasion, Mr Domboşi used the longitudinal parallel channels to carve several transversal grooves (at 30 cm intervals), and used a modern iron axe to cut out large lumps of raw salt. The pieces of salt rock thus extracted measured around 30 x 30 x 30 cm. This method of raw salt extraction, using metal axes, proved to be efficient, but it required an intensive effort (cf the observations made at Dumitra: DEÁK 2008).

Conclusions

These first results obtained during the 2010 experiments from Băile Figa have shown the unexpectedly high efficiency of the trough as an implement for boring into the salt rock by means of fresh water. Even though there is still much more to do, we can state that it would appear to have been strikingly complex, sophisticated and efficient in terms of process and productivity. If we also take into account the facilities for draining off the water, as well as those for storing the extracted salt, it is evident that the extraction activity at Băile Figa was highly complex.

Given the fact that several such 'troughs' were discovered along the course of the Sărat stream, at relatively short distances from each other, we can assume that these too were part of an installation designed to transport water from the point of collection to the point of utilisation, such that the fresh or slightly briny water could be saved and re-used as it passed from trough to trough.

In the future we plan to continue experiments on the functioning of the wooden troughs by testing other possible methods for extracting salt:

- Through the evaporation of brine in wooden troughs;
- By crystallising the salt on the hemp cord during the summer months (CHINTĂUAN 2005, 75–78);
- Concentrating brine by allowing it to trickle through the cords into another container beneath (HARDING 2009).

Chapter 8
Prehistoric technologies of salt production in the Carpathian zone

ANTHONY HARDING – VALERII KAVRUK

The material presented in the preceding pages has given an indication of where, when and how salt was produced in the countries that lie in the Carpathian zone of Europe. At the same time it has raised many questions about production and usage, both for the wooden installations we have described, and for other methods of producing salt.

There were four principal means of obtaining salt in the prehistoric world (with variations, these include the main methods still in use today). The most obvious, and the one requiring least equipment, is through solar evaporation of brine in salt lagoons and lakes, whether of sea water, of inland salt lake water, or of brine from springs. This method is that found throughout the Mediterranean at the present day, and known to have been used in antiquity (for instance on the coast north of Rome). It is (potentially, at least) what happens around the Great Salt Lake (Tuz Gölü) of Anatolia, and other salt lakes in the world. Where the sun shines hotly enough for long enough, the evaporated brine forms crystals which can simply be collected; though for human consumption one would today expect further purification to take place. This method is not generally applicable to the Carpathian zone, though in hot spells in the summer it is noticeable that brine streams can dry up to a trickle, and salt crystals form on the stream bed. This would not be a very effective way of obtaining salt, however, as the crystals are quite small and not easily collected except with the pebbles on which they form.

The second method in general use is that of quarrying rock salt. This method is used by peasants in Romania today, though (as discussed in Chapters 3 and 7) it is arduous and requires the use of heavy tools and/or the application of fresh water to break up the rock. Modern salt mining is of course highly automated and can cut or blast through the rock relatively easily. The blocks thus obtained must then be processed, i.e. ground up small (where intended for human consumption, very small, in fact into powder); itself an arduous process. In much of Europe today, mined rock salt is the principal source of salt, but it may be questioned how much mining went on in prehistoric times.

The other two methods are represented by the briquetage technique, in which brine is boiled in coarse ceramic containers, and by what we may term the trough technique, as represented by the installations seen at Figa and elsewhere.

Briquetage

Briquetage is widely known in Iron Age Europe, but in the Neolithic and Bronze Age it is occurs much less frequently. In a Bronze Age context, that round Halle (Saale) in central Germany is perhaps the best known, but it is found in other parts of eastern Germany, in western France and in England. As more finds keep coming to light, and it is often so abraded as to be almost unrecognisable, it is likely that what is known at present is only a small fraction of what was once present. In the context of the Carpathian zone, the finds of briquetage near Kraków in southern Poland are important, certainly in Bronze Age contexts and probably also in Neolithic ones as well (Chapter 6); and the Neolithic finds from Moldavia are highly suggestive. The find contexts of the Polish pieces are, however, relatively uninformative and the quantities small by comparison with what is found on the great briquetage sites of the Iron Age (for instance in the Seille valley in eastern France). It is striking that there is no certain briquetage on Bronze

Age sites in Romania, which is perhaps surprising given that the technology seems to have been well established in the Neolithic.

Interestingly, BODI (2007) found that producing salt by means of briquetage was more difficult than collecting salt crystals that had formed on wood after brine was sprinkled onto a fire; in fact, that it took much longer to crystallize salt in clay containers, and producing a salt cake was never achieved.

We have argued elsewhere that in the Bronze Age, at least, the briquetage technique was used for small-scale production, probably for domestic use and perhaps for special purposes, for instance in ritual performance. The few exceptions apart, the briquetage technique does not seem to have played a major role after the Neolithic in the Carpathian zone.

Troughs

The trough technique is represented, so far, on four sites in Romania and one (possibly two) in Ukraine. At present we have no indication that it is present elsewhere, but given that until recently only two certain examples were known in total and we now have twelve possible pieces, it would be surprising if more examples did not turn up. Admittedly their geographical spread is rather small, which might suggest this was a trick peculiar to Transylvania and immediately adjacent areas. This remains to be elucidated.

With radiocarbon dates now available for eight troughs, or parts thereof, it is possible to suggest some aspects of their form and development.

Băile Figa: four troughs

1. Trough found by I. Chintăuan: length 3.22 m, square holes in base (*Fig. 4.11*)
2. Trough in Trench 1, eastern sector: preserved length 1.33 m, round holes (Plate 17.1)
3. Trough in Trench 1, western sector: preserved length 1.72 m, round holes (Plate 10.7)
4. Trough in Trench XV, near the Chintăuan trough: length unknown (2 m visible in excavation), square holes (*Fig. 4.29*)

Caila: 1 or possibly 2 troughs (damaged end of one embedded in stream side, length unknown (*Fig. 4.81*); trough base found in stream bed (*Fig. 4.82*).

Ocna Dej: one trough; preserved length 0.605 m, original length unknown (sawn off); round holes. (CAVRUC – CHIRICESCU 2006, 237 no. 195, where it is wrongly stated to come from Valea Florilor; this information was provided by the Muzeul Etnografic al Transilvaniei, Cluj, where the piece is held; WOLLMANN 1996, 246, 417 Plate XVI, 4 lower, citing MUREŞAN 1964, though the article does not appear to mention the trough; on p. 417 footnote 331 a reference to p. 372 in the same volume is given (not seen); WOLLMANN – CIUGUDEAN 2005, 101 Plate III, 4). Not dated.

Valea Florilor: probably three troughs: only one originally published, length about 2 m (MAXIM 1971); part of two as found in the house of Professor Maxim in Cluj in 1994 illustrated by Wollmann and Ciugudean (WOLLMANN 1996, 246 Plate XVI, 3; WOLLMANN – CIUGUDEAN 2005, 98 Plate III, 2–3). From the scales on the published photographs, one appears to be about 90 cm long, the other about 83 cm. Round holes (*Fig. 4.1*).

Királyvölgy (Tisolovo): trough, preserved length 1.36 m, original length unknown (sawn off) (HARDING 2011). Square holes (*Fig. 5.8*). Trough base, fragment, square holes (*Fig. 5.11*).

Solotvino (Aknaszlatina): one trough, known only from a historical account and not certainly different from the preceding (SZATHMÁRY 1867). Apparently complete, with both ends present and ten pegs in the base (shape not known) (*Fig. 5.2*).

Form

Unfortunately no surviving trough is complete. The piece found at Aknaszlatina (Solotvino) is shown by Szathmáry as being complete, though as mentioned previously, it is uncertain if this is different from the Királyvölgy piece; and even if it is, we have no independent account of it. According to the rather inadequate drawing of the find from Valea Florilor, that trough was indeed complete, about 2 m long, enclosed at both ends and having 14 or 15 holes in the base (the text says 15, but the drawing shows only 14). The surviving fragments studied by Wollmann do not give any clue as to the original state. Trough 1 from Figa is 3.22 m long, one end undamaged, the other damaged so that it is unclear whether it was originally enclosed or not; there are 16 preserved holes out of an estimated original 20.

The troughs from Királyvölgy and Ocna Dej are sawn off, only one end being preserved. Troughs 2 and 3 from Figa are incomplete, with one end only surviving. Trough 4 is still sunk in the ground; the end that protrudes is broken, and the same is true for the Caila trough.

It is thus unclear whether the troughs were originally enclosed at both ends or not. No surviving piece has two ends present, but the drawing of the Valea Florilor piece shows both being present. Given the status of that trough, it remains uncertain what the original form was.

Holes and pegs

The holes, and thus the pegs that fit in them are of two shapes: round and square. Troughs 2 and 3 from Figa, that from Dej, and apparently that from Valea Florilor, have round holes and pegs, while troughs 1 and 4 from Figa, and those from Caila and Királyvölgy, have square holes, with pegs whose cross-section is square on the body and more round at the neck, where a collar is present to stop the peg slipping through the trough. All pegs are perforated. Trough 1 at Figa had twisted cord in at least one of the pegs. Trough 3 at Figa had a needle-like object formed of a wood splinter wrapped in ?bark inserted into one of the peg holes (Catalogue no. 7). Neither method would have been watertight. This may have been intentional (to allow a slow drip of water through the peg) or it may, in the case of the wooden needle, have been intended to stop up a peg at certain moments.

Pairs of troughs?

There is some evidence that the troughs did not function alone but in groups of two or more. Troughs 2 and 3 at Figa were found only a metre or so apart though their original position is unknown. Troughs 1 and 4 lay very close together, only 1–2 metres apart. At Valea Florilor parts of no less than three troughs were apparently present in the same shaft or pit. There is no evidence from the other sites.

This might suggest that the troughs were intended to function in series or in parallel, either side by side to increase the output, or with the water from one pouring into the next.

Date

Interestingly, the radiocarbon dates on troughs and pegs indicate a chronological development, with the round holes/pegs lying earlier than the square (Figa Troughs 2 and 3 lie in the 16th and 15th centuries BC, while troughs 1 and 4 lie in the 11th–9th centuries, Királyvölgy and Caila in the 12th–11th centuries BC; the broad error terms of the Valea Florilor trough give a range of three centuries or more). The big difference between the dates suggests that this is a real technological and/or typological development. One reason might be that round pegs in round holes were too likely to slip through the bottom of the trough – though the addition of a collar would seem to obviate the problem. Square pegs could not of course be twisted around like round ones, which might make for more stability and a more watertight sealing.

Table 8.1. Trough dates in relation to peg shape.

Site	Date cal BC 1σ	Date cal BC 2σ	Peg shape
Figa 3 (combined dates)	1607–1516	1612–1502	Round
Figa 2 (combined dates)	1496–1452	1501–1438	Round
Valea Florilor	1378–1128	1425–1012	?Round
Királyvölgy trough	1191–1000	1259–927	Square
Királyvölgy base	1112–948	1130–919	Square
Caila	1187–1041	1211–1008	Square
Figa 1 (combined dates)	1050–981	1112–938	Square
Figa 4	895–812	904–806	Square

Figure 8.1 shows this material diagrammatically.

Method of operation

A number of methods have been suggested for how the trough technique would have operated. This discussion goes back to Preisig's original publication of the Királyvölgy discoveries, when he suggested

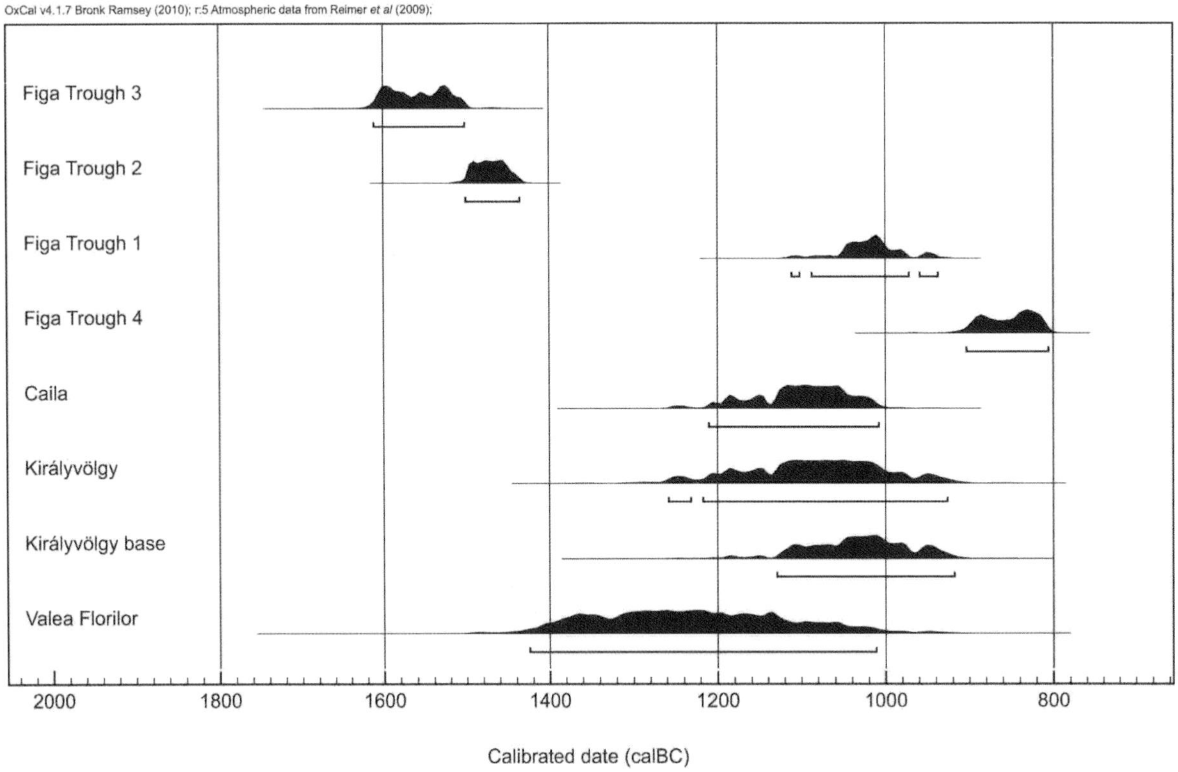

Fig. 8.1. Diagram showing the relative dates of troughs from different locations.

that water was introduced into the trough and allowed to drip slowly onto the rock salt, enabling it to be broken up more easily. This is the method explored in the experiments described in Chapter 7, and which seems on present evidence to work effectively.

One can probably dismiss any idea that the troughs served as containers in which brine was boiled, for instance by means of the addition of heated stones. If that were the case, we would surely have found evidence in the form of burnt or cracked stones on the sites; as far as we are aware, such evidence has never been forthcoming.

The fact that perforated pegs were inserted into the holes at the bottom of the troughs, and filled with cord or wood needles, suggests that the intention was to allow water to percolate slowly through the apertures. In theory the percolation could have taken place upwards by capillary action, to enable concentrated brine to be collected in the troughs, but in practice this seems unlikely. It is most likely, as suggested in Chapter 7 above, that the brine percolated slowly downwards.

What happened after that is also a matter of speculation. Preisig suggested, and Buzea follows this suggestion, that fresh water dripped onto the salt rock. But there are other possibilities. Harding has previously described a small experiment carried out by a former student at Bristol University, Katie Killick, who followed up a suggestion that brine might drip slowly down from the trough into a second container beneath (HARDING 2009). In this small reconstruction, a mixture of earth and salt water was put in a plastic container, with a hole in the bottom. This was filled with a cork, itself perforated and the perforation filled with twine. The container was placed over a second container (*Fig. 8.2*), and the whole left for several hours, being stirred occasionally. After several hours a clear liquid had run down

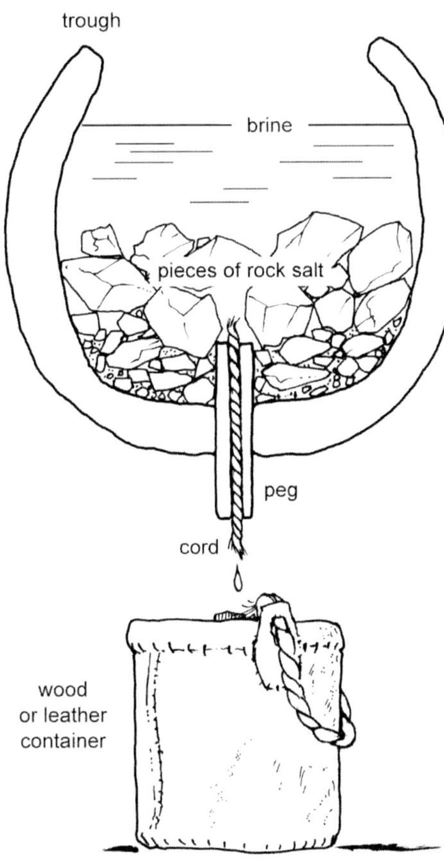

Fig. 8.2. A possible reconstruction of the method of concentrating brine using two containers.

the twine and collected in the lower container. The earth remained in the upper container. The liquid in the lower container was tasted and found to be very salty.

This leads to the following alternative suggestion about the way the troughs functioned:

1. The troughs are supported on a wooden framework, or suspended using bindings. This is supported by the fact that a binding was found around Trough 2 at Figa, and a large groove on the underside was clearly created for the binding to fit into;
2. Lumps of rock salt are brought from whichever surface was being worked and put into the trough; briny water is added;
3. The water, in proximity to the rock salt, becomes salty;
4. The resulting brine drips slowly out, and is either collected in containers (wood or ceramic), run along channels to other processing equipment, or drips onto a textile surface;
5. If the brine is channelled into a second trough, the process might be repeated in order to achieve a higher concentration;
6. If the brine drips onto textile, in sunny weather crystallization might occur quite fast and the resulting crystals could be picked off;
7. The cords might hang down from the trough a considerable way below the pegs, and salt crystals would form on them. By pulling the cord out the crystals might fall off and be collected; or they could simply be collected by hand.

What is not clear from this discussion is the function of the other Bronze Age installations on the sites, in particular at Figa. It will be noted from Chapter 4 that the northern excavation area in particular (Trench III) contained a series of enclosed constructions, some consisting of plank walls, some of wattle fences. Some of the constructions seem to have been roofed. The wattle constructions form oval or roundish enclosures (traces were also found in Trench I). We suppose that these enclosures were storage ponds for brine, perhaps brine waiting to be evaporated, or places where lumps of rock salt were accumulated and introduced water allowed to become salty prior to evaporation. Such fences would not have been watertight, unless caulked with moss or some similar substance, so any possible reconstruction needs to take this into account.

Wattle fences or walls are present at Săsarm, probably too at Caila, and are known from Cucuieți and other sites in Moldavia. At Săsarm some of them appear to be revetment for adits going into the stream sides. Since it is easy to construct such fences, one can imagine that they may have had multiple uses.

The Iron Age

In the Iron Age it seems that the technology employed was quite different. While it cannot be excluded that the troughs continued to be used, and obviously they were still lying around on site and potentially available for use, there is no evidence that they were in fact used. Instead, from Trench I at Figa we can see that shafts and pits were dug down to access the rock salt. One of these (the so-called box construction) was timber-lined, the timbers being braced by cross-members, presumably to provide safe and (relatively) easy access for the miners, given that a depth of four or more metres had to be descended. We may note that the ladder recovered from this area was well over 5 m long.

At the rock face iron hammers may have been used to break up the rock salt, though none are present at Figa. The single iron pick found in the northern excavation area seems quite flimsy, and far removed from the heavy sledgehammers used by present-day workmen to break up the salt (cf Chapter 3). In the Roman period it appears that a range of mechanical devices were available (WOLLMANN 1996), though none has so far turned up in a certain salt mine in the Dacian province and none is known from the preceding Iron Age.

Conclusions

While much is still uncertain about the details of the technology employed for the prehistoric extraction of salt in the Carpathian region, the available evidence has allowed a series of suggestions to be made. These revolve particularly around the way the wooden troughs were used, but other aspects, such as the fenced enclosures, also remain to be fully elucidated. We are confident that the progress of research, in both Transylvania and Moldavia – and potentially in other neighbouring countries – will see a resolution of at least some of these difficulties. Since our own work has only taken place in the last eight years, prior to which almost nothing was known for certain, it is highly likely that the renewed interest in prehistoric salt production will see further dramatic advances in the coming years.

Chapter 9
Tracing 6th–5th millennium BC salt exploitation in the Carpathian Basin

ESZTER BÁNFFY

Introduction

In the Neolithic, salt exploitation has been detected south and east of the Carpathian ranges, above all in Moldavia, Bulgaria, Serbia and the Adriatic Karst (GAYDARSKA – CHAPMAN 2007; MONAH 2002; MONTAGNARI KOKELJ 2007; NIKOLOV 2008; TASIĆ 2000; URSULESCU 1977). This short paper attempts to give an overview of Neolithic salt research within the Carpathian Basin. Archaeological traces and research on Neolithic salt production is rather uneven in this area. While in several cases it is highly probable that a site specialized on salt exploitation, or certain regions served as the home of exchange networks, there are hardly any archaeologically tangible finds to prove it. Tracking salt exploitation through archaeological finds and sites cannot yet be observed before the early 5th millennium. I therefore begin by summarising the results from this period, the Carpathian Late Neolithic.

Direct evidence from earlier Neolithic phases of the central Carpathian Basin is hardly more than hypothetical. But given that salt is a necessity for the maintenance of human life, and becomes even more essential when the proportion of meat and fish protein in the diet decreases as reliance on plant foods increases, salt procurement must certainly have taken place. Along with the appearance of the first domesticated plants, at the onset of the Neolithic transition (in the Carpathian Basin: 6000–5800 cal BC), consumption of hunted and foraged food gradually changed to a diet that relied more on cereals. The Neolithic lifestyle also means animal domestication. Sheep, goat and cattle were brought to the Carpathian Basin in an already domesticated form, so supplying them with salt was essential. These physiological needs are complemented by the need for salt in the preservation of food, the production of leather, or the consumption of festive meals within the community; in other words salt also has a social function (CHAPMAN – GAYDARSKA 2003).

Thus, where there is no local salt source, salt must have been brought in from elsewhere. In order to make a first attempt at reconstructing possible early salt networks, it would seem sensible to collect data on settlement patterns and other data from a range of different areas, especially salt regions. Such networks may turn out to be one of the major keys to understanding the different forms of social and cultural development in the 6th and early 5th millennia cal BC. In this sense it is worth trying to trace them back to the earliest Neolithic centuries, and specifically to the Neolithic transition. In the second part of this paper I shall try to reconstruct a possible new salt route which may have played a role in the neolithization process from Transdanubia to the whole of Central Europe.

In order to avoid confusion by looking for finds that are unlikely to be present, it is necessary to distinguish three types of activity which can be connected to salt in the Neolithic. First, it is only worth looking for finds, mainly potsherds, that were used for evaporating brine in the immediate vicinity of salt springs. The second concerns the transport to salt-poor regions. In order to detect such routes, the transport network needs to be traced back with the help of finds that were imported or exported, or other indicators such as local commodities (e.g. raw materials). The way the salt was transported is unlikely to be visible as pottery vessels may not have been used, more probably bags made of leather or textile; like salt, these are archaeologically invisible. Thirdly, there is no direct archaeological signal for the consumption of salt. Logically, this should be explored in regions that have no salt springs, for instance in the centre of the Carpathian Basin. The present summary, therefore, tries to infer salt use and salt routes by means of this threefold approach.

Evidence for salt production and trade in the Late Neolithic

Salt production techniques in the Neolithic may simply have consisted of heating brine in pots. More complicated techniques probably only developed in the Middle or Late Bronze Age in the Carpathian Basin.

In the Carpathian Late Neolithic, the first half of the 5th millennium BC, the Lengyel group covers a large area from Transdanubia to the Munich Basin, Moravia and Małopolska (south-east Poland). The latter is one of the major salt regions of Europe, and these are the first sites of the Lengyel culture with pottery that can be connected to salt production, in the form of briquetage (Fries-Knoblach 2001; Kaczanowska 2006, 104–5). Given the well-documented cultural connections within the Lengyel area in central Europe, it probably indicates that the communities living in the huge Lengyel-Stichband-Moravian Painted and Münchshöfen cultural area were connected with the Małopolska salt production, and that the Lengyel communities also traded in salt. Other commodities that reveal the existence of dense cultural and exchange contacts can be followed by tracking finds like marble armrings (Zápotocká 1984). Similarly, in the East Carpathian salt area, pottery associated with brine evaporation occurs in the Cucuteni A culture (Cavruc – Chiricescu 2006, 202).

The contacts, seen from an archaeological point of view, were traditionally seen as having gone through Moravia. In the past two decades, however, it has become clear that as well as the Moravian valleys another route can be reconstructed. Nándor Kalicz pointed to a series of Lengyel sites along the north-east Hungarian mountains that are linked with related East Slovakian sites and finds (Kalicz 1994). In this way, the Lengyel settlement of western Hungary expanded through eastern Slovakia as far as the Wieliczka salt region.

By contrast, further south, the east Hungarian Alföld groups do not seem to be connected with the Małopolska area. It is much more likely that by means of successful long-distance networks with the East Carpathian mountainous region they obtained salt from the springs in that region (Sófalvi 2005, 22–3). The existing links and exchanges between the Alföld late Neolithic cultural groups and their contemporary Transylvanian neighbours is proven by many features that appear in both areas, and by imported finds, as recent research has shown (Ignat 1998; Kalicz – Raczky 1984; Paul 1992).

The beginnings of sedentary life, food production – and access to salt

In the centuries before the Late Neolithic Lengyel, Tisza and related cultures, and even before the long-lived LBK groups, the Starčevo, Körös, and Criş cultures occupy roughly the southern half of the whole Basin (*Fig. 9.1*).

Earlier research on the East Carpathian area was mainly targeted on identifying salt springs, and only secondarily on working out when they may have been in use. Today, there is firm evidence that brine springs started to be exploited in the Eastern Carpathians already at around 6000 BC (Danu *et al.* 2010; Munteanu *et al.* 2007; Weller *et al.* 2008; Weller – Dumitroaia 2005). Pottery of the Criş culture occurs at Lunca-Poiana Slatinei, Calabatoaia and Cucuieți, at or near salt springs (Cavruc – Chiricescu 2006, 195). There are early Neolithic archaeological sites, both in Transylvania and in Moldavia, which specifically reflect a close correlation with salt springs (Ursulescu 1984, 41; 2001). Further, Neolithic sites were investigated further west in Transylvania, which have been considered possibly to have been linked with salt (Lazarovici – Lazarovici 2011).

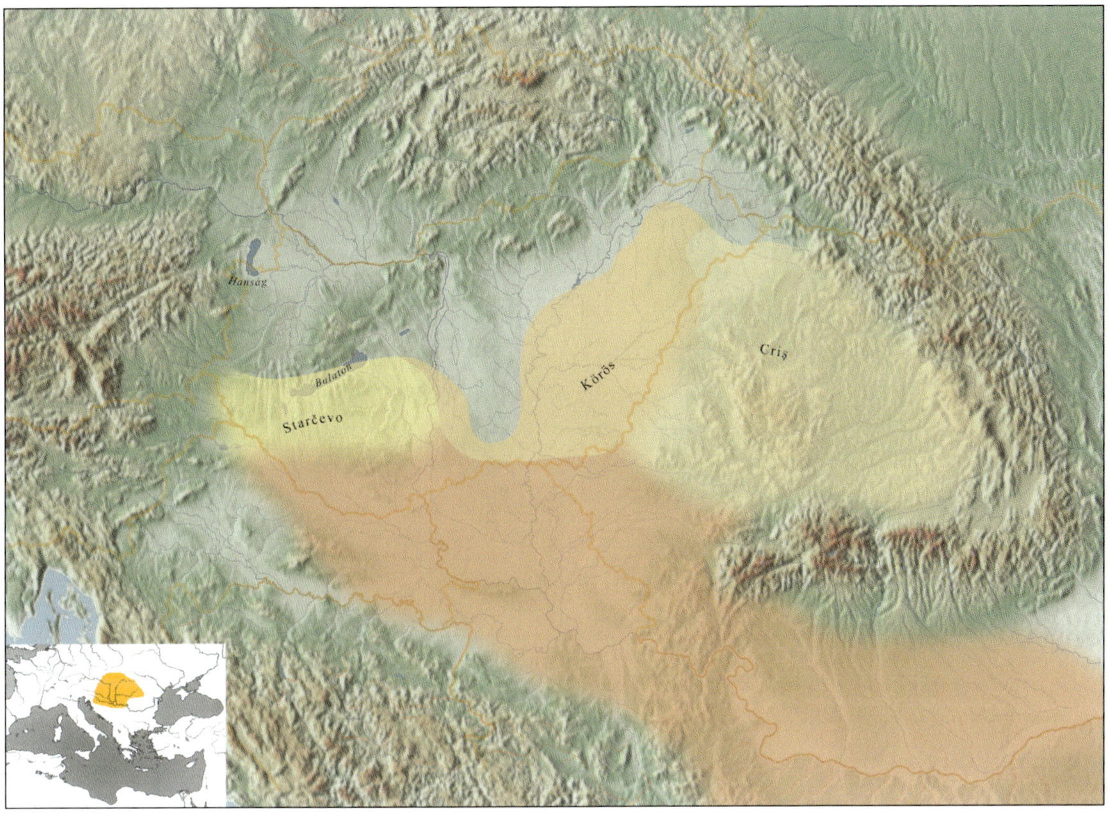

Fig. 9.1. Map showing Early Neolithic cultural formations in the Carpathian Basin (author).

One such settlement is Gura Baciului near Cluj (LAZAROVICI – MAXIM 1995). The site is located in the immediate vicinity of salt springs (MAXIM, Z. 1999), and was inhabited for the whole of the Early Neolithic Criş culture, with broad contacts within the East Carpathian Early Neolithic (LAZAROVICI – MAXIM 1995, 346–352). The radiocarbon dates indicate a long sequence of some 800 years (SPATARO 2008, 92). The site was considered to play a role in salt exploitation and trade by the excavators and also by Nenad Tasić (TASIĆ 2000, 40).

The most recent discussion of salt concerns an area further west still, near the present Romanian-Hungarian border. The publication of Méhtelek and the Méhtelek group is seen as part of a regional sub-group within the Körös-Criş culture. The contact routes, supported by a number of finds including the special flat „slab" figurines that are typical of the Méhtelek group, can be reconstructed convincingly between the East Hungarian and West Romanian Körös-Criş and the Transylvanian and Moldavian salt regions. Kalicz himself presumed that – apart from participating in the long-distance obsidian network – salt may have played a crucial role within these contacts, with Méhtelek communities playing a part in the salt trade (KALICZ 2011; 2012, 121). Within the Körös culture, a certain funnel-shaped, coarse pottery type (as at Ibrány-Nagyerdő: KALICZ 2012, Fig. 3/10) is sometimes assumed to have been used for evaporating brine (FRIES-KNOBLACH 2001, Taf. 6/1). Similar types from later periods are thought to have been used as briquetage. However, if one assumes that one will only find briquetage near brine springs, there is no point in arguing for an export of such pots from salt-poor regions.

The implication that the salt network may have continued into early LBK is only supported by the reviving connections between the Northern Alföld and Transylvania when late LBK groups formed, i.e. in the last centuries of the 6th millennium cal BC. The Esztár group in the eastern part of the Alföld, with its painted pottery, which is closely related to the Lumea Noua group in Western Transylvania, needs to be mentioned here (GLIGOR 2009; GOLDMAN – SZÉNÁSZKY 1994).

After drafting possible long-distance connections between Transylvanian salt and the East Hungarian Körös culture, we need to consider how communities of the Körös culture in southern Hungary might have supplied themselves with salt.

According to the maps showing the main salt recources in Europe, the central part of the Carpathian Basin (present-day Hungary) is entirely lacking in salt springs. The only possibility might have the exploitation of small alkali ponds in the lowlands, e.g. those close to Szarvas in the Alföld, but these were barely enough to fulfil the needs of even one village community (*Fig. 9.2*).These ponds occur even more scarcely close to the narrow settlement area along the Danube, where a dense settlement pattern of the Körös culture was found, probably traces of a smaller group of people who stayed there for some generations, then withdrew from the entire region. Possible reasons for the Körös groups abandoning the Danube might include climatic conditions (BÁNFFY forthcoming 2013). In the marshy oxbows of the Danube even a slight rise in the water table placed the small arable islets under water or impossible to access. Social problems were perhaps equally important: these groups were left in an isolated position with apparently no contacts to nearby Starčevo groups to the west, with a rather unsettled area to the north and south, and a sandy area to the east. After attempting to establish life on the Danube for some generations, this "enclave" probably had good reason to return to the Alföld, using the same routes by which they had arrived, the stream valleys crossing the sand back to the valleys of the Tisza and Körös. In this way, they may have taken advantage not only of larger community identities and bigger kinship structures, but also easier access to commodities essential for life, like salt.

Fig. 9.2. Alkali ponds in the Danube-Tisza interfluve region (photo R. Balázs).

Salt and the Neolithic transition in western Hungary and central Europe?

It is a fact that the western Carpathian Basin, Transdanubia, is a region virtually without salt. Even the nearest salt deposits, at Tuzla in Bosnia, are located far away, in a region with which the first Transdanubian farmers (the people of the Starčevo culture) did not have particularly close contacts. It has also become an accepted fact that the communities of the first farmers of central Europe, the people of the LBK culture, formed in the western part of the Carpathian Basin. These communities spread rapidly to present-day Austria, Moravia and southern Germany along the Danube, to the Halle/Saale region to the North and towards southern Poland. After some 150 years, a second wave of expansion followed. As a result, vast regions in Europe, from the Paris Basin to Ukraine turned to the LBK type of sedentary and farming lifestyle (*Fig. 9.3*).

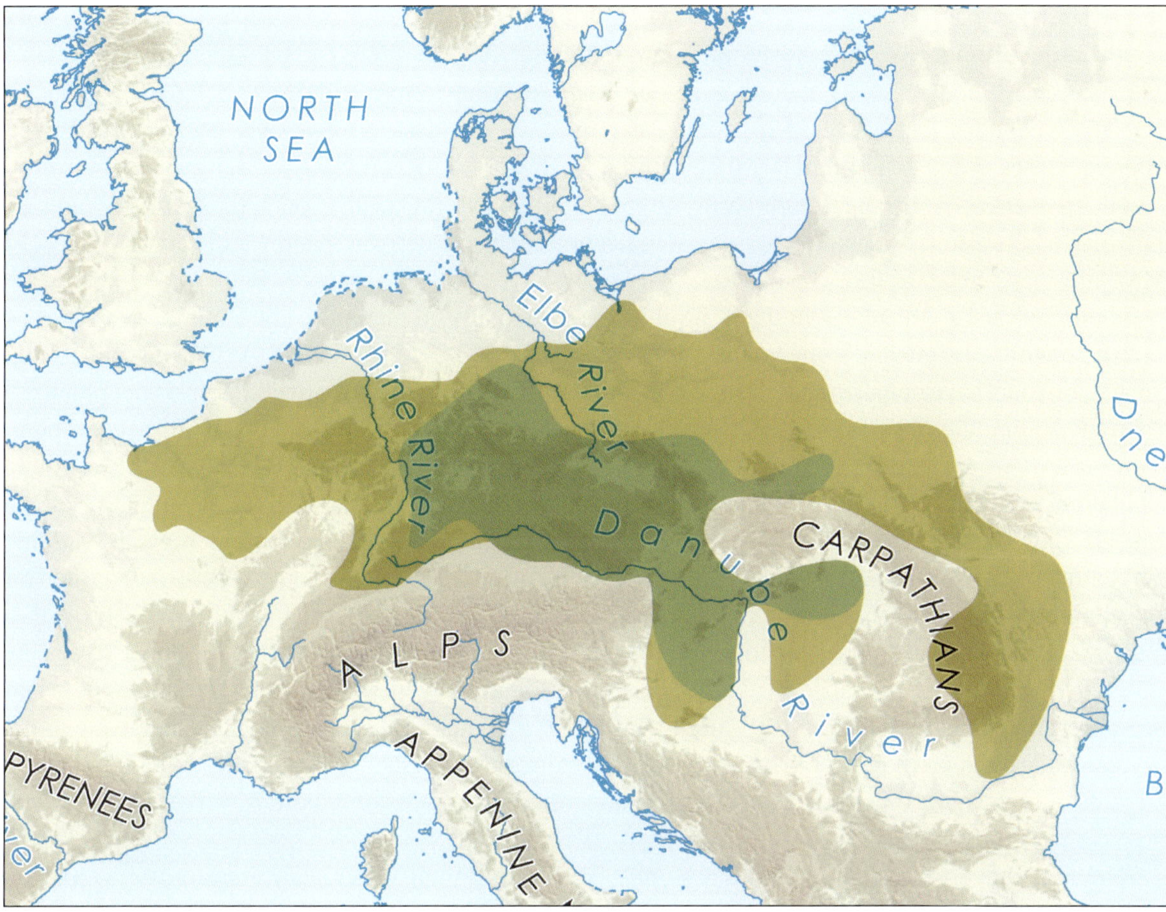

Fig. 9.3. Map showing the distribution of the first farmers (LBK culture) in central Europe (shaded area) (modified after HANSEN *2007).*

The first farmers of North Balkan origin (the people of the Starčevo culture), crossed the Drava and occupied a wooded, hilly landscape until they reached the marshy Balaton region, where they must have met and mingled with local foragers. The expansion to the heartland of central Europe was so rapid that it left no typological differences in the archaeological record (LÜNING 1988; QUITTA 1960) and – at least for the beginnings of the expansion – successive phases cannot be be pinned down by means of radiocarbon dates (GLÄSER 1991; LENNEIS *et al.* 1996; LÜNING 2005, 60–62). Flint provenance studies also bolster the idea of migration and direct contact (LENNEIS – LÜNING 2001; PAVLŮ – VOKOLEK 1992; RAMMINGER 2011). The LBK migration from Transdanubia has recently been supported by other methods: ancient DNA and stable isotope analyses

In his discussion of the possible causes triggering migration, David Anthony offers several explanations. For me, his most important observation was that negative 'push' and positive 'pull' forces can be distinguished among the causes leading to migration (ANTHONY 1990, 898). Consequently, some of the causes triggering migration arise in the area of origin: overpopulation, deterioration of the climate, drought, famine and social tension. Several examples can be quoted for the 'pull' type from history: one common feature is preliminary contact with the target area. A low population density, fertile soil, proximity to water, good climate, and possible raw material sources are all factors that make a given new area attractive. If exchange relations can also be created and maintained, an area of this type usually attracts settlers.

'Push' forces in terms of an overpopulation or climatic deterioration, can be rejected in the case of the LBK expansion. In contrast, there is evidence for almost each of the 'pull' forces. The main emphasis is on the incentives that triggered the largest expansion in prehistory. Obviously, these were several different causes, since "migration is a social strategy" (ANTHONY 1997, 22).

The presence of Szentgál radiolarite in preneolithic Moravia, in Southern Germany and of Danubian shells also in Germany are modest indications of these networks in the archaeological record (MATEICIUCOVÁ 2004; 2008).

In order to justify the idea that the Central European expansion may (also) be explained in terms of contact networks, let me cite some arguments for similar prehistoric phenomena. As Curtis Runnels and Tjeerd van Andel have noted, the Neolithic expansion can, together with its innovations, be conceptualized as a trade commodity forming the basis of wealth, whose acquisition was probably an attractive option (RUNNELS – VAN ANDEL 1988, 102). In Germany, the earlier, western contact network of the Mesolithic population was also exploited by the early farmers, and judging from the growing intensity of the contacts, they probably improved and expanded it. Andreas Zimmermann has convincingly argued that the central places (*Zentrale Orte*), whose emergence can hardly be dissociated from the settlement concentrations (*Siedlungskammer*) (ZIMMERMANN 1995, 61–2), were the main settings for down-the-line exchanges. The communal identity, the remarkably uniform material culture and most likely the similar social structure remained virtually unchanged for many generations. This would suggest that the contact networks remained in force until the time of the later LBK groups in Transdanubia and the Flomborn phase in Germany, perhaps even for some time afterwards. What remains to be explored is the mutual interest forming the basis of these contact relations. What commodities were traded between these distant groups?

The real range of traded commodities must have been much wider than appears today, and no doubt included wares that leave no trace in the archaeological record, such as furs, textiles, leather, and foodstuffs. This is not to mention the possible exchange of craftsmen, of individuals introducing a new technology to a particular region, or marriage and kinship ties between groups. These contacts will perhaps never be detected unless many more LBK cemeteries containing well-preserved skeletal remains are uncovered and analyzed.

The idea that the commodity received in exchange for productions of Transdanubian origin may have been salt first occurred to me when I visited the Bad Nauheim–Niedermörlen settlement. The finds from this site reflects surprisingly strong ties with Transdanubia. The Mörlener Bucht area is rich in haematite deposits (SCHADE-LINDIG – SCHWITALLA 1999, 28), but this red mineral is not so rare as to form the basis of exchange relations. However, Bad Nauheim lies in an area rich in salt. Excavations in the Niedermörlen area have uncovered extensive Iron Age and Roman salt mining. Wendelin Leidinger, who studied the remains of Neolithic salt production in more northerly regions (Westphalia), has described in detail how the earliest Linear Pottery communities in that area produced salt (LEIDINGER 1983; 1996). It is therefore possible that this easily transportable commodity, so essential for diet, food preservation and animal health, was exchanged for commodities from the Danube valley. Tasić considers salt and access to salt deposits (regions with soils with a high salt content and brine springs) a major factor in the neolithization of the Balkans (TASIĆ 2000). Trade in salt also played an important role in the cultural development of this region and in its contacts with other areas in later periods (MONAH 1991).

One negative factor must be introduced into this discussion. The Hallstatt region in the Salzkammergut, rich in salt deposits, lies closer to Transdanubia than the other sources considered here – so why would the first farmers have gone so far for salt? The distribution of Early Neolithic sites indicates that the migration route went along the north side of the Danube, through the Munich Basin to south-west Germany. Although they were closer, the Hallstatt salt mines were unknown in the 6th millennium BC and can thus be rejected as a possible source.

It may become possible to detect salt (sodium chloride) in the matrix of pots that were used for salt production (HORIUCHI *et al.* 2011). But in the Neolithic briquetage would only be found where salt production occurred, i.e. near salt springs. How salt was transported from salt-rich areas to the salt-poor central Carpathian Basin is hard to say, since it is not only the salt that is archaeologically invisible but also any bags or sacks used for carrying it.

If we consider the distribution map of the first farmers in central Europe, the darker patch on *Fig. 9.4* shows how far the first wave of the LBK reached. It becomes apparent that the three furthest points – western Germany, the Elbe-Saale region, and Małopolska – are major salt regions. Was this pure chance?

To connect salt and the central European migration at the start of the Neolithic is, in the current state of knowledge, hardly more than speculation. This is hardly surprising, given the scarcity of data on the Mesolithic-Neolithic transition and the process of neolithization. In the future, new methods may be able to test the hypothesis of a connection between the Neolithic transition and access to salt.

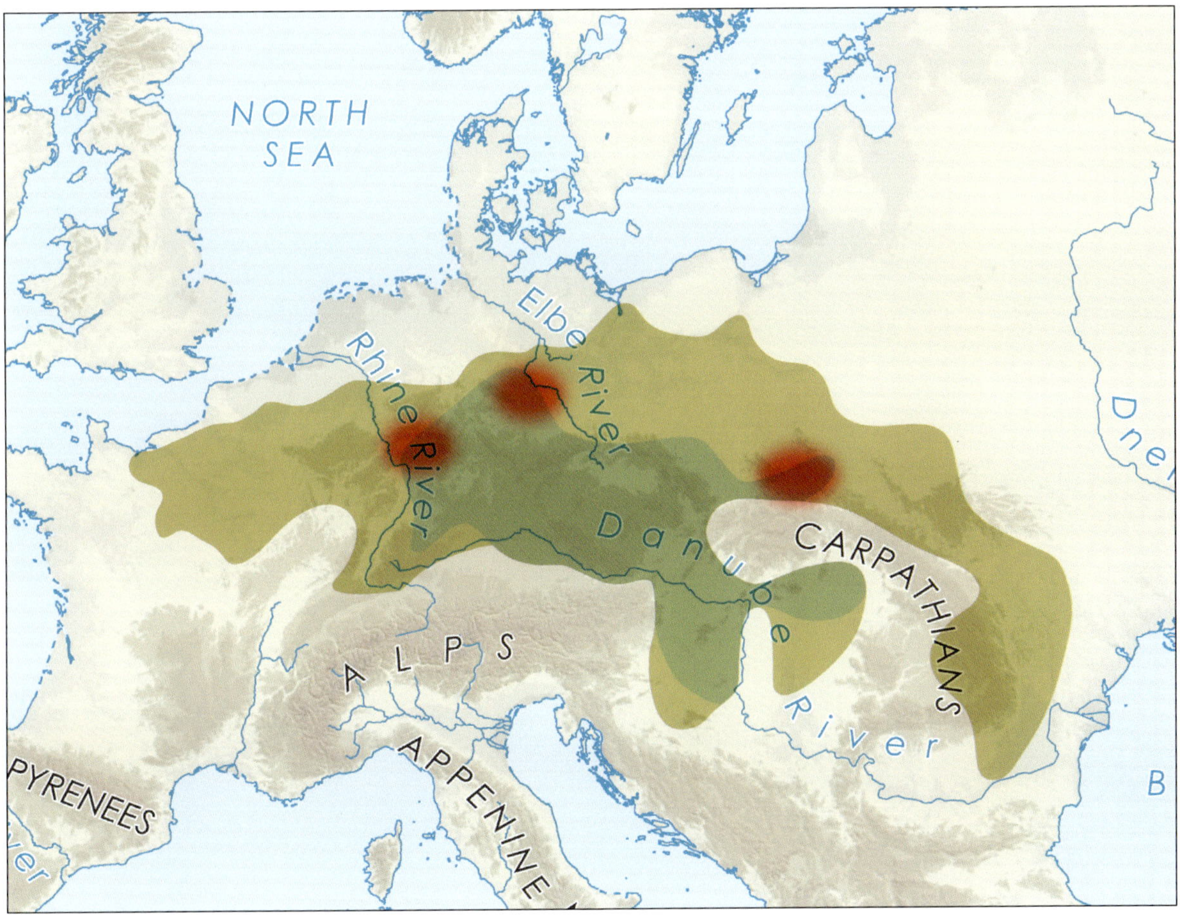

Fig. 9.4. Map showing the distribution of the first farmers (LBK culture) in central Europe, with three major salt regions marked by dark shading (modified after HANSEN *2007).*

Chapter 10
Prehistoric production and exchange of salt in the Carpathian-Danube Region

ANTHONY HARDING – VALERII KAVRUK

State of research

Research on the prehistoric exploitation of salt in the Carpathian-Danube-Pontic region has sometimes included consideration of salt exchange. Thus, Katalin Biró thought it possible that salt from the Carpathians was 'exported' to the Hungarian Plain from the Neolithic onwards (BIRÓ 2004); we have both separately suggested that salt from Transylvania and Maramureş was exported to the Hungarian Plain during the Middle and Late Bronze Ages (CAVRUC 2010; HARDING 2011). John Chapman and Bisserka Gaydarska (2003) assumed that the Tripolye mega-settlements were supplied with salt obtained by brine evaporation in sub-Carpathian Moldavia. On the other hand, Dan MONAH (2008, 22–3) considers that salt cakes obtained by brine evaporation in 'briquetage' were prestige objects and not intended for mass consumption and use. He also believes that it was rock salt from the Buzău salt-rich zone that was brought to the salt-poor Tripolye areas for mass consumption and use (MONAH 2012). Recently, Vassil Nikolov, in publishing the Provadia-Solnitsata excavations, suggested that this was a centre that supplied salt to a large part of the east Balkan area during the Neolithic and Eneolithic (NIKOLOV 2008, 47, 339).

These suggestions have attracted some criticism. Thus, Magda Mircea and Marius Alexianu contest the Chapman-Gaydarska hypothesis by showing that, contrary to one of its premises, there were local salt sources in the North Pontic area rich enough to ensure the supply of the mineral to Tripolye groups (MIRCEA – ALEXIANU 2007). Kavruk (Cavruc) asserted that as long as the character and the scale of prehistoric salt production in the Carpathian–Danube–Pontic region remain unresolved, it is premature to talk about salt exchange (CAVRUC 2008, 79).

All these authors – with the notable exception of Monah – approach the issue of prehistoric salt exchange without making any distinction between its various forms in ancient societies. At the same time, both viewpoints indicate the difficulty (even the impossibility) of finding positive indications of prehistoric salt exchange.

This chapter aims to define the essential features of the various salt production and exchange systems in the prehistoric societies of the Carpathian-Danube-Pontic region; to identify the archaeological indications that correspond to each type of salt production and exchange, and to integrate the prehistoric archaeological evidence available in the Carpathian-Danube-Pontic area into a typology of salt production and exchange.

Methodological considerations

The exchange of materials and goods implies at least four components

- the origin point (winning and production area) of the material
- the destination
- the distance between the two
- the benefits to each side in the exchange.

however, the main difficulty in finding evidence for prehistoric salt exchange lies in the fact that this soluble and consumable mineral is archaeologically 'invisible'. Beyond the few prehistoric salt mines that are currently known, finds of prehistoric processed salt are very rare. In most cases the processed salt is found near the extraction sites, and only in one case has it been found in the place where it was used. This is the case of the marine salt found in a cave near Zakros in Crete inside a Middle Minoan pot, together with residues of animal origin, probably salted food (Kopaka – Chaniotakis 2003).

Archaeological evidence for salt exploitation

Since only certain elements are archaeologically visible, we consider here only those aspects which can be certainly documented. These fall into a number of categories:

- The type of source exploited (brine, rock salt, crystalline products collected from brine springs or streams)
- The salt content of the source
- Production method (evaporation in large pottery vessels, in small vessels (briquetage), evaporation without pottery vessels, mining at shallow depths, mining at considerable depth
- Production complexity (equipment, other aspects of the technology)
- Scale of production (local or distant consumption needs)
- Site location (in relation to settlement and to routes to other areas)
- Processing (further concentration, removal of impurities)
- The final product (purity, form)

Geographical, historical and ethnographical considerations of the salt trade in the Carpathian-Danube-Pontic region

Exchange routes have at least two ends: origin and destination. Every indication that salt was produced in quantities exceeding those of local needs suggests an industry orientated towards exchange; the point of origin may thus be recognised relatively easily. It is much more difficult to discern the destination points. For this reason it is helpful to consider the geographical situation, as well as written and ethnographic evidence for salt exchange in historic times and in traditional culture.

*Natural routes connecting salt-rich and salt-poor areas (*Fig. 10.1*)*

The salt-rich, salt-poor and salt-less areas of the Danube-Carpathian region are often connected by a network of natural communication routes. The best opportunities for long-distance exchange of salt between these areas are given by the navigable rivers. Thus, the best options for connecting Transylvania and Maramureş to the Hungarian Plain are the rivers Someş (including its tributaries Someşul Mic and Someşul Mare), Mureş (including its tributaries Târnava Mare and Târnava Mică), and Tisza. The Tisza, as a tributary of the Danube, connects the Carpathian and northern Balkan zones. The rivers Morava and Drina, which are southern tributaries of the Danube near the junction of the Tisza with the Danube, provide good conditions to transport salt further south.

From a geographical point of view the saline areas of Transylvania, as well as those of Maramureş, seem ideally suited to have connections with regions to the west, thanks to the landscape and the fact that the most important rivers (with the exception of the Olt) flow in that direction. On the other hand, the Olt assures excellent connections between the salt deposits of Ocna Sibiului and the Lower Danube.

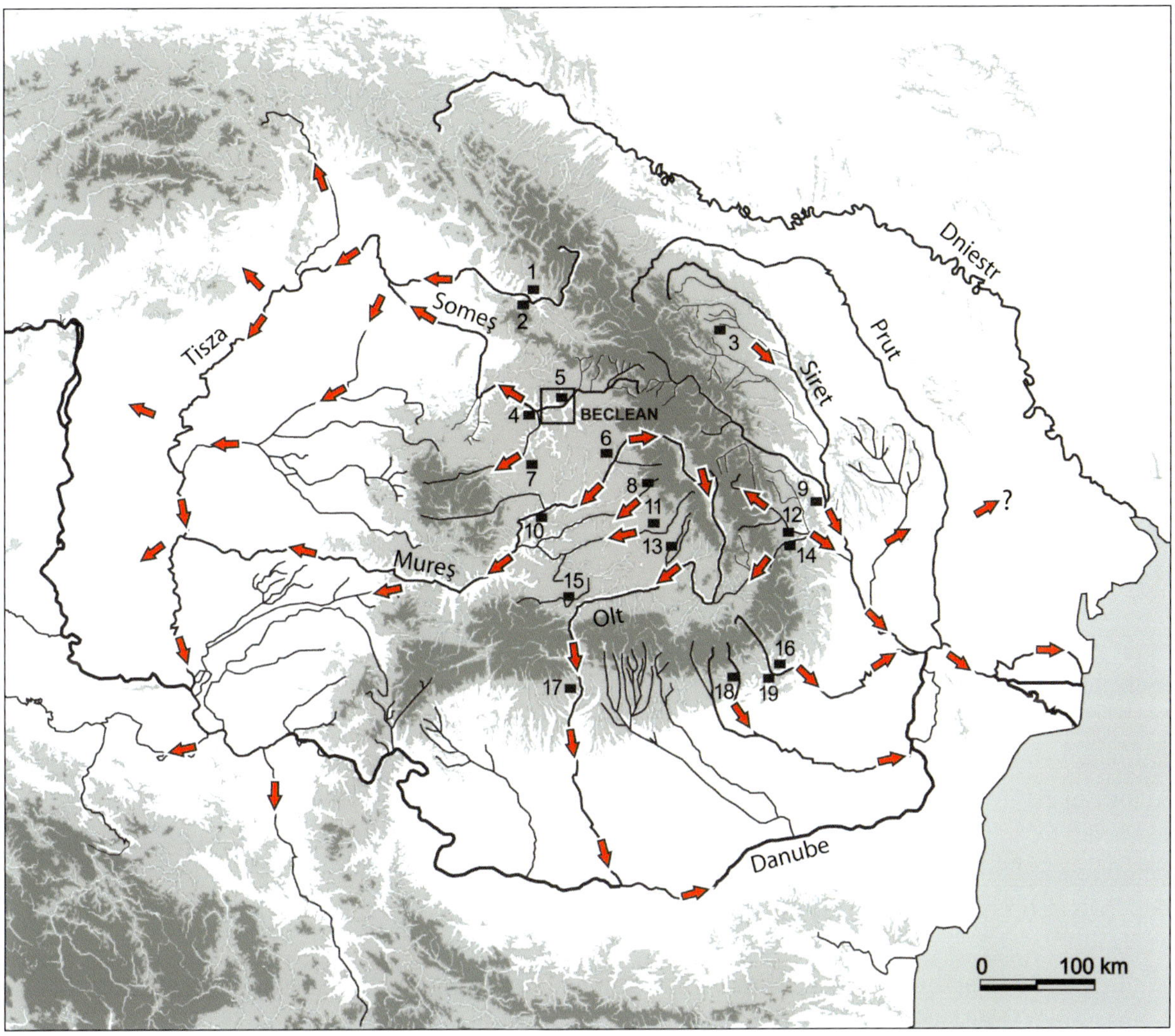

Fig. 10.1. Major rock salt deposits and main salt trade routes in the Carpathian Basin and adjacent territories (after Stoica – Gherasie *1981;* Medeleţ *1995;* Rustoiu *2005;* Cavruc *2005;* Marc *2006). Arrows show potential movement routes. Black squares: rock salt deposits: 1. Solotvino; 2. Coştiui; 3. Cacica; 4. Ocna Dejului; 5. Dumitra; 6. Ideciu; 7. Cojocna; 8. Praid-Sovata; 9. Sărata; 10. Ocna Mureş; 11. Corund; 12. Slănic Moldova; 13. Sânpaul; 14. Târgu Ocna; 15. Ocna Sibiului; 16. Lopătari; 17. Ocnele Mari; 18. Slănic Prahova; 19. Sărata Monteoru.*

By contrast, the natural communication routes with saltless eastern Transylvania (the Giurgeu, Bilbor, Borsec, Ciuc depressions, and the Tulgheş gorge) and salt-poor south-eastern Transylvania (Braşov, Întorsura Buzăului, and Baraolt Depressions) are more difficult of access. These regions are separated from the rest of Transylvania by the western chain of the Eastern Carpathians. Nevertheless, several gorges and passes offer traffic routes from the saline areas of Transylvania to its eastern and south-eastern salt-poor areas. One of the best ways to reach the salt-rich eastern area of the Transylvanian Plain from the salt-less Giurgeu Depression is to follow the flow of the Upper Mureş, through the Topliţa-Deda gorge. Once here the easiest communication route is along the middle stretch of the Mureş that leads to the Reghin – Solovăstru – Ideciu de Jos – Jabeniţa area, which is very rich in salt sources. This route was turned to good account as part of the local traditional economy; there is a peasant market place at Deda where one can still buy lumps of rock salt (Marc 2006). Another natural communication route which ties the salt-less Giurgeu Depression to the salt-rich Reghin area crosses directly through Ditrău village, over the Gurghiu Mountains, following the flow of the Gurghiu river. Until recently, peasants brought rock salt from Reghin to Ditrău by cart; they first followed the flow of Gurghiu river and then crossed the mountains on a forest path at Bătrâna Peak.[5] At the same time the Giurgeu Depression is connected to the Praid – Sovata saline area (one of the richest salt areas in Europe). The easiest way to get there from the Giurgeu Depression is to follow the route that connects Gheorgheni and Praid, along the Barzontul Mic river, through the Bucin Pass, and then along the Creangă Mare river and the valley of the Târnava Mică. It is hardly a coincidence that along this route a road called the 'salt road' was built and maintained from the middle of the 17th century (Horváth 1998, 49).

The massive occurrence of salt sources along the eastern slope of the Eastern Carpathians and along the southern slope of the Southern Carpathians made any transport of salt from Transylvania to Moldavia, Wallachia and Oltenia unnecessary.

Written and ethnographic evidence for the exchange of salt in the Carpathian-Danube-North-Pontic region

Since geographical constraints affect salt transport in recent times as well as in the past, historical and ethnographic evidence on production and exchange of salt may be a helpful analogy for understanding prehistoric production and exchange of salt.

Salt production locations

Medieval written records indicate the main mines in the Carpathian area from which rock salt was extracted for trade: Turda, Dej, Sic, Cojocna, Ocna Sibiului, Praid, Sânpaul, as well as the areas around Gurghiu and Bistriţa in Transylvania; Coştiui and Rona in Maramureş (Simon 2006). The ethnographic data add also several smaller sources of salt exploited for home consumption and/or for short-distance exchange.

In Moldavia and Bessarabia written sources mention salt mines in Cacica, Târgu Ocna, Grozeşti (Oituz Valley), Valea Sării (Vrancea area), Hârlău, Gura Slănicului (in the Trotuş river valley), Ibăneşti (Dorohoi areal), Poduri, Muntiori, Ireşti, and Ocniţa (Ichim 2006).

Six major mines are attested in Ţara Românească (Wallachia and Oltenia) as being in use in the Middle Ages: Ocnele Mari (Vâlcea), Telega, Slănic, Ghitioara, Teişani (west of Vălenii de Munte) and Ocna Mică (Târgovişte) (Ciobanu 2006).

Destinations of salt exchange and the main routes involved

Medieval written records attest the salt trade between Transylvania and Maramureş with the Hungarian Plain and Moravia. The main shipping routes of the rock salt from Transylvania towards Hungary were the Someş, Mureş and Tisza. Nicolaus Olahus, for example, in the middle of the 16th century, mentions that the salt from Transylvania was transported by boats on the Mureş and Someş to the Tisza, whence it was distributed and sold throughout Hungary. In Transylvania, another salt road was also important during the medieval period, coming down to the Mureş from Uioara to Alba Iulia and Deva, and from there to Banat. From the Ocna Sibiu area towards Oltenia and the Danube valley salt was transported along the Olt (Marc 2006).

According to some written sources, during the medieval period Carpathian salt was exported also to the Balkans, through Bosnia and Serbia. Thus, there was a salt warehouse at Salankemen near the confluence of the Tisza and Danube, south of the Danube; salt was transported to that point from

[5] According to Lenuţa Uştea, born at Subcetate near Ditrău.

Transylvania via the Mureş and Tisza (CIOBANU 2006, 90). According to Simon, between August 1514 and March 1516, some 226–266 tonnes of salt was exported from Hungary through the Smederovo custom house near Belgrade to the Balkans (SIMON 2006).

For Moldavia, in the mid-17th century salt was exported from there to Russia, Podolia, Ukraine and Turkey. From the port of Galaţi on the Danube, salt among other goods, was exported to Crimea, Constantinople and Egypt. According to medieval documents, the land route along which salt was transported for commercial purposes (so called 'salt roads'), was the Ocna road following the Trotuş valley, and then along the Siret down to Galaţi. Among the destinations that Moldavian salt reached by land, the documents indicate Braşov, Ghimeş and Ciuc. The residents of the Seat of Ciuc preferred to buy salt from Moldavia and not Transylvania (ICHIM 2006).

From Wallachia and Oltenia salt was traded mainly to the Lower Danube area. Thus, from Ocnele Mari salt was sold to Caracal, Bechet, Corabia and Islaz, Craiova, Calafat, Zimnicea and Giurgiu. From the Teleajen basin and Prahova valley salt mines, salt was transported down to the Danube: to Giurgiu, Slon, Greaca, Olteniţa, Călăraşi and Brăila (MARC 2006, 152). Also in the Middle Ages, salt from Ţara Românească reached Constantinople and Braşov. Even nowadays salt lumps, extracted from the Teleajen basin and Prahova zone deposits, are often brought to the Întorsura Buzăului depression in the south-east corner of Transylvania and exchanged.[6]

The forms of traded salt

In the Middle Ages, salt was usually transported as cubes of rock salt, of predetermined size, which helped to quantify the volume of salt sold. More rarely the salt was carried as shredded salt, in the form of pinches of salt or very small pieces (SIMON 2006). Ethnographic evidence shows that salt was traded only in the form of lumps. There are no proven cases of moving brine over long distances. According to recent ethnographic research carried out in eastern Transylvania (as part of the joint British-Romanian Project), local peasants obtained sometimes cakes of salt by boiling brine in ceramic vessels. After the cakes are created in this way, they are removed by breaking the vessels. The peasants then sell the cakes to people in neighbouring saltless settlements (CHIRICESCU 2013).

Means of transport

The means of transport used for the long-distance exchange of salt were riverine, as well as overland. The vessels used in the Middle Ages, as well as in recent times, were different types of canoe, boat and raft (SIMON 2006, 92). Rafting was one of the most effective means for transporting salt. In Transylvania this is attested since the Roman period (the documents mention the College of rafters – *dendrophorum*). In some places, rafts were used until the mid-20th century (MARC 2006). Ethnographers have noted the particular type of raft for transporting salt. These had a suspended floor made of beams placed at least half a metre above the raft base so that water does not affect the salt (MARC 2006). Orbán Balázs mentions that in the Topliţa Valley there was a special sawmill for cutting boards '12-arms' long meant for the 'salt-vessels' (cited by MARC 2006).

Among means of land transportation, medieval sources mention ox-drawn carts. The medieval documents also mention 'salt riders'. They carried salt lumps on horseback, in pouches, wicker baskets and leashes, or rope nets, by the mountain tracks or paths. The salt transported in this way was covered with blankets or other fabrics, to protect it against rain, drizzle, fog or dew (MARC 2006). This latter means of transportation was rather used for short-distance transport.

[6] Information from Dr. Dan Buzea, born in Întorsura Buzăului.

Storage of salt intended for trade

Authors reveal a constant concern for storing salt in good conditions before transportation. Close to the salt production places the salt was kept in fenced rooms, covered by leather, straw or other means, since salt exposed in unprotected areas deteriorated, mainly because of rain (MARC 2006).

Goods and benefits obtained in exchange for salt

Everywhere during ancient and medieval times and later, salt resources were a state monopoly. The state would sometimes allow private bodies, and especially churches, to exploit salt mines, in exchange for loyalty, services and rents.

During the Middle Ages salt production and trade produced significant revenues. Thus, in Hungary salt was the Crown's second income source, after taxes. In the time of Matthias Corvinus (1458–1490), for instance, the income obtained from salt constituted about 15% of the total budget of the Hungarian Crown (SIMON 2006, 92).

Ethnographic evidence attests to various forms of bartering (for sandstone, timber, potatoes, corn, etc).

Taking these facts into account, we can try to establish the type of salt production attested by the available evidence in Transylvania and Maramureş, and to suggest how it was integrated into the south-east European exchange network.

Site position in relation to settlement and external connections

In the present project, in no case were traces of habitation in close proximity to salt production places found. This suggests that the producers paid regular visits to the production sites, but did not live there.

All the sites and finds discussed in this book lie near navigable rivers: the Someşul Mare (Caila, Săsarm, Băile Figa, Ocna Dej), the Arieş (a tributary of the Mureş) (Valea Florilor), Tisza (Királyvölgy; Solotvino). In every case access to the rivers is easy (some sites lie right next to one of the rivers, while others are directly connected to them by natural routeways).

The final product: purity and form

In some parts of the Băile Figa site, both in the excavation and in pits dug by the developer for the spa construction, it was possible to examine the rock salt. In some places the salt deposit is very 'pure', being of an immaculate white colour, while elsewhere the deposit contains impurities and is grey in colour. It would hardly have been possible to 'clean' the salt. In any case, there is no evidence that the blocks and lumps of salt detached from the bedrock were processed in any way.

The image of the Királyvölgy finds (*Fig. 5.3*) shows a block of salt with depressions on its upper surface. Another such find may have come from Solotvino/Aknaszlatina (SZATHMÁRY 1867). The salt block has straight sides. Thus it seems possible that at least a part of the produced salt might have been formed of roughly cube-shaped blocks, what might have been suitable for long-distance transportation. However, some of the salt might have been in the form of irregular lumps and even shredded salt (shovels and wooden trays seem suggestive in that respect). Different forms might have been intended for different destinations, according to the difficulty of the route to be taken.

Taken together with the experimental evidence presented above, this suggests that production at the sites described in this book can be considered industrial production of salt intended for long-distance exchange. Historical and ethnographic analogies, and the cultural context of the Carpathian Basin and south-east Europe in the 17th–9th centuries BC, suggest that the main destination of this exchange was the salt-less territory of the Great Hungarian Plain and the northern Balkans. Both of these zones were among the most advanced cultural areas of Europe during the Middle and Late Bronze Age, yet were for the most part without salt resources of their own. We do not know what might have been exchanged in return for the salt – if anything.

Production and exchange of salt: conclusions

Who were the owners and the producers of the salt from Maramureş and Transylvania in the Bronze Age? Could it be local people, settlers or itinerant craftsmen? When one compares the distribution maps of finds studied here with those for the distribution of archaeological cultures in the time in question (about 1620–850 cal BC), a variety of groups were present. At the end of the Middle Bronze Age and at the beginning of the Late Bronze Age (approximately 17th–14th centuries BC) Transylvania was occupied first by the Wietenberg culture (Boroffka 1994) and then by the Noua culture (Andriţoiu – Vasiliev 1993), with the Suciu de Sus culture in Maramureş (Bader 1979; Kacsó 2003). From the most recent map (Kacsó 2008, Fig. 10), Solotvino and Királyvölgy are situated close to the northern 'border' of the Suciu de Sus culture area, while Ocna Dej, Băile Figa, Caila and Săsarm area lie on the northern periphery of the Wietenberg and Noua cultures, close to the border with Suciu de Sus-Lăpuş area. Valea Florilor lies in the Wietenberg and Noua area. These finds are thus situated in different cultural environments. At the beginning of the Late Bronze Age (about 15th/14th – 12th/11th centuries BC) Maramureş was occupied by the Lăpuş group, Transylvania by the Noua culture. In the next period (approximately 12th–9th centuries BC), the northern part of Transylvania and Maramureş are covered by the Gáva culture, encompassing an area much larger than that of the salt mines. In such circumstances we can conclude that the specific phenomenon of salt exploitation was extra- or supra-cultural. Other interpretations are of course also possible. However, there is a clear discrepancy between the technological continuity shown by our finds on the one hand, and the discontinuity in archaeological cultures, on the other.

Determining what goods might have been moved in exchange for salt is a matter of speculation, but one may consider the range of imported goods, visible and invisible, which have been suggested. Raw materials (wood, salt, gold, silver, copper and possibly tin) are readily available in Transylvania, but elaborate bronzes and other exotic goods might have been brought in from afar. It has often been suggested that spiral decoration on bronzes and bone objects indicates connections with the Mycenaean world, and there are now increasing signs that trade goods came up to the Black Sea from the Aegean.

One interesting possibility is that tin came to Transylvania in exchange for salt. Boroffka (2006) has drawn attention to the presence of tin-bearing ores in Romania, but there is no direct evidence that these were exploited in prehistory. Given the enormous volume of bronze in Late Bronze Age hoards in Transylvania, for which a large input of tin was necessary, one must imagine a very substantial tin trade across Europe. Whether such tin came from western Serbia, where tin is present (Đurman 1997), from the Erzgebirge (believed by many to have been inaccessible to Bronze Age technology), or from western or southern Europe (even as far as Cornwall), it is impossible to say. New methods for determining the origin of tin (Haustein *et al.* 2010) may enable more definite conclusions in the future.

While this remains speculative, what remains certain is that the evidence from northern Transylvania indicates a scale of production that we see as industrial in nature, with all that this implies for the movement of salt across significant distances in Bronze Age Europe.

Chapter 11
Conclusions and prospects

ANTHONY HARDING – VALERII KAVRUK

The preceding pages have presented field evidence and ethnographic and historical source material for an understanding of ancient salt production and exploitation in the Carpathian zone of Europe. Original fieldwork in Transylvania was supplemented by field visits to other regions and other countries in the area with a view to building up a comprehensive picture of how, when and where salt was exploited in ancient times in the region. These attempts have been successful as far as was achievable given the time and resources available; more remains to be done in all the countries involved if this aim is to be achieved.

The period which has been the major beneficiary of our work is the Bronze Age, since the bulk of the dated material from Transylvania, as well as the objects from Ukraine now in Sopron, Hungary, belong to that period. When we began our work, however, we had no means of knowing that this would be the case. Indeed, at the first site where we spent some time (Sânpaul, Harghita county) we expected evidence for Roman exploitation to be mainly in evidence, given that an altar recording a *conductor salinarum* was known from that locality, also the site of a Roman fort and *vicus*. The extraction of the first trough from Băile Figa in 2005 and our visit to the Bistriţa area a few weeks later, led to the obtaining of the first radiocarbon dates that indicated a prehistoric date. This was followed shortly afterwards by the publication of a date for the Valea Florilor trough, also falling in the Bronze Age. After that, more and more dates were obtained to confirm this dating, from Caila and Săsarm in Transylvania as well as from Királyvölgy (Tisolovo) in Ukraine.

Not all material fell in the Bronze Age, however. A significant number of dates from Figa belong to the pre-Roman Iron Age ('Dacian'), which has also been shown to have been a major period of salt extraction at that site. A somewhat earlier date was obtained on a timber from Sânpaul. Subsequently we were able to show that early medieval activity also occurred at Figa, while medieval and early modern dates were obtained from Sânpaul and from Királyvölgy and Tereblya in Ukraine. Finally, a date on a timber from a stream in Slovakia falls in all probability in the 18th century AD. Thus the dating evidence, taken as a whole, suggests a long timespan for salt extraction in the Carpathian zone; something that comes as no surprise, given the richness of the salt deposits, and the other evidence (e.g. medieval and early modern documents) that indicate extensive exploitation over the centuries.

If the Bronze Age dating of much of the material was of great interest and importance, of no less interest was the evidence for how the wooden installations were used to obtain salt. It must be admitted that at present this matter is not definitively resolved, though there are strong indications that the method suggested long ago by Eduard Preisig, that of letting fresh water drip through the holes in the trough base onto rock salt in order to create cracks and depressions, is correct. Alternative methods have been suggested, and may indeed work. But this is to consider only the wooden troughs, which have figured so largely in the discussion. Many of the other elements found at Figa remain mysterious – the wattle-fenced enclosures, for instance, and the plank-walled constructions, as well as the numerous wooden artefacts whose function can only be a matter of speculation. It seems highly likely from the situation of the Iron Age wooden constructions in Figa Trench I that the aim was to get at the underlying rock salt, presumably so as to break chunks of rock off and bring them to the surface for grinding up into usable pieces. Since in Figa Trench III the rock salt appeared close to the surface, and actually had Bronze Age timbers resting on it, it is most likely that something of the sort happened there too. How this relates to the use of the troughs, however, is uncertain, as no troughs have yet been found in Trench III (though in the nearby eastern tributary stream, structures similar to those in Trench III, along with a trough, have been found). The level of the water-table must also have caused problems for ancient miners, as it does for modern archaeologists; at present, it is only with constant pumping of water that excavation is at all possible in those areas where the wooden constructions are sunk into the stream bed. There is circumstantial evidence that the stream has altered course since ancient times, which might make the problem a little less acute, but surely not much.

The same problems would no doubt attend excavation in the other sites where there is direct evidence for prehistoric salt working, for instance Caila, where the land is relatively flat (the stream bank rises steeply only on the western side). Any excavation in the area where the piles and trough are visible would quickly reach the water table and require pumping.

This means that future work on these sites will require a very considerable input of funds, to enable even a moderate amount of excavation. There is the further point that the wood recovered needs immediate reburial or immediate conservation if it is to survive. We have removed relatively little wood from Figa, and all of it has either been conserved or (in the case of the troughs) is still undergoing conservation.

A short history of salt production in the Carpathian zone

Salt production began in the Carpathian zone in the early Neolithic, as the finds from Cacica and Solca in Moldavia demonstrate; abundant pottery of the Criş culture was found at these sites along with vessels that can realistically only have been used as briquetage (Andronic 1989; Ursulescu 1977). There is no evidence that exploitation started any earlier than this, though it is unclear what form such evidence from Mesolithic contexts might take. Investigations at Lunca-Poiana Slatinei also suggest that salt was the reason for the existence of this site (Weller – Dumitroaia 2005). On a European level, the evidence for Neolithic salt production has come from France and Spain, where the association of sherds and stonework with salt sources gives a strong indication of early exploitation (Weller 2002a; b).

Somewhat later, the pointed-base vessels of the Lengyel culture from Wieliczka seem to indicate that the briquetage technique was used in southern Poland, while at one site at least (Barycz), ground installations seem designed to assist in the evaporation of brine through the use of fire (Jodłowski 1968).

All this suggests that obtaining salt was indeed a concern in early farming times; in the Hungarian context there are strong grounds for supposing that the areas to the east and west of the Danube had different procurement opportunities, and thus different development trajectories (full discussion in Chapter 9).

What happened between the Lengyel and the Middle Bronze Age is unclear, but the few radiocarbon dates from Figa that fall in the Eneolithic and Early Bronze Age, along with the assemblage of Early Bronze Age pottery, suggest that this source at least was known and used. The available dates really begin, however, in the sixteenth century cal BC, and follow a continuing pattern down to the ninth century. We have presented above our reasons for believing that there is a sequence at Figa from south to north on the site; certainly we have demonstrated that salt production took place at Figa during the Middle and Late Bronze Age even if direct associations with cultural groups have been harder to ascertain. What is more, the available radiocarbon dates indicate that other sites in the area were similarly active in the Late Bronze Age (Caila, Săsarm and Királyvölgy), as well as – less certainly – Valea Florilor, which has just one date with wide error terms.

Perhaps most important, in the Middle and Late Bronze Age, the technique used was that involving troughs and plank or wattle enclosures, in contrast to the briquetage technique used in the Neolithic. The earliest dates for the use of troughs are those from Trench I, and belong to the sixteenth century cal BC. Presumably the idea may have been in existence earlier than this, but there is no evidence to inform us on this point. The technology continued for some 800 years, on present knowledge; the latest dates fall in the ninth century and could be as late as 800 cal BC. After that, it appears that things changed; there is a gap in demonstrable exploitation for probably 400 years.

When production resumed – at a time which cannot be specified precisely but which lay somewhere in the fourth to second centuries cal BC – the technology used was different. There is no evidence that the trough technique was still used, though the troughs were evidently still lying around. There is no sign of wattle or plank enclosures. Instead, shafts and pits were dug down to the rock, in at least one instance lined with split timbers and the rock reached by means of ladders. We do not yet know how the rock was broken up, but with iron tools now available perhaps iron picks and hammers were used to chip large or small pieces off the parent blocks. Such shafts were found at Figa, but there is also an Iron Age radiocarbon date from Sânpaul, where we know there was also a Roman presence.

We have no direct evidence for the technology used to extract salt in Roman times in the Carpathian Basin. Wollmann presented a range of materials from mining sites that may belong to this period and may give some indication of the techniques, but this remains quite uncertain (WOLLMANN 1996). So far no field project has specifically examined the question of Roman salt mining in Romania, at least in our area.

In the late Roman or early medieval period, however, there are signs at Figa of renewed interest in the site. The indications are somewhat vague and ambiguous, but five radiocarbon dates fall in the fifth to the sixth centuries cal AD, and intriguingly, some of the dated timbers come from the same archaeological context as those of the Bronze Age, more than 1000 years earlier. Presumably this activity revolved around the exploitation of the same rock salt surfaces as had been worked previously, and perhaps by the same technique.

There is some evidence a little later for activity in other areas as well: at Királyvölgy there is one radiocarbon date centring on the ninth century cal AD; at Tereblya two widely separated dates span the medieval to early modern period, while at Sânpaul three dates fall in the fifteenth to sixteenth centuries cal AD. None of this can be firmly attached to any particular phase or type of activity, beyond indicating a continuing interest in the rich salt resources of the Carpathian area.

Salt in the Carpathian Basin in its European context

We have considered in the pages above not only the ways in which salt was produced but also the role which it might have played in trade networks in central and eastern Europe in prehistoric and early historic times. There can be no possible doubt that the volume of production is comparable with that in other parts of Europe and with that produced by other means (deep mining, coastal lagoons, briquetage). We have suggested that each of these methods – which would in principle have been available much more widely than was actually the case – is likely to have served different purposes, with the salt produced destined for different ends. Although much remains speculative, we have argued that the volumes of salt produced, using the trough technique, would mean that the production can be regarded as industrial in scale, that is to say, far beyond the means and needs of peasant household requirements. In any case, when one looks at prehistoric Europe as a whole, it is not until the Iron Age that one sees the great batteries of ovens that were used for brine evaporation using briquetage: in the Bronze Age, and presumably also in the Neolithic, briquetage installations are modest in scale and number – in fact, it is usually only the ceramic containers themselves that survive. By contrast, the trough technique had the potential to obtain large amounts of salt rock, if not also brine crystals (depending on how one imagines the troughs were used). Probably it was only the mining or quarrying of rock salt at outcrops that could have produced quantities comparable to what was produced with the trough technique.

Taking the continent as a whole, therefore, the situation regarding salt in the Carpathian Basin is something that cannot be overestimated. The deposits are very large, extensive, and in many cases close to the earth's surface, so there is no geological reason why large amounts of salt could not be extracted. In such a situation, one can speculate on the possible destination of the traded salt. The obvious answer as far as Transylvania is concerned is the saltless area to the west occupied by present-day Hungary, which was home to rich cultural groups throughout much of prehistory, and especially during the Bronze Age.

Much has been made in the past of a possible link between salt resources and the appearance of large bronze hoards, especially in Transylvania. The analysis conducted here, which only covers Bistrița-Năsăud county in northern Transylvania, showed no such correlation, but it must be admitted that for any conclusive answer to be obtained in this matter a much fuller analysis is needed, notably for those areas (such as Alba and Mureş counties) where the largest hoards occur. Realistically one might expect that hoards would turn up most prolifically in areas with rich metal resources. While this is true to some extent for Transylvania, it does not explain the large amounts of metal found on the Hungarian Plain, where there are neither salt nor copper deposits.

In terms of human actors, nothing is yet known about who conducted the salt extraction business – status, gender, or age. Some recent work has shown that females as well as males were present in the Hallstatt mines (PANY – TESCHLER-NICOLA 2007) but the implications of this have yet to be fully explored. It might be suggested that salt processing was a 'soft' activity, in contrast to 'hard' activities like metallurgy; but if we are correct that salt production at Figa was actually a form of quarrying, involving the use of heavy mallets and sledgehammers, then very considerable physical strength is needed – hardly very 'soft', at least by comparison with the briquetage technique, where the manipulation of wood for fires might be the most labour-intensive activity. It is thus currently impossible to quantify amounts of salt, numbers of workers involved, or numbers of extraction areas open at one time, but the latter at least is a question that could theoretically be answered, given sufficient resources and access to the relevant parts of the right sites.

Salt and tin

One possibility worth considering is that salt from Transylvania and Maramureş was exchanged for tin. The role of tin in the Carpathian Basin shows an incremental development. In the third millennium BC tin bronzes were produced irregularly and accidentally from mixed ores. In the first half of the second millennium exotic goods made of tin-bronze (prestige weapons and jewellery) were produced by intentional alloying, and in the second half of the second millennium and beginning of first, tin was widely used for alloying. We do not enter here into the disputed question of the availability of tin in the Carpathian Basin and the Balkans; a number of deposits have been identified (BOROFFKA 2006). but most appear to be small.

As we have seen, salt production in north-eastern Transylvania appears to fall into a number of different periods: around 3000, 2400–2000 and 1600–850 cal BC. In the third millennium the rock salt was extracted irregularly and without specialised equipment or tools. By contrast, between 1600 and 850 cal BC salt production became a large-scale, highly organized industry oriented towards external consumers. Taking these two sets of dates together, one could argue that the phases of tin use in the

Carpathian Basin correspond to a large extent to the dynamics of salt production in Transylvania and Maramureş. In particular, the period of 'industrial' salt production corresponds to the period of maximum use of tin in the bronze industry.

The most important ways by which Carpathian salt supplied the Hungarian Plain, and perhaps parts of the Balkans, were along the Tisza, Someş/Samos and Mureş/Maros. The fortified settlements, rich hoards and elite weapons concentrated in key strategic positions along these rivers suggest that during the Middle and Late Bronze Ages they were intensively used for the transport of goods. For instance, along the Someş/Samos three groups of Middle and Late Bronze Age fortified settlements, hoards and tumuli are known: Coldău-Uriu, Medieşul Aurit-Apa, and Ópályi. The Coldău-Uriu group is in the Beclean area, near the three salt production sites presented above. This group includes the Coldău-Varbă fortified settlement (late Wietenberg), the Coldău tumulus cemetery (presumably Suciu de Sus), and the Uriu hoard (Br D). From its position and topography, the Coldău-Varbă fortified site might have acted as a port settlement connected to the Băile Figa, Săsarm and Caila salt production sites. From this port salt could be transported along the Someş/Samos westwards to the salt-less Hungarian Plain. One might compare the situation in the Banat in the same period (ca 1400–1250 BC), where a group of five huge fortified settlements (including Corneşti and Sântana), had the potential to control the lower Mureş, the middle and lower Tisza, as well as the Bega and Timiş rivers, connecting Transylvania and Banat with the northern Balkans.

The many long-lived Bronze Age settlements and the important role of stock-breeding in the Hungarian Plain and adjacent territories required much salt. The high level of social complexity and organization, as well as the complex network of regional and interregional connections of this area, would imply that the local Middle and Late Bronze Age communities needed to ensure a regular supply of salt – most obviously from Transylvania and Maramureş. What is more, they might have moved some of the salt further south, supplying the north-western part of the Balkan peninsula.

Although direct evidence is lacking, we can surmise that tin circulated widely and by various routes across Europe. It might have come to the Danube-Tisza interfluve from different sources and by different routes as part of a complex exchange system. In that case, Transylvania and Maramureş could have been supplied from the west via this route, though any suggestion of direct exchange of Carpathian salt for Bohemian-Saxon tin seems rather simplistic – even if we could demonstrate that this latter source was actually exploited. With analytical work now able to identify the origin of tin with some certainty, we expect some of these questions to be solved in the next few years.

Prospects

The future of salt research in the Carpathian zone depends on a number of things: sufficient surviving evidence on site, most obviously, but also resources and site protection measures. We have been able to show in this account how, unexpectedly, an extraordinary degree of preservation was present on a few sites in northern Transylvania, including some that we were able to investigate. We believe it is very likely that more such sites remain to be discovered, though with increased awareness of our work the possibilities have become more limited (perhaps surprisingly, nothing comparable has yet turned up in Moldavia, for instance). Our preliminary work in Ukraine and Slovakia, along with the material made available to us from south-east Poland, did not reveal any more sites similar to the Beclean group, but since it took several years for us to find these last localities, it could hardly be expected that the problem could be solved in other areas in a few days. Systematic work in Poland and Ukraine does, we believe, hold considerable potential; the situation in Slovakia is more uncertain, as most salt-bearing areas have already been explored by local archaeologists. Even there, however, our discovery of a pile site in a stream at Cigel'ka shows what might yet turn up.

Site protection is perhaps the biggest threat to future research along the lines we have explored. None of the sites was protected when we first visited, and most are still not registered officially as sites of archaeological importance (Figa is an exception). Registering a site, however, does not mean that it is effectively protected, at least in Romania. We have indicated how the Figa site has been encroached upon by modern development, which has removed any possibility of further exploring certain areas (though we hope and believe that the most sensitive and informative areas will survive somewhat longer). Săsarm is subject to strong erosional forces, while wood in the northernmost part of the site has already been removed since our survey in 2006. Caila is eroding fast; the trough first noted in 2008 is now severely damaged, simply through erosion, while twisted bindings observed the same year have completely vanished. Sites in other areas, such as Valea Florilor or Királyvölgy, no longer exist in any visible form.

In order to explore a site like Figa further, large resources will be necessary, since much of the important material lies under the water table, so that pumping will be necessary. Timbers, as exposed or recovered, will need immediate conservation to prevent their drying out, cracking, and eventual disintegration. Skilled excavators will be needed to cope with the delicate conditions on the ground; and a range of laboratory techniques will be needed in order to maximise the information recovered. Any project will therefore need both a dedicated and skilled team, and a lot of time and money. We are optimistic that these problems will be solved, and a fuller programme of work undertaken in all the areas discussed above.

While we count ourselves extraordinarily fortunate in having had the opportunity to begin exploration on these highly important sites, we are only too aware that we have just scratched the surface, and much remains to be done – even at Figa, where the combined efforts of archaeologists and developers have already exposed or destroyed much of what the site has to offer. We have worked with the facilities and techniques available to us; the next generation will find new and better ways of carrying out salt research in the Carpathian Basin. We are convinced that matters now obscure will be made clear in the coming years, and sites and artefacts now unfound will delight and amaze our successors.

Appendix 1
Catalogue of wood and stone

ADELA KOVÁCS
with contributions by Marion Uckelmann and Seán Goddard

Wood

Trench I

1. Trough 2 (*Pl. 9.1*)
Trench I, to the east of the stream; sq. A1; depth below surface: 1.21–1.51 m
Preserved L: 133 cm; maximum W: 37.5 cm; maximum H: 38.5 cm; size of the holes in the base from closed end: 2.2 cm diam., 2.3 x 2.5 cm, 2.5 x 2.5 cm oval, 2.2 x 2.5 cm, 2.5 x 2.5 cm, 2.5 x 3 cm, distance between the holes 10 to 12 cm. Additional hole in base near one edge 15 mm diam.
Quercus spp. (oak with sapwood, with the pegs of elder)
Bottom rounded on the outer side. At the median line of the bottom of the trough a row of round and square holes, in which axially perforated pegs were inserted. At the closed end a deep channel or groove 4.1 cm deep and 9.5 cm wide, is cut around the body, tapering out at the top. Base of trough 2.5 cm to 4.3 cm thick, sides 26 cm high. Base of trough at closed end carved deeper around the first hole. Four pegs were found in the bottom of the trough inside the holes. The numbering of the pegs starts from the closed end of the trough.
Preservation state: good
Museum Inv. No. 14959

2. Peg (*Pl. 9.2*)
Peg from Trough 2
L: 12.7 cm; D: 2.3 cm
Sambucus nigra (elder)
Cylindrical shape, circular in section, made from whole roundwood with some bark on the surface. Axially perforated by 1.1 cm diam. hole. Broken in two equal halves and cut across at both ends.
Preservation state: fair
Museum Inv. No. 14959_1

3. Peg (*Pl. 9.3*)
Peg from Trough 2
L: 11.5 cm; D: 2.2 cm
Sambucus nigra (elder)
Cylindrical shape, circular in section, made from whole roundwood. Axially perforated by 1.1 cm diam hole. The distal end broken in ancient times, other end cut across.
Preservation state: fair
Museum Inv. No. 14959_2

4. Peg (*Pl. 9.4*)
Peg from Trough 2
L: 8.7 cm; D: 2.5 cm
Sambucus nigra (elder)
Cylindrical shape, circular in section made from whole roundwood. Axially perforated by 1.4 cm diam hole, the distal end slightly fractured. Other end cut across at top. Broken into four pieces.
Preservation state: fair
Museum Inv. No. 14959_3

5. Peg with ivy inside (*Pl. 9.5*)
L: 9 cm; D: 1.8 cm; perforation D: 1.2 cm
Sambucus nigra (elder)
Cylindrical shape, made from whole roundwood, cut horizontally at both ends. Axial perforation. On the upper surface are traces of bark. Inside the perforation is a twisted ivy stem in the form of a knot.
Preservation state: good
Museum Inv. No. 14965

6. Roundwood branches bound together with ivy stem (*Pl. 9.6*)
Trench I; sq. A/1; depth below surface: 1.40–1.60 m
L: 150 cm; W: 10 cm; thickness: 1 cm
Quercus spp. (oak)
Two large branches, used to support Trough 2, bound together with an ivy stem in the shape of an "X". Bark partially preserved. Both have axe marks at bottom end.
Preservation state: fair
Museum Inv. No. 14848

7. Trough 3 (*Pl. 10.7*)
Trench I, to the west of the stream; sq. A–B / 2; depth below surface: 0.85–1.55 m
L: 1.72 m; maximum W: 34.5 cm; maximum H: 34.5 cm; sides of trough 2.6 to 4.4 cm wide; diameter of the holes, except the central one, between 2 and 2.5 cm, diameter of the central hole: 3.5 cm; distance between holes: varies from 6 to 9 cm
Alnus spp. (alder)
Bottom straightened by adzing. At the median line on the bottom of the trough a row of round holes, in which axially perforated pegs were inserted. In the perforation of one of the pegs a wooden needle had been inserted, wrapped in a thin, narrow wooden band, to provide a tight fit in the hole. A shallow channel 5.2 to 7 cm wide and 2.2 to 2.9 cm deep runs along the interior of the trough joining up the holes. At the closed end a deep channel or groove is cut round the body up to 2.5 cm deep and 7.5 cm wide. The numbering of the pegs starts from the closed end of the trough, from "a" to "m".
Preservation state: good
Museum Inv. No. 14960

8. Peg (*Pl. 10.8*)
Peg from Trough 3
L: 9.4 cm; D. max: 2.6 cm; D. perforation: 0.9 cm
Sambucus nigra (elder)
Cylindrical shape, circular in section, axially perforated. The upper end has traces of cut marks. The surface has many such marks along its length. Traces of bark in the distal part. The upper part has a slight collar for fixing. At both ends are traces of friction from rubbing against a cord.
Preservation state: good
Museum Inv. No. 14960_a

9. Peg (*Pl. 10.9*)
Peg from Trough 3
L: 8.5 cn; D. max: 2.4 cm; D. perforation: 1.2 cm
Sambucus nigra (elder)
Cylindrical shape, circular in section, axially perforated. Made from whole roundwood with bark removed. The upper end has traces of cut marks. At both ends are traces of friction from rubbing against a cord.
Preservation state: good
Museum Inv. No. 14960_b

10. Peg (*Pl. 10.10*)
Peg from Trough 3
L: 9.5 cm; D. max: 1.8 cm; D. perforation: 0.7 cm
Sambucus nigra (elder)

Cylindrical shape, circular in section, axially perforated. Bark removed. Both ends have traces of cut marks around the perforation. At both ends are traces of friction from rubbing against a cord. Preservation state: good
Museum Inv. No. 14960_c

11. Peg (*Pl. 10.11*)
Peg from Trough 3
L: 8.6 cm; D. max: 2 cm; D. perforation: 1.1 cm
Sambucus nigra (elder)
Cylindrical shape, circular in section, axially perforated. Bark removed. Both ends have traces of cut marks around the perforation, tapering the ends. Also cut on all sides over 6 cm length at one end. Trace of bung near interior end.
Preservation state: good
Museum Inv. No. 14960_d

12. Peg (*Pl. 10.12*)
Peg from Trough 3
L: 8.6 cm; D. max: 2.6 cm; D. perforation: 0.6 cm
Sambucus nigra (elder)
Cylindrical shape, circular in section, whole roundwood, axially perforated. At distal end traces of bark but torn at one side over 3.2 cm. The upper end has cut marks around the perforation. On the surface are longitudinal cut marks so that in the upper part a slight collar is formed. At the lower end traces of friction from rubbing against a cord.
Preservation state: good
Museum Inv. No. 14960_e

13. Peg (*Pl. 10.13*)
Peg from Trough 3
L: 9.8 cm; D. max: 2.8 cm; D. perforation: 0.9 cm
Sambucus nigra (elder)
Cylindrical shape, circular in section, whole roundwood, axially perforated. Bark removed. The upper end with traces of several cut marks and bottom end cut to create bevelled edge. The inferior end has traces of friction from rubbing against a cord.
Preservation state: good
Museum Inv. No. 14960_f

14. Peg (*Pl. 10.14*)
Peg from Trough 3
L: 11.3 cm; D: 1.9 cm; D. perforation: 1.3 cm
Sambucus nigra (elder)
Cylindrical shape, circular in section, axially perforated. Bark removed. Two knots from former branches, cut away. Both ends have several traces of cut marks around the perforation.
Preservation state: good
Museum Inv. No. 14960_g

15. Peg (*Pl. 10.15*)
Peg from Trough 3
L: 7.8 cm; D: 2.9 cm; D. perforation: 1.1 cm
Sambucus nigra (elder)
Cylindrical shape, circular in section, whole roundwood, axially perforated. Bark removed. The upper end with traces of cut marks around the perforation. The lower end broken.
Preservation state: good
Museum Inv. No. 14960_h

16. Peg (*Pl. 10.16*)
Peg from Trough 3
L: 9.2 cm; D: 2.8 cm; D. perforation: 1.2 cm
Sambucus nigra (elder)

Cylindrical shape, circular in section, whole roundwood axially perforated. The bark removed. Two knots from former branches, cut away. Both ends have traces of cut marks around the perforation, at bottom creating a bevelled end. Cut on outside circumference over bottom 5 cm.
Preservation state: good
Museum Inv. No. 14960_i

17. Peg (*Pl. 10.17*)
L: 10.2 cm; D: 2.6 cm at the top 2 cm at the bottom; D. perforation: 1.1 cm
Sambucus nigra (elder)
Cylindrical shape, circular in section, axially perforated. Bark removed. The upper end has traces of cut marks around the perforation. On the body are several longitudinal cut marks over the bottom 5.6 cm, leaving a wider collar at the top. At both ends are traces of friction from rubbing against a cord.
Preservation state: good
Museum Inv. No. 14960_j

18. Peg (*Pl. 10.18*)
Peg from Trough 3
L: 7.2 cm; D: 1.9 cm; D. perforation: 1 cm
Sambucus nigra (elder)
Cylindrical shape, circular in section, whole roundwood, axially perforated. Bark mostly removed, longitudinal cuts in the lower part. The upper part has traces of cut marks around the perforation with both ends bevelled and traces of friction from rubbing against a cord.
Preservation state: good
Museum Inv. No. 14960_k

19. Peg (*Pl. 10.19*)
L: 12.2 cm; D: 2.8 cm; D. perforation: 1 cm
Sambucus nigra (elder)
Cylindrical shape, circular in section, whole roundwood, axially perforated. Bark mostly removed, traces in the lower part. Two knots from former branches, cut away. Both ends have cut marks around the perforation.
Preservation state: good
Museum Inv. No. 14960_l

20. Peg (*Pl. 10.20*)
Peg from Trough 3
L: 11.2 cm ; D: 2.8 cm; D. perforation: 1.1 cm
Sambucus nigra (elder)
Cylindrical shape, circular in section, whole roundwood, axially perforated. Bark removed. Both ends have traces of cut marks around the perforation. Longitudinal cuts on outside over 6.9 cm from bottom end.
Preservation state: good
Museum Inv. No. 14960_m

21. Ladder (*Pl. 11.21*)
Trench I. Overall surviving length: 5.31 m [1047] with seven mortise holes, the shorter side 3.85 m [1141], with five mortise holes. Width: 54 cm.
Ulmus spp. (elm)
Formed of two long side pieces and six rungs surviving in position, with a further two recovered separately. Distance between the two side pieces (i.e. width of step): 40 cm. The distance between the rungs varies from 50 to 57 cm, and the distance from the bottom to the first rung is 79 cm. The side pieces are formed from whole roundwood that had been squared off and carefully chamfered at the corners. They were thickest at the bottom rung (7.8 cm square), the narrowest section (5.5 by 5.8 cm) at the break in the longer side. Mortise holes 2.8–3 cm wide and 5.5–6.4 cm tall. The rungs are rectangular with a stopped tenon. The step has been reduced to a roughly octagonal shope and is thinnest at the centre, probably worn from extensive use. Two types of tenon are used: alternately pegged and unpegged. In the first, the ends of the rung are squared off and simply pushed into the perforation (Rungs b (23), c (24) and e (26)). In the other, the ends project beyond the ladder sides and have a

rectangular perforation, into which a wedge is inserted. The working of the ladder is clearly visible and the whole is very well preserved, apart from the breakage to rung and sides.
Preservation state: good
Museum Inv. No. 15580

22. Rung (*Pl. 12.22*)
Trench I., Sq: C–D / 3; Abs. Ht.: 295.94 m
L: 64 cm; W: 4.7 cm; Th: 1.9 cm; middle D: 3.5 cm; perforations 3x3 cm
Ulmus spp. (elm)
Rectangular shape, with the middle part circular. Axe marks along both tangential faces. Perforated ends, fixed with wedges. Both wedges preserved, one a tapering radial piece 7.2 cm long and 1.5 cm square at widest end, the other 6.2 cm long and 1.6 x 1.5 cm at widest end. One of the ends is rounded, the other squared.
Preservation state: good
Field no. 1452
Museum Inv. No. 15580_a

23. Rung (*Pl. 12.23*)
Trench I, Sq: C–D / 3; Abs. Ht.: 295.94 m
L: 57 cm; W: 5 cm; Th: 2 cm; middle D: 3.5 cm
Ulmus spp. (elm)
Rectangular shape of tangential conversion, rounded in the middle. Both ends are cut straight, projecting 8.5 and 8.8 cm with ends 4.9 x 1.5 cm and 4 x 1.9 cm in size. One end is slightly bevelled. No perforated ends. Cut marks on the body of the rung. One axe blade mark 4.5 cm wide and 0.3 cm deep across the profile.
Preservation state: good
Field no. 1451
Museum Inv. No. 15580_b

24. Rung (*Pl. 12.24*)
Trench I, Sq: C–D / 3; Abs. Ht.: 295.94 m
L: 56.2 cm; W: 5.5 cm; Th: 2 cm; middle D. 3.4 cm
Ulmus spp. (elm)
Rectangular shape of tangential conversion with sapwood surviving on one side, rounded in the middle. Both ends are cut straight and slightly rounded. Ends, 4.4 x 2.1 cm and 4.5 x 2.2 cm not perforated.
Preservation state: good
Field no.1402
Museum Inv. No. 15580_c

25. Rung
Trench I, Sq: C–D / 3; Abs. Ht.: 295.94 m
L: 63 cm; W: 4 cm; Th: 1.9 cm; perforations: 3 x 3 cm
Ulmus spp. (elm)
Rectangular shape, of tangential conversion, becoming rounded in the middle with neatly chamfered edges. Both ends cut straight and slightly rounded. Perforated ends 1.2 cm long and in end section 4.5 x ,8 cm and 4.8 x 2.4 cm, fixed with wedges. Wedges are radial conversions, 5.1 cm long and 1.6 x 0.09 cm and 1.2 x 0.09 cm at wider ends.
Preservation state: good
Field no. 1150
Museum Inv. No. 15580_d

26. Rung (*Pl. 12.26*)
Trench I, Sq: C–D / 3; Abs. Ht.: 295.94 m
L: 61 cm; W: 5.4 cm; Th: 2 cm
Ulmus spp. (elm)
Rectangular shape, tangential conversion, rectangular in section. Both ends cut straight and slightly rounded, projecting 9.7 and 10 cm, with end sections of 3.8 x 1.7 cm and 3 x 0.9 cm. Ends not perforated. One axe blade profile 4 cm wide and 0.1 cm deep across the profile.

Rectangular shape, rectangular in section, straight profile. One end cut off straight from several attempts. The other end chopped from two sides, forming an oblique edge. On the upper surface exfoliating layer, traces of burning.
Preservation state: fair
Museum Inv. No. 16759

39. Large wedge (*Pl. 15.39*)
Trench I, without context
L: 33.8 cm; W: 9.5 cm; Th: 4.4 cm
Quercus spp. (oak)
Made from a tree root, fairly rectangular shape, oval in section, straight profile. A slight extension on one of the ends. Short handle, with the ends cut oblique.
Preservation state: fair
Museum Inv. No. 16760

40. Large wedge (*Pl. 15.40*)
Trench I, Sq: E/2; Abs. Ht.: 295.82 m
L: 36.8 cm; W: 4.6 cm; Th: 2.5 cm; D: 4.1cm
Wood species not determined
Cylindrical shape, circular in section. One end chopped from two sides, forming an edge. The other end round, chopped from several directions.
Preservation state: good
Field number: 1440
Museum Inv. No. 17335

41. Large worked piece (*Pl. 15.41*)
Trench I, Sq: D/2; Abs. Ht.: 296.47 m
L: 93 cm; D: 16 cm
Quercus spp. (oak)
Roundwood, with bark surviving. On the surface knots from five cut off small branches. One end cut across moderately flat with three surviving axe facets, max. 4.1 cm wide. The other end is cut across flat at the top, 5 x 3.6 cm, the sides then tapering outwards over c. 4 cm to produce a width of 13 cm. The fourth side is cut vertically over 9 cm to produce a right-angle shoulder utilising a side branch. The branch is cut across to produce a flat upper surface over 13 cm and is cut flat at the end where it is c. 4.5 cm in diameter.
Preservation state: good
Field number: 1022
Museum Inv. No. 15579

42. Beam / worked piece of large wood (*Pl. 15.42*)
Trench I , Sq: D/2–3; Abs. Ht.: 296.30 m
L: 117 cm; D: 24 cm
Quercus spp. (oak)
Roundwood with bark largely intact. One end chopped from two sides over 17 and 13 cm, forming a blunt wedge-shaped end. The other end is cut from one direction. On the body a 20 cm long curved notch, 0.7 cm deep, was cut on one side.
Preservation state: good
Field number: 1021
Museum Inv. No. 15576

43. Roundwood baton (*Pl. 16.43*)
Trench I , Sq: D–E/2; Abs. Ht.: 295.44 m
L: 26.5 cm; D: 2.2 cm
Wood species not determined
Roundwood with bark surviving Both ends cut from several directions, forming edges.
Preservation state: good
Field number: 1012
Museum Inv. No. 17336

44. Short broken board (*Pl. 16.44*)
Trench I, Sq: E/III; Abs. Ht.: 296.80 m
L: 34.5 cm; W: 12.6 cm; Th: 2.1 cm; handle L: 10.6 cm
Wood species not determined
Oval asymmetric shape, oval in section, straight profile. Projecting part rectangular in section, cut straight. Probably a board for use in a muddy environment, broken on short grain section leaving only half at one end.
Preservation state: good
Field number: 1436
Museum Inv. No. 17337

45. Board (*Pl. 16.45*)
Trench I, Sq: A/4 ; Abs. Ht.: 295.50 m
L: 21 cm; W: 3.5 cm; Th: 1.7 cm
Quercus spp. (oak)
Radially split piece, rectangular in section. Chopping traces on both sides. On one surface are several cut marks forming small steps. One end broken, the other cut to a tapering point.
Preservation state: fair
Museum Inv. No. 14845

46. Small splinter (*Pl. 16.46*)
Trench I , Sq: E/III; Abs. Ht.: 296.80 m
L: 15.2 cm; W: 6.6 cm; Th: 1.5 cm
Wood species not determined
Quite rectangular shape, rectangular in section, curved profile. One end cut straight, the other chopped from one side, forming an edge. The product of an axe cut to a roundwood stem, with a squared end (axe off-cut).
Preservation state: good
Museum Inv. No. 17338

47. Small splinter, maybe wedge (*Pl. 16.47*)
Trench I , Sq: D/2; Abs. Ht.: 295.45 m
L: 15.5 cm; W: 4.4 cm; Th: 1.5 cm
Wood species not determined
Rectangular shape, rectangular in section, straight profile. One end cut straight, the other chopped from one side, forming an edge.
Preservation state: good
Museum Inv. No. 17339

Trench III

48. Trough (channeled piece) (*Pl. 16.48*)
Trench III; Sq. A / 1; depth below surface: 2.44 m; Abs. Ht.: 286.12 m
L: 35 cm; W: 19.5 cm, Th: 18.3 cm; hollow: 7 cm
Wood species not determined
Object made from a trunk split lengthwise, hollowed inside. Has the shape of a "saddle". The inside has a channeled shape. Both ends broken.
Preservation state: fair
Museum Inv. No. 17271

49. Small trough (*Pl. 17.49*)
Trench III; Sq. E / 5; depth below surface: 1.25 m; Abs. Ht.: 288.25 m
L: 53 cm; W: 23 cm; Th: 4.5 cm
Quercus spp. (oak)
Small trough, hollowed, made from tree trunk that had been quarter split. The bowl is oval in shape, carved, with curved profile and a interior depth of c. 5.3 cm. The ends are rounded and slightly damaged, as is one side. Numerous tool marks, 4.2 cm wide, show that the interior of the bowl was created by cutting towards the middle

from both ends. The ends of the bowl are a maximum of 3.2 cm thick, while the thickness of the main body varies from 1.7 cm (base) to 2.9 cm (side). The damaged side is only 1.1 cm thick in places.
Preservation state: fair
Museum Inv. No. 15263

50. Small bowl (*Pl. 17.50*)
Trench III; Sq. H / 6; depth below surface: 0.50 m; Abs. Ht.: 289.00 m
L: 44 cm; W: 16.5 cm; Th: 3.5 cm
Populus spp. (poplar)
Small trough made from a quarter split log. Oval bowl with rounded edges but both edges and one end are incomplete so original dimensions are unknown. Both the outer and inner surfaces have been roughly shaped with tool marks still clearly visible. The interior has been adzed from many different directions. One blade profile survives on the interior, 4.4 cm long and completely straight. Another jam curve at the broken end has a full curving blade profile of 4 cm. The most complete end is 3.9 cm thick, the body varying between 0.6 cm and 1.6 cm in thickness. The bowl has a maximum interior depth of 6 cm.
Preservation state: fair
Museum Inv. No. 15264

51. Small bowl (*Pl. 17.51*)
Trench III; Sq. D / 4; depth below surface: 1.05 m; Abs. Ht.: 288.45 m
L: 50 cm; W: 19 cm; Th: 2.5 cm
Populus spp. (poplar)
Made from a half split log. One end and both sides broken so the original dimensions are uncertain. At the more complete end is a 5 cm long handle that is broken along one side. The concave interior retains numerous adze marks from several different directions, the widest toolmarks being 5.2 cm wide. The interior has a maximum depth of 9.8 cm along the profile. The end 3.4 cm thick and the centre of the bowl 1.6–2.1 cm thick.
Preservation state: good
Museum Inv. No. 14839

52. Piece of small trough (*Pl. 18.52*)
Trench III; without context
L: 31 cm; W: 19 cm; Th: 3.9 cm
Wood species not determined
Trough made of half a tree trunk. The bowl is fragmentary, of (possibly) oval shape. The outside and inside of the bowl have traces of repeated chops made with a sharp tool. Damaged edges.
Preservation state: fair
Museum Inv. No. 17267

53. Small channeled piece (*Pl. 18.53*)
Trench III; Sq. A / 7; depth below surface: 0.50 m; Abs. Ht.: 289 m
L: 13.5 cm; W: 5 cm; Th: 1 cm
Sambucus nigra (elder)
Carved object of wood cut in half and hollowed inside. The outer surface has been cut with a knife and the interior has been hollowed out with a spoon chisel with a 3.5 cm wide curved end. The toolmarks on the inside are cut from the pointed end. One end is cut straight across with a slight step carved on the outside 1.3–1.6 cm from the end. The other end is cut on two sides to leave a projecting end 1.9 cm wide. The sides of the object area 0.5 and 0.7 cm thick, increasing to 1.1 cm in the middle.
Preservation state: good
Museum Inv. No. 14729

54. Small channeled piece (?) (*Pl. 18.54*)
Trench III; Sq. B / 5; depth below surface: 0.70 m; Abs. Ht.: 288.80 m
L: 13 cm; W: 5 cm; Th: 5.8 cm
Wood species not determined
Carved object of wood cut in half, longitudinally, hollowed inside. At the ends and inside are traces of shaping done with a sharp tool. The outside is smooth. Completely preserved.

Preservation state: good
Museum Inv. No. 14730

55. Hollowed wood fragment (vessel?) (*Pl. 18.55*)
Trench III; Sq. B / 2; depth below surface: 1.08 m; Abs. Ht.: 288.42 m
D: 14 cm; W: 1.8 cm; Th: 3 cm
Quercus spp. (oak)
Tangentially split piece of wood worked from the outer part of a log. It is semicircular in shape, oval in section, with a curved profile. The edges are all broken, apart from one end, so its original dimensions are unknown. It may represent a bowl fragment, the most complete end possibly representing part of a handle. A 1.4 cm diameter groove cuts across the thickest end.
Preservation state: fair
Museum Inv. No. 14972

56. Peg (*Pl. 18.56*)
Trench III; Sq. A / 1; depth below surface: 1.30 m; Abs. Ht.: 288.20 m
L: 12 cm; max. D: 3.2 cm; perforation D: 1 cm
Sambucus nigra (elder)
Cone-shaped peg, rectangular body, made from roundwood. Longitudinal central hole. A 2.1 cm long collar exists at the top end, neatly bevelled on its bottom edge. Below the collar the width immediately decreases to 2.1 cm (from 3.1 cm) and gradually tapers to 1.5 cm at the bottom end. Around the perforation, at both ends, and on the body are traces of cutting with a sharp tool, probably a knife. Completely preserved.
Preservation state: good
Museum Inv. No. 14967

57. Peg (*Pl. 18.57*)
Trench III; Sq. D / 3; depth below surface: 1.26 m; Abs. Ht.: 288.24 m
L: 16.5 cm; max. D: 4.2 cm; perforation D: 1 cm
Sambucus nigra (elder)
Cone-shaped peg, circular body, both ends with circular section, made from complete roundwood. Longitudinal central hole, in which a string was inserted. In the upper end is a 2.4 cm long collar 4.1 x 3.9 cm, with bevelled edge, decreasing to 3 x 2.5 cm below the collar. The bottom 7.8 cm has tapering sides leaving a width of 2.4 cm at the bottom end. Around the perforation, at both ends, and on the lower part of the body, there are traces of cutting with a sharp tool. Completely preserved.
Preservation state: good
Museum Inv. No. 14968

58. Peg (*Pl. 18.58*)
Trench III; Sq. B / 2; depth below surface: 1.00 m; Abs. Ht.: 288.50 m
L: 16 cm; max. D: 3.5 cm; perforation D: 1.2 cm
Sambucus nigra (elder)
Cone-shaped peg, rather circular body, both ends with circular section. Longitudinal central hole. In the upper part were two thin branches, removed by a straight cut. In the upper part a slight, 1.4 cm long, collar, 3.5 cm in diameter, below which the body tapers to 1.8 cm diameter at the bottom end. The body has traces of cutting with a knife along all the surfaces. Completely preserved.
Preservation state: good
Museum Inv. No. 14969

59. Peg (*Pl. 18.59*)
Trench III; Sq. C / 2; depth below surface: 1.40 m; Abs. Ht.: 288.10 m
L: 13 cm; max. D: 3 cm; perforation D: 1 cm
Sambucus nigra (elder)
Cone-shaped peg, rather rectangular body, both ends with circular cross section. Longitudinal central hole. The upper end is oval in shape, 3 x 2.8 cm, forming a 2.1 cm long collar, bevelled on top and bottom edges. Below the collar the dimensions are reduced to 2.5 x 2.1 cm and form a rectangular cross section. Cut down over last 2.7 cm

to produce a roughly circular 1.6 x 1.3 cm tip. Inside, a 3.9 cm long and 0.8 cm wide oak wedge was inserted from the bottom. Completely preserved.
Preservation state: fair
Museum Inv. No. 14970

60. Peg (*Pl. 18.60*)
Trench III; Sq. D / 5; depth below surface: 0.55 m; Abs. Ht.: 288.95 m
L: 14 cm; max. D: 3.2 cm; perforation D: 1.2 cm
Sambucus nigra (elder)
Cone-shaped peg, circular body, made from whole roundwood. Diameter decreases from 3.1 cm at the top to 2 cm at the bottom. Longitudinal central hole, that is slightly curved along the length of the peg. The upper end has traces of cut marks leaving a bevelled top. Completely preserved.
Preservation state: good
Museum Inv. No. 14974

61. Peg (*Pl. 18.61*)
Trench III; Sq. E / 5; depth below surface: 1.20 m; Abs. Ht.: 288.30 m
L: 13 cm; max. D: 3.5 cm; perforation D: 1.2 cm
Quercus spp. (oak)
Cone-shaped peg, rather rectangular body, both ends with circular section. Top cut flat, almost circular (4 x 3.8 cm). Below the 1.2 cm long collar it is roughly rectangular, 2.6 x 2.2 cm, and tapers over the last 3 cm to form a rounded end 1.9 x 1.7 cm. Clear knife marks exist along the sides and around the top. Completely preserved.
Preservation state: good
Museum Inv. No. 15254

62. Peg (*Pl. 18.62*)
Trench III; Sq. D / 9; depth below surface: 1.10 m; Abs. Ht.: 288.40 m
L: 13 cm; max. D: 3.5 cm; perforation D: 1.2 cm
Sambucus nigra (elder)
Cone-shaped peg made from whole roundwood. Roughly circular top of c. 3 cm, below which is a 0.8 cm long collar. Beneath the collar the peg is rectangular, 2.4 x 2.3 cm but over the last 5 cm it tapers on all sides to create a roughly circular tip of 1.7 cm diameter. Longitudinal central hole. On the body and around the perforation are knife marks. The tip has a small groove, possibly made by friction against a small cord, or the edge may just be broken as it is only 0.2 cm thick on that side. Completely preserved.
Preservation state: good
Museum Inv. No. 15256

63. Peg (*Pl. 19.63*)
Trench III; Sq. A / 2; depth below surface: 1.50 m; Abs. Ht.: 287.06 m
L: 14.2 cm; max. D: 3.6 cm; perforation D: 1.5 cm
Quercus spp. (oak)
Cone-shaped peg, oval in the upper part, circular tip. Longitudinal central hole. A slight collar fashioned from two sides. On the body and at the ends are carvings made with a sharp tool. Completely preserved.
Preservation state: good
Museum Inv. No. 16655

64. Peg (*Pl. 19.64*)
Trench III; Sq. A / 4; depth below surface: 2.50 m; Abs. Ht.: 286.06 m
L: 10.5 cm; max. D: 5.3 cm; perforation D: 3.9 cm
Quercus spp. (oak)
Cone-shaped peg, rather rectangular body, at the ends circular in section. Longitudinal central hole. A slight collar cut from two sides. On the body and at the ends are carvings made with a sharp tool. Completely preserved.
Preservation state: good
Museum Inv. No. 16665

65. Peg (*Pl. 19.65*)
Trench III; without context
L: 15.5 cm; max. D: 3.6 cm; perforation D: 1 cm
Sambucus nigra (elder)
Rather cone-shaped peg, body rectangular in section, circular at both ends. Longitudinal central hole. Collar cut from several directions. On the body and at the ends are carvings made with a sharp tool. Completely preserved.
Preservation state: good
Museum Inv. No. 16712

66. Peg (*Pl. 19.66*)
Trench III; Sq. A / 9; depth below surface: 0.95 m; Abs. Ht.: 288.50 m
L: 12.5 cm; max. D: 2.8 cm; perforation D: 1.2 cm
Sambucus nigra (elder)
Cone-shaped peg made from whole roundwood. The top is roughly square, 2.8 x 2.8 cm, cut down vertically to form a 1.9 cm long collar. Below the collar it is roughly square, 1.7 x 1.7 cm, but over the last c. 3 cm it is reduced on all sides to form a circular cross section, 1.5 x 1.3 cm at the tip. Longitudinal central hole. Around the perforation, at both ends, and on the body, are traces of cutting with a knife. Completely preserved.
Preservation state: good
Museum Inv. No. 14966

67. Peg (*Pl. 19.67*)
Trench III; without context
L: 11.6 cm; max. D: 3.2 cm; perforation D: 0.7 cm
Wood species not determined
Rather cone-shaped peg, body rather rectangular in section, circular at both ends. Longitudinal central hole. Collar in the upper part, cut from two sides. Signs of wear around the holes. On the body and at the ends are cut marks made with a sharp tool. A small crack in the upper part. Completely preserved.
Preservation state: fair
Museum Inv. No. 17305

68. Hub (?) (*Pl. 19.68*)
Trench III; Sq: C/3; depth below surface: 0.82 m; Abs. Ht.: 288.68 m
L: 10.7 cm; max. D: 5.8 cm, perforation L: 5 cm; perforation W: 1 cm
Quercus spp. (oak)
Made from a straight branch or the trunk of a young tree. Round section, at either end sawn off. At about 1/3 of its length the hub is slightly thinner and has a polished surface. The thicker part is perforated by two criss-crossed holes. One of the holes is rectangular in shape, and was apparently made for the insertion of a plank-shaped piece. Another hole is square and much narrower, and a wooden wedge, square in section, was inserted in it. The object is evidently some kind of compound device, possibly a wheel hub. Completely preserved.
Preservation state: good
Museum Inv. No. 16656

69. Mallet (*Pl. 19.69*)
Trench III; Sq. A / 3; depth below surface: 1.30 m; Abs. Ht.: 287.26 m
L: 25.8 cm; W: 12.3 cm; Th: 3.6 cm; handle L: 52.4 cm; handle D: 3.5 cm
Wood species not determined
Mallet made of three pieces. The use part is rectangular, oval in section, with flat profile. In the middle a transversal rectangular peforation in which the handle was fixed. The handle cylindrical, slightly degraded at the end, fixed in the use part by means of a rectangular wooden wedge, forcibly inserted. Bark removed. Completely preserved.
Preservation state: good
Museum Inv. No. 16662

70. Mallet (*Pl. 20.70*)
Trench III; Sq. A / 15; depth below surface: 2.70 m; Abs. Ht.: 286.80 m
L: 58 cm; body W: 24 cm, body Th: 27 cm; handle D: 5 cm
Quercus spp. (oak)

Tool made from a trunk with a branch attached. The body is rather triangular in section, broken in ancient times. At one end are traces of use. The handle was cut obliquely, from two sides.
Preservation state: fair
Museum Inv. No. 16708

71. Tool haft (*Pl. 20.71*)
Trench III; Sq. A / 5; depth below surface: 1.42 m; Abs. Ht.: 288.08 m
L: 56 cm; D: 4 cm
Quercus spp. (oak)
Haft carved from a branch, with the long part used as handle. The short part has a small step where the axe was fixed, polished due to intensive use. Completely preserved.
Preservation state: good
Museum Inv. No. 14837

72. Tool haft (*Pl. 20.72*)
Trench III; Sq. E / 5; depth below surface: 0.85 m; Abs. Ht.: 288.65 m
L: 37 cm; D: 3 cm
Quercus spp. (oak)
Haft made from a forked branch, with the long part used as handle. The handle is 28.4 cm long to the junction and is 2.9 x 2.2 cm across at that point. Over the last 10 cm of the handle it is only cut down on the outer face as the branch that it is formed of begins to naturally curve outwards in that direction. The apparent tapered end of the handle is formed by curving edge of the branch on the inner face over the last 9 cm. Above this the handle is cut down along the inner face. The bulbous head is cut down steeply from the direction of the shorter branch. The short part has a small step where the adze was fixed. It only slightly tapers from 1.7 cm diameter by the junction to 1.6 cm diameter at the tip. Bark is partially preserved around the junction and along the shorter branch. Completely preserved.
Preservation state: good
Museum Inv. No. 14838

73. Tool haft (*Pl. 20.73*)
Trench III; Sq. α / 1; depth below surface: 2.37 m; Abs. Ht.: 286.13 m
Max. L: 71.5 cm; D: 6 cm; handle D: 4.5 cm
Wood species not determined
Hook made from a branch with one side stem, in an asymmetrical "V" shape. The handle was made from the thinner branch. The end was cut straight and then rounded. The head, about half the length of the handle, was made from the larger branch. There is grooving and wear traces on the top. The tip was sharpened by paring with a knife. On the inside part are traces of friction (use wear). Completely preserved.
Preservation state: fair
Museum Inv. No. 17272
Field no. 762/2010

74. Shovel (*Pl. 21.74*)
Trench III; Sq. A / 5; depth below surface: 1.25 m; Abs. Ht.: 288.77 m
L: 47 cm; W: 23 cm; Th: 3.5 cm
Quercus spp. (oak)
Shovel with a slightly hollowed bowl and handle, fashioned from a tangentially split trunk. The shovel blade is sub-rectangular with curved edges, 20 cm long and 11.3 cm wide, with a small piece missing on its shoulder. It gradually tapers from 2.8 cm at the top to 1 cm at the tip. The back of the shovel has numerous axe cuts that form the gently curving back of the implement. The internal section across the width of the shovel has a depth of 0.6 cm. The handle, slightly oval in section, is 4.1 x 3.4 at the end and 3.6 x 2.8 cm just before the shovel blade. The shovel appears quite worn from long usage. Nearly the whole object is preserved.
Preservation state: good
Museum Inv. No. 14840

75. Shovel (*Pl. 21.75*)
Trench III; Sq. C / 10; depth below surface: 1.14 m; Abs. Ht.: 288.36 m

L: 47 cm; W: 8.5 cm; Th: 3 cm
Wood species not determined
Narrow shovel made from tangentially split timber. The bowl is an elongated oval shape with rounded edges, forming about half of the object. It gradually narrows in thickness from 2.4 cm to 0.9 cm at the tip. The bowl has a break on a side. The handle, oval in section (2.8 x 2.6 cm), tapers gradually towards the end.
Preservation state: good
Museum Inv. No. 15259

76. Shovel (*Pl. 21.76*)
Trench III; Sq. K / 6; depth below surface: 1.05 m; Abs. Ht.: 288.45 m
L: 58 cm; W: 18 cm; Th: 3.5 cm
Quercus spp. (oak)
Shovel made from a tangentially split board. The 34 cm long handle is broken at the top and is roughly rectangualar in shape with the edges bevelled. It is 5.5 x 4.3 cm just above the bowl but tapers towards the top. The bowl is just under 30 cm long and 18.5 cm at its widest point. The inner face of the bowl shows no toolmarks, is quite flat and may represent the original split face. The shovel is fairly blunt with the thickness reducing from 2.5 cm to 1.3 cm only near the tip. The outer face is covered by numerous toolmarks, the widest being 4.2 cm in width, leaving a concave cross-sectional profile. The bowl has breaks on one side and at the top.
Preservation state: good
Museum Inv. No. 15262

77. Shovel (?) (*Pl. 21.77*)
Trench III; Sq. A / 1; depth below surface: 2.44 m; Abs. Ht.: 286.12 m
L: 26.8 cm; W: 10.2 cm; Th: 2.7 cm; handle D: 2 cm; handle L: 6.5 cm
Wood species not determined
Shovel (?) with bowl and handle. Rather rectangular shaped bowl, oval section, with ancient breaks. Short handle, rather conical shape, oval in section.
Preservation state: fair
Museum Inv. No. 17269

78. V-shaped object (*Pl. 22.78*)
Trench III; Sq. D / 6; depth below surface: 0.70 m; Abs. Ht.: 288.80 m
L: 22 cm; D: 3.5 cm; perforation L: 3.5 cm; perforation W: 2 cm
Quercus spp. (oak)
V-shaped object, one half preserved, broken at curving end. Thickest at curved end (5.6 x 4.7 cm) and c. 3.7 cm in diameter at complete end. Traces of bark on the outside. at 5.9 cm from the end is a transversal rectangular perforation, with slightly sloping edges across the width of the hole, showing that it had been carefully chiselled on those edges to help form the hole. The surface has a strong sheen due to prolonged use. Some damage to the outer face, c. 1 cm deep. A 0.5 cm wide and 0.3 cm deep scar exists on one side 1.9 cm from the complete end, possibly from a binding.
These objects are highly unusual. It seems likely that they were used as springs to force constructional elements into place or apart.
Preservation state: good
Museum Inv. No. 14963

79. V-shaped object (*Pl. 22.79*)
Trench III; Sq. E / 4; depth below surface: 1.10 m
L: 22 cm; W: 4 cm; Th: 4.5 cm; perforation L: 3.5 cm; perforation W: 2 cm
Quercus spp. (oak)
V-shaped object, only one half preserved, broken at curved end. The body is rectangular in section, with rounded edges. Towards the complete end a transversal rectangular perforation with slightly sloping edges across the width of the hole, showing that it had been carefully chiselled on those edges to help form the hole. The perforation goes at a slight angle through the piece, the two top edges being 5.2 and 6.1 cm from the complete end.
Preservation state: medium to poor
Museum Inv. No. 14964

80. V-shaped object (*Pl. 22.80*)
Trench III; Sq. D / 10; depth below surface: 0.96 m; Abs. Ht.: 288.54 m
L: 19.6 cm; W: 4.4 m; Th: 2.5; perforation L: 1.9 cm; perforation W: 1 cm
Fagus sylvatica (beech)
V-shaped object made from a branch, broken at the curving end. The preserved branch is carved, of cylindrical shape, c. 3.5 cm diameter at the complete end and 5 cm diameter at the broken end. Near the complete end is a rectangular perforation in which a small (7 x 2.2 x 1 cm) wedge was found. The curved end has been cut down on its outer face. Outside surface has a marked sheen. No bark surviving. Half of the object preserved.
Preservation state: fair
Museum Inv. No. 15255

81. V-shaped object (*Pl. 22.81*)
Trench III; Sq. B / 3; depth below surface: 2.40 m; Abs. Ht.: 286.162 m
L: 24.5 cm; max. D: 3.8 cm; min. D: 2.9 cm; perforation L: 4.4 cm; perforation W: 1.8 cm
Quercus spp. (oak)
Symmetrical V-shaped object made from a branch with two stems, cut where the two connect. Both branches cylindrical. The ends are chopped from several directions. Near the ends, there are transversal rectangular perforations. Curved end is cut with a sharp tool and then polished. On the outside is strong sheen. Traces of bark partially preserved. Nearly half of the object preserved.
Preservation state: fair
Museum Inv. No. 16663

82. Binding (*Pl. 22.82*)
Trench III; Sq. E / 5; depth below surface: 1.20 m; Abs. Ht.: 288.30 m
L: 31 cm; W 14 cm; max. D: 5.5 cm; handle D: 3.5 cm
Corylus avelana (hazel)
Oval binding achieved by weaving and entwining a stem with many side stems. The small (all under 2 cm diameter) stems are twisted and braided, and finish in a knot. The main branch has an obliquely cut end, cleared of bark. It may have been a handle. Completely preserved.
Preservation state: good
Museum Inv. No. 15258

83. Binding (*Pl. 22.83*)
Trench III; Sq. B / 4; depth below surface: 3.00 m; Abs. Ht.: 286.45 m
D: 27 cm; Th: 4 cm
Salix spp. (willow)
Oval binding achieved by weaving and entwining a stem with many side stems. The small branches are twisted and braided, and then bound. It was found in the upper part of a vertical pole stuck in the ground. Completely preserved.
Preservation state: good
Museum Inv. No. 16713

84. Perforated beam (*Pl. 23.84*)
Trench III; Sq. D / 5; depth below surface: 0.60 m; Abs. Ht.: 288.90 m
L: 88 cm; W: 10 cm; Th: 6.5 cm; perforation L: 4.6. 4.2 and 4.2 and 5 cm; perforation W: 3.3, 3.7 and 3.4 cm
Quercus spp. (oak)
Perforated beam, rectangular in section, made from quarter split log. Three roughly oval transversal perforations. The first hole is 16.5 cm from the narrow end. The second hole is 27 cm from the first and the third 12 cm from the second. The third hole is 17 cm from the other end. Broken at both ends in ancient times.
Preservation state: fair
Museum Inv. No. 14843

85. Perforated piece (*Pl. 23.85*)
Trench III; Sq. A / 5; depth below surface: 1.10 m; Abs. Ht.: 288.40 m
L: 20 cm W: 8.5 cm; perforation D: 4 cm; Th: 3.8–4.1 cm
Quercus spp. (oak)

Broken perforated circular disc, made from tangentially split timber. Broken in antiquity along the rays. Half-circular shape, with one side flattened over 10.5 cm, other edges gently curved. The perforation is not exactly in the middle, the edges being 8.2 and 7.7 cm from the edge along the broken side. Half of the object preserved.
Preservation state: fair
Museum Inv. No. 14973

86. Perforated piece (*Pl. 23.86*)
Trench III; Sq. F / 8; depth below surface: 0.70 m; Abs. Ht.: 288.80 m
L: 25 cm; W: 6.8 cm; Th: 7.2 cm; perforation L: 3.2 cm; perforation W: 6 cm
Quercus spp. (oak)
Broken perforated object made from radial split timber subsequently split tangentially. Sub-rectangular in section widest at flat end where it is a maximum of 8.7 x 7.2 cm. That flat end appears to be the product of a modern cut. The other end is broken and torn across a notch. Another notch exists 6 cm from the other end. It is roughly rectangular 3.7 cm across and 3.6 cm deep. The long side has traces of polish. The surfaces have traces of repeated chopping, with facets of 4.7 cm width made by a slightly curved blade. Half of the object is preserved.
Preservation state: good
Museum Inv. No. 15260

87. Perforated piece (handle ?) (*Pl. 23.87*)
Trench III; Sq. C / 2; depth below surface: 0.78 m; Abs. Ht.: 288.72 m
L: 105 cm; max. D: 6.5–8.5cm, perforation L: 6.5 cm; perforation W: 3.5 cm
Quercus spp. (oak)
Slightly curving perforated roundwood object. Largest end 8.6 cm diameter, other end 4.9 cm diam. Both ends show traces of wear. Cuts visible on larger end. Rectangular perforation 51 cm from smaller end. On one face the surface either side of the hole has been cut down towards the hole over 6.2 cm (to wider end) and 10 cm. Bark removed. Completely preserved.
Preservation state: good
Museum Inv. No. 15265
Field no. 317/2008

88. Half of a perforated cradle-like object (spacer for stretcher? horse-pannier?) (*Pl. 23.88*)
Trench III; Sq. C / 2; depth below surface: 0.82 m; Abs. Ht.: 288.68 m
L: 65 cm; W: 6.5 cm; Th: 6.5 cm;
Quercus spp. (oak)
Rectangular object, with a body of rectangular section, made from a quarter split timber. Curves in one plane following the natural curvature of the parent trunk. Rounded edges; at the ends are traces of carving and polishing. The body has five transversal indentations, three rectangular; and two semi-circular at the ends. The circular indentations (6.1 across by 4.8 cm deep and 5.7 cm across by 3.6 cm deep) are not perfectly round. The rectangular indentations are slightly wider at the mouth than at the base, varying from 6.2–7.1 cm wide and 1.1–2.4 cm deep. The spacing of the indentations is not consistent (8.8; 9.9; 9.9 and 10.8 cm). The surfaces are smooth. Half of the original tool is preserved, broken longitudinally.
Preservation state: fair
Museum Inv. No. 15266
Field no. 318/2008

89. Perforated cradle-like object (*Pl. 23.89*)
Trench III; Sq. D / 5; depth below surface: 1.30 m; Abs. Ht.: 287.70 m
L: 65 cm; W: 8.5 cm max; Th: 6.5 cm;
Quercus spp. (oak)
Rectangular object, the body rectangular in section, the ends oval in section. Tangentially split from a naturally curving branch, with sapwood surviving along the inner face of the curve. The body has four transversal perforations, two rectangular and two near the ends circular. The two circular perforations are 6.2 cm in diameter, the rectangular ones 5.8–6.4 cm long and 2.8–2.9 cm wide. The corners of the notches are slightly curved suggesting that they were cored before being chiselled out by 1 cm wide tools with a slightly curved cutting edge. Axe marks are

present on the flat surfaces, max. 3.5 cm wide. Traces of friction around the perforations. One end is broken on the centre line of a perforation. Almost entirely preserved.
Preservation state: good
Museum Inv. No. 15267
Field no. 434/2008

90. Small wedge (*Pl. 23.90*)
Trench III; Sq. A / 4; depth below surface: 2.50 m; Abs. Ht.: 286.06 m
L: 21.8 cm; W: 4.4 cm; Th: 1.8 cm
Quercus spp. (oak)
Rectangular chip, rectangular in section. Both surfaces carved. One end carved in two directions, forming an edge cut off straight, like a tip. The other end has ancient breaks.
Preservation state: fair
Museum Inv. No. 16666

91. Large wedge (*Pl. 23.91*)
Trench III; Sq. α / 6; depth below surface: 2.00 m; Abs. Ht.: 287.51 m
L: 26.5 cm; max. D: 9.1 cm; Th: 5 cm
Quercus spp. (oak)
Rectangular chip, half-circular in section. Both ends were obliquely cut off in the same direction. In the middle is a rectangular carving. The surfaces are smooth.
Preservation state: good
Museum Inv. No. 16657

92. Large wedge or chip (*Pl. 23.92*)
Trench III; Sq. A / 4; depth below surface: 1.40 m; Abs. Ht.: 287.16 m
L: 37 cm; max. D: 6.5 cm; tip W: 6 cm; tip W: 1.4 cm
Quercus spp. (oak)
Cylindrical chip made of a thick branch. Upper end cut off straight with a sharp tool, after several attempts. The lower end is carved starting from the middle area of the wedge from two sides, forming two edges. Bark removed. Almost completely preserved.
Preservation state: good
Museum Inv. No. 16660

93. Large wedge (*Pl. 23.93*)
Trench III; Sq. D / 6; depth below surface: 0.73 m
L: 31 cm; W: 7.5 cm; Th: 1.1–2.3 cm
Quercus spp. (oak)
Oval-shaped cleaver made from a tangentially split timber, with a relative short (8.1 x 4.5 x 1.1 cm) handle. Maximum thickness in the middle is 2.3 cm. The use edge is slightly curved, and is cut on one face over 11.5 cm to produce an edge that is only 1.1 cm thick. Almost entirely preserved.
Preservation state: good
Museum Inv. No. 14841
Field no. 189/2007

94. Large wedge (*Pl. 23.94*)
Trench III; Sq. α / 3; depth below surface: 0.90 m; Abs. Ht.: 288.60 m
L: 33 cm; max. D: 9.9 cm; tip W: 11 cm; tip Th: 5 cm
Quercus spp. (oak)
Cylindrical wedge made from a thick branch. Upper end cut off straight with a sharp tool, after several attempts. The lower end is carved starting from the middle area of the wedge from two sides, forming two edges. Bark removed. Almost completely preserved.
Preservation state: good
Museum Inv. No. 16661

95. Large chip (*Pl. 23.95*)
Trench III; Sq. E / 8; depth below surface: 1.22 m; Abs. Ht.: 288.28 m
L: 31 cm; W: 8 cm; Th: 2.5 cm
Quercus spp. (oak)
Rectangular chip, rectangular in section, with an oval tip, with rounded edges. Made from a radially split timber with some sapwood present along one edge of the main body and handle. The tip has wear traces, some breaks. It was created by cutting down the wider faces over 7.3 and 3.7 cm from the end, reducing the thickness from 1.6 cm to 0.3 cm. The 13.3 cm long handle has stepped shoulders, 1.1 cm deep and is cut on one side along the top 5.3 cm, reducing the thickness from 1.5 cm to 0.6 cm. Numerous axe marks are visible on the flat faces and the ends and notched shoulders, the maximum width of the facets being 3.5 cm, possibly representing the full blade width. Almost completely preserved.
Preservation state: good
Museum Inv. No. 15261

96. Small wedge (*Pl. 24.96*)
Trench III; without context
L.: 16.4 cm; LA.: 4.1 cm; TH: 2.7 cm
Wood species not determined
Trapezoidal wedge, rounded edges, triangular in section. Both ends cut off oblique, with traces of wear. Completely preserved.
Preservation state: good
Museum Inv. No. 17264

97. Small wedge (*Pl. 24.97*)
Trench III; Sq. γ / 1`; depth below surface: 2.89 m; Abs. Ht.: 285.67 m
L.: 12.9 cm; L.: 1.8 cm; TH: 0.7 cm
Wood species not determined
Rather rectangular wedge, rectangular in section, with straight edges. Slightly curved profile. Both ends have carving traces, with a sharp tool, mainly on the sides. Completely preserved.
Preservation state: fair
Museum Inv. No. 17266

98. Small wedge (*Pl. 24.98*)
Trench III; Sq. α / 1`; depth below surface: 2.65 m; Abs. Ht.: 285.91 m
L: 7.7 cm; W: 2.4 cm; Th: 1.6 cm
Wood species not determined
Tapering wedge with rounded edges, half-circular in section. The upper end is broken. The tip is rounded, with use traces.
Preservation state: good
Museum Inv. No. 17345

99. Peg (*Pl. 24.99*)
Trench III; Sq. E / 4; depth below surface: 0.55 m; Abs. Ht.: 288.95 m
L: 16.5 cm; D: 3.5 cm
Wood species not determined
Cylindrical shape at one end but rectangular in cross section lower down, one end cut off straight, the other slightly broken. Bark removed. At one end is a blind hole 3.1 cm deep and 0.9 cm in diameter. On one surface are traces of cuts, made with a sharp tool. This may be a peg in the process of manufacture.
Preservation state: fair
Museum Inv. No. 14971

100. Peg (*Pl. 24.100*)
Trench III; without context
L: 9.7 cm; W: 2.3 cm; Th: 1.3 cm; D: 1 cm
Quercus spp. (oak)

Rectangular peg, rectangular in section, with a cylindrical tip. The rectangular end is cut straight, the circular end, broken. Almost completely preserved.
Preservation state: good
Museum Inv. No. 16658

101. Worked roundwood (*Pl. 24.101*)
Trench III; without context
L: 13.3 cm; D: 3.4 cm
Wood species not determined
Cylindrical chip, made from a branch. One end cut straight, the other oblique. On the outside are cuts made with a sharp tool. Bark removed. Completely preserved.
Preservation state: good
Museum Inv. No. 17263

102. Worked roundwood, maybe wedge (*Pl. 24.102*)
Trench III; Sq. A / 1`; depth below surface: 2.12 m; Abs. Ht.: 286.44 m
L.: 6.8 cm; L.: 2.2 cm; TH: 1.4 cm
Wood species not determined
Tapering wedge, made from a branch. One end thicker, broken in ancient times. The tip is rounded, with traces of use. Almost completely preserved.
Preservation state: good
Museum Inv. No. 17265

103. Worked roundwood (*Pl. 24.103*)
Trench III; Sq. C / 5; depth below surface: 0.40 m; Abs. Ht.: 289.10 m
L: 8.5 cm; D: 4 cm
Populus spp. (poplar)
Worked wood (2 pieces) made from a branch. It has a triangular split along its whole length, going to the off-centre pith. The middle was cut off straight (separating the pieces), the ends are broken.
Preservation state: good
Museum Inv. No. 14731

104. Worked roundwood (*Pl. 24.104*)
Trench III; Sq. C / 3; depth below surface: 0.80 m
L: 50 cm; D: 1.5 cm
Corylus avelana (hazel)
Small roundwood sharpened to a point at one end. The tip is worn and smoothed with no traces of tool marks. At 6.2 cm from the tip, is a small 1 cm long cut on one side. On the surface traces of bark. Completely preserved.
Preservation state: good
Museum Inv. No. 14842

105. Worked piece of wood (*Pl. 24.105*)
Trench III; Sq. A / 1; depth below surface: 2.45 m; Abs. Ht.: 286.11 m
L: 24.5 cm; W: 12.3 cm; Th: 6.3 cm
Wood species not determined
Chip, triangular shape, rectangular in section. One end has tool marks.
Preservation state: good
Museum Inv. No. 17268
Field no. 766/2010

106. Worked piece of wood (*Pl. 24.106*)
Trench III; Sq. A / 1; depth below surface: 2.44 m; Abs. Ht.: 286.12 m
L: 20.6 cm; W: 7.7 cm; Th: 2.5 cm
Wood species not determined
Chip, irregular shape, with rather straight edges. On one side are several cuts made with a sharp tool. Almost completely preserved.

Preservation state: fair
Museum Inv. No. 17270

107. Worked piece of wood (*Pl. 25.107*)
Trench III; Sq. α / 3; depth below surface: 1.70 m; Abs. Ht.: 286.286 m
L: 17.4 m; W: 3.6 m; Th: 3.7 m
Quercus spp. (oak)
Object made from a tree root. Curved profile, rectangular in section. Broken or torn at one end, oblique at the other.
Preservation state: good
Museum Inv. No. 16654

108. Worked piece of wood (*Pl. 25.108*)
Trench III; Sq. A / 2; depth below surface: 2.36 m; Abs. Ht.: 286.20 m
L: 45 cm; max. D: 16 cm
Wood species not determined
Wedge made from a trunk fragment. The body is short, one end is rather cylindrical, cut off straight. The tip is tapering, slightly curved. On the tip are burning traces. On the body is the mark of a removed branch. Bark removed. Completely preserved.
Preservation state: good
Museum Inv. No. 17277
Field no. 771/2010

109. Prop/worked piece of wood (*Pl. 25.109*)
Trench III; Sq. F / 8; depth below surface: 0.80 m; Abs. Ht.: 288.70 m
L: 28.3 cm; max. D: 11.2 cm
Quercus spp. (oak)
Log cut obliquely at both ends, cutting angles 30 and 52 degrees. Bottom side cut flat. Several axe marks including one blade jam curve 5 cm wide with a depth across the blade profile of 0.9 cm. Bark removed. Completely preserved.
Preservation state: fair
Museum Inv. No. 15332

110. Worked piece of wood (*Pl. 25.110*)
Trench III; Sq. α / 3; depth below surface: 2.27 m; Abs. Ht.: 286.29 m
L: 37.5 cm; W: 16.8 cm; Th: 9.9 cm; handle L: 11.3 cm; handle W: 7.6 cm; handle Th: 2.9 cm
Quercus spp. (oak)
Tool made from cutting across a tree trunk. The section approximately half-circular. The body has V shaped carvings. One end extends to form a handle. Completely preserved.
Preservation state: fair
Museum Inv. No. 16664

111. Worked piece of wood (*Pl. 25.111*)
Trench III; Sq. α / 1; depth below surface: 2.81 m; Abs. Ht.: 285.75 m
L: 18 cm; W: 16 cm; Th: 15 cm
Wood species not determined
Natural burr, with axe marks from the removal from the tree. On one of the surfaces are traces of repeated tool marks. Fragmented in ancient times. The outer bark partially preserved. Only a small part of the object (about 25%) preserved.
Preservation state: good
Museum Inv. No. 17276

112. Small splinter (*Pl. 25.112*)
Trench III; Sq. A / 1`; depth below surface: 2.12 m; Abs. Ht.: 286.44 m
L: 9.6 cm; W: 5.3 cm; Th: 1.6 cm
Wood species not determined

Rectangular splinter, rectangular in section. One end cut off oblique, the other with carving traces, broken. Almost completely preserved.
Preservation state: fair
Museum Inv. No. 17261

113. Small splinter (*Pl. 25.113*)
Trench III; Sq. A / 1`; depth below surface: 2.12 m ; Abs. Ht.: 286.44 m
L: 16.8 cm; W: 4.3 cm; Th: 1.8 cm
Wood species not determined
Rectangular splinter, rectangular in section, with straight edges. Both ends have carving traces. Completely preserved.
Preservation state: good
Museum Inv. No. 17262

Pieces from wooden structures

114. Pole (*Pl. 26.114*)
Trench III; Sq. A / 1; depth below surface: 0.57 m; Abs. Ht.: 287.99 m
L: 220 cm; max. D: 18 cm; min. D: 7 cm; upper part D: 8 cm
Wood species not determined
Long tapering pole with a notch close to the top. The taper starts just below the notch with the last 50 cm worked more finely to a point. Working traces very visible. The upper end has two branches in a V shape, cut off straight. Several branches cut away from the body. The bark partially preserved. Completely preserved.
Preservation state: good
Museum Inv. No. 17273
Field no. 155/2010

115. Pole (*Pl. 26.115*)
Trench III; Sq. A / 1; depth below surface: 0.59 m; Abs. Ht.: 287.97 m
L: 190 cm; max. D: 22 cm; min. D: 11 cm
Wood species not determined
Stout pole. The tip is carved 1/3 of its length. Circular in section in the upper part, rather more rectangular towards the tip. Towards the tip are multiple traces of carving made with a sharp tool. The upper end is cut off straight. The bark is well preserved on the body. Completely preserved.
Preservation state: good
Field no. 154/2010
Museum Inv. No. 17274

116. Wood piece (*Pl. 27.116*)
Trench III; Sq. A B / 3; depth below surface: 2.82 m; Abs. Ht.: 268.68 m
L: 114.3 cm; W: 36 cm; Th: 4.5 cm
Quercus spp. (oak)
Timber (possibly a beam) made from a half-split trunk, half-circular in section. Both ends were cut off straight with a sharp tool. Bark removed. Completely preserved.
Preservation state: good
Museum Inv. No. 16758

117. Timber (*Pl. 27.117*)
Trench III; Sq. A / 1; depth below surface: 1.65 m; Abs. Ht.: 286.91 m
L: 50.5 cm; W: 1.5 cm; Th: 7.5 cm
Wood species not determined
Timber (possibly a beam) made from a half-split trunk. Rectangular shape, half-circular section. One end cut oblique after several chops. The other end is cut straight, with cutting traces on it. On the body are traces of removed branches. Fallen bark. Completely preserved.

Preservation state: good
Museum Inv. No. 17275
Field no. 755/2010

118. Plank / Beam (*Pl. 27.118*)
Trench III; Sq. A / 1; depth below surface: 2.55 m; Abs. Ht.: 286.01 m
L: 119 cm; W: 13 cm; max. Th: 8 cm
Wood species not determined
Plank (pole?) made by vertically splitting a tree trunk. Rectangular shape, trapezoidal in section. The upper end is cut straight (with a modern saw for dendrochronological sampling). The tip is sharpened by means of several chops. Used to hold up part of the wattle structure (Complex 3).
Preservation state: good
Museum Inv. No. 17278
Field no. 158/2010

119. Large worked piece of wood (*Pl. 27.119*)
Trench III; Sq. A–B / 1–2; depth below surface: 2.00 m; Abs. Ht.: 286.56 m
L: 164.5 cm; W: 53 cm; Th: 4 cm
Quercus spp. (oak)
Carved object with chopping traces on both surfaces. Triangular shape, asymmetrical, rectangular section, curved profile. The edges are straight, with traces of shaping. Peaked end made by chopping the two sides. In the upper part a vertical crack.
Preservation state: good
Museum Inv. No. 16723

120. Plank (*Pl. 27.120*)
Trench III; Sq. A / 1; depth below surface: 2.02 m; Abs. Ht.: 287.48 m
L: 114 cm; W: 18 cm; Th: 7 cm
Wood species not determined
Plank made by vertically splitting a tree trunk. Trapezoidal, rectangular in section. The upper end cut off straight. The tip is chopped over about half of its length. Both surfaces are well smoothed. One edge is carved straight; the other is cut oblique, to form a tip. Completely preserved.
Preservation state: good
Museum Inv. No. 17279
Field no. 161/2010

Trench XV

Note: the most important pieces in this trench, the trough and ladder, were not extracted and so could not be studied in detail for inclusion here.

121. Half part of a V-shaped object (*Pl. 28.121*)
Trench XV; Sq. A /1; depth below surface: 0.20 m; Abs. Ht.: 293.24 m
L: 22 cm; W: 4 cm, wedge L: 9 cm; wedge W: 4 cm; wedge Th: 1.5 cm
Quercus spp. (oak)
V-shaped object made from a branch, broken at the curved end with the other half of the object broken off. The preserved branch is carved, of cylindrical shape. Near one end is a rectangular (3.7 x 2.2 cm) perforation in which a small wedge was found. It is cut down on the outside of the curve over a length of 13 cm and is also cut across on the inside of the curve. On the outside a marked sheen. No bark survives. Half of the object is preserved. It has a 0.4 cm wide crack down one side and a portion has been split off down one side next to the hole on the inside of the curve. The peg is formed from a half split oak roundwood that had been squared off.
Preservation state: good
Museum Inv. No. 15582
Field no. 5000/200

122. Small wedge (*Pl. 28.122*)
Trench XV; Sq. B / 1; depth below surface: 0.40 m; Abs. Ht.: 293.04 m
L: 15.6 cm; W: 2.8 cm; Th: 1.4 cm
Quercus spp. (oak)
Rectangular shape, radially split, trapezoidal in section, the profile slightly curved. Carving traces on both surfaces. One of the ends chopped symmetrically from two sides, forming an edge, cut straight. The other end cut off straight. Completely preserved.
Preservation state: good
Museum Inv. No. 15581
Field no. 5019/2008

123. Small wedge (*Pl. 28.123*)
Trench XV; Sq. I / 1; depth below surface: 0.30 m; Abs. Ht.: 293.14 m
L: 15.3 cm; W: 2.6 cm; Th: 2 cm
Quercus spp. (oak)
Rectangular wedge, rectangular in section, made from a small tangentially split piece of wood. One end chopped and cut off straight from several attempts, the other end broken. The tip was cut from two sides to form a wedge-shaped point 2.1 cm wide and 0.4–0.6 cm thick. Completely preserved.
Preservation state: good
Museum Inv. No. 15584
Field no. 5061/2008

124. Small splinter, possible wedge (*Pl. 28.124*)
Trench XV; Sq. I / 1; depth below surface: 0.30 m; Abs. Ht.: 293.14 m
L: 12.5 cm; W: 3.4 cm; Th: 1.7 cm
Quercus spp. (oak)
Wedge made from a longitudinal half-split branch, half-circular in section. On the body traces of cut-off branches and a simple axe cut, 2.4 cm wide. One end is chopped across obliquely at 40 and 50 degrees by two axe blows, creating a stepped end. The other end is cut from the same direction at 30 degrees, forming a chisel-shaped point.
Preservation state: good
Museum Inv. No. 15585
Field no. 5062/2008

125. Worked roundwood (*Pl. 28.125*)
Trench XV; Sq. A / 2; depth below surface: 0.55 m; Abs. Ht.: 292.89 m
L: 11.5 cm; max. D: 5.1 cm
Quercus spp. (oak)
Roundwood with no surviving bark. On one face is a small notch, 3.3 cm long and 1.2 cm deep. On the same face is a vertical axe cut 3.3 cm wide. Both ends broken. The object has deteriorated and has many cracks.
Preservation state: medium to poor
Museum Inv. No. 15583
Field no. 5006/2008

Trench XXII

126. Peg (*Pl. 28.126*)
Trench XXII; without context
L: 14 cm; max. D: 3 cm; perforation D: 0.7 cm
Sambucus nigra (elder)
Tapering peg, circular in section. Longitudinal central hole. Very slight collar, barely visible. Around the perforation, at both ends, and on the body, cutting traces with a sharp tool. Completely preserved.
Preservation state: good
Museum Inv. No. 16706

127. Peg (*Pl. 28.127*)
Trench XXII; without context
L: 13.2 cm; max. D: 2.5 cm; perforation D: 0.8 cm
Sambucus nigra (elder)
Tapering peg, oval in section in the upper part, circular in section at the tip. Longitudinal central hole. Very visible collar in the upper part. On the body traces of carving with a sharp tool. Completely preserved.
Preservation state: good
Museum Inv. No. 16715

128. Peg (*Pl. 28.128*)
Trench XXII; Sq. L / 12; depth below surface: 0.30 m; Abs. Ht.: 288.22 m
L: 13.2 cm; max. D: 2.8 cm; perforation D: 0.8 cm
Sambucus nigra (elder)
Tapering peg, oval in section in the upper part, circular in section at the tip. Longitudinal central hole. In the upper part are several breaks. Traces of rubbing against a cord. On the body there are several longitudinal chop marks. Collar in the upper part. Completely preserved.
Preservation state: good
Museum Inv. No. 16717

129. Peg (*Pl. 28.129*)
Trench XXII; Sq. J/ 12; depth below surface: 0.30 m; Abs. Ht.: 288.22 m
L: 14.2 cm; max. D: 3.6 cm; perforation D: 0.7 cm
Sambucus nigra (elder)
Tapering peg, circular in section. Longitudinal central hole. On the body are several carvings, more visible in the lower part, towards the tip. In the upper part, around the perforation there are traces of rubbing from a cord. Slight collar, fashioned from two sides. Completely preserved.
Preservation state: good
Museum Inv. No. 16719

130. Peg (*Pl. 28.130*)
Trench XXII; Sq. I / 16; depth below surface: 0.30 m; Abs. Ht.: 288.22 m
L: 11.5 cm; max. D: 3.5 cm; perforation D: 0.5 cm
Quercus spp. (oak)
Tapering peg, oval in section. Longitudinal central hole. No collar. Partially preserved, the tip broken longitudinally, at the half-way point.
Preservation state: fair
Museum Inv. No. 16718

131. Mallet (*Pl. 29.131*)
Trench XXII; Sq. K /15; depth below surface: 0.76 m; Abs. Ht.: 287.77 m
L: 53 cm; body D: 19 cm; handle L: 60.5 cm; handle D: 6 cm
Quercus spp. (oak)
Tool made from a tree trunk with branch attached. Rather cylindrical in shape, with several carvings on the body. Both ends cut off straight. The use end has wear traces. The handle is cylindrical, slightly degraded at the end. The bark partially present. Completely preserved.
Preservation state: good
Museum Inv. No. 16722
Field no. XXII/36

132. Shovel (*Pl. 29.132*)
Trench XXII; Sq. K / 15; depth below surface: 0.80 m; Abs. Ht.: 287.27 m
L: 41 cm; W: 9 cm; Th: 1.9 cm; handle D: 2.2 cm
Quercus spp. (oak)
Small shovel made from a board. The bowl is rather oval, with rounded edges, flattened; about half of the object survives. The handle is tapering, oval in section, thinner towards the end. The bowl has a small break on one side.

155. Grooved hammer (*Pl. 33.155*)
Trench I; sq. B/4; depth below surface: -0.40 m
L: 25.6 cm; W: 11.6 cm; Th: 8.4 cm; Weight: 4.268 kg
Rectangular shape, rectangular in section. One flat end. One surface and one end battered.
Museum Inv. No. 14835

156. Grooved hammer (*Pl. 33.156*)
Trench I, Without context
L: 20 cm; W: 9.1 cm; Th: 7 cm; Weight: 2.867 kg
Rectangular shape, rectangular in section, at one end and almost triangular at the other. All three edges from the distal end have grooves.
Museum Inv. No. 15586

157. Grooved hammer (*Pl. 33.157*)
Trench I, Without context
L: 21.8 cm; W: 12.3cm; Th: 6.2 cm; Weight: 2.399 kg
Trapezoidal shape, triangular in section with rounded edges. The distal end broken.
Museum Inv. No. 15587

158. Grooved hammer (*Pl. 33.158*)
Trench I, Without context
L: 15.1 cm; W: 8.8 cm; Th: 6.1 cm; Weight: 1.535 kg
Rectangular shape, rectangular in section with rounded edges. One end with traces of wear, the other end battered.
Museum Inv. No. 15588

159. Grooved hammer (*Pl. 33.159*)
Trench I; sq. D/4; depth below surface: -0.30 m
L: 29.4 cm; W: 13.9 cm; Th: 8.9 cm; Weight: 4.619 kg
Triangular shape, triangular in section. Rounded end with traces of wear. One end battered.
Museum Inv. No. 14836

160. Painted grooved hammer (*Pl. 33.160*)
Trench III; sq. A/3; Abs. Ht.: 288.30 m
L: 12.3 cm; W: 9.1 cm; Th: 4.3 cm; Weight: 0.761 kg
Rectangular shape, rectangular in section with rounded edges. Three of the four edges are broken. Two small channels on the upper surface bear traces of red ochre.
Museum Inv. No. 15268

161. Grooved hammer (*Pl. 34.161*)
Trench III, Without context
L: 22.5 cm; W: 18.7 cm; Th: 5.7 cm; Weight: 2.370 kg
Trapezoidal shape, triangular in section with rounded edges.
Museum Inv. No. 16671

162. Grooved hammer (*Pl. 34.162*)
Trench III, Without context
L: 12.3 cm; W: 10.2 cm; Th: 4.9 cm; Weight: 1.187
Trapezoidal shape, oval section. Grooves on two lateral edges.
Museum Inv. No.17259

163. Grooved hammer (*Pl. 34.163*)
Trench III; sq. E/6; Abs. Ht: 288.45 m
L: 26.8 cm; W: 11.9 cm; Th: 6.7 cm; Weight: 4.827 kg
Rectangular shape, triangular in section, with rounded edges. One end rounded, the other flat.
Museum Inv. No. 15326

164. Grooved hammer (*Pl. 34.164*)
Trench III; sq. J/3; Abs. Ht: 288.45 m
L: 26.9 cm; W: 13.9 cm; Th: 8.1 cm; Weight: 4.989 kg
Rectangular shape, rectangular in section, with rounded edges. One side and one surface broken. Grooves on three of the four edges.
Museum Inv. No. 15331

165. Grooved hammer (*Pl. 34.165*)
Trench III
L: 19.2 cm; W: 13.1 cm; Th: 11.2 cm; Weight: 4.867 kg
Rectangular shape, rectangular in section. One flat end, the other broken. Groove on one edge.
Museum Inv. No. 15001

166. Grooved hammer (*Pl. 34.166*)
Trench III
L: 33.8cm; W: 12.4 cm; Th: 8.4 cm; Weight: 4.487 kg
Pyramidal shape, triangular in section at one end and rectangular at the other. The active end is rounded, with utilisation traces. Grooves on two edges.
Museum Inv. No. 15002

167. Grooved hammer (*Pl. 34.167*)
Trench III; sq. D/8; Abs. Ht.: 289.10 m
L: 25 cm; W: 10.8 cm; Th: 5.5 cm; Weight: 2.590 kg
Rectangular shape, rectangular in section. One end pointed. On the edges slight grooves.
Museum Inv. No. 15161

168. Grooved hammer (?) (*Pl. 34.168*)
Trench III; sq. α /3; Abs. Ht: 288.05 m
L: 19.6 cm; W: 12.5 cm; Th: 5.5 cm; Weight: 2.125 kg
Oval shape, oval in section. Both surfaces flattened. One end round, the other broken. Two possible grooves.
Museum Inv. No. 15327

169. Grooved hammer (*Pl. 35.169*)
Trench III; sq. A/1; Abs. Ht: 287.83 m
L: 21.6 cm; W: 13.3 cm; Th: 8.9 cm; Weight: 3.208 kg
Circular shape, trapezoidal in section. Groove on one side (?). Traces of use hard to see.
Museum Inv. No. 15330

170. Grooved hammer (*Pl. 35.170*)
Trench III; sq. A/3; Abs. Ht: 286.289 m
L: 24 cm; W: 8 cm; Th: 3.2 cm; Weight: 0.998 kg
Oval shape, half circle in section. Profile slightly curved. Battered along the whole length, one surface preserved.
Museum Inv. No. 16669

171. Grooved hammer (*Pl. 35.171*)
Trench III; sq. A/3; depth below surface: 0.70 m
L: 14 cm; W: 7.6 cm; Th: 4.1 cm; Weight: 0.453 kg
Trapezoidal shape, rectangular in section. One end broken, one with battering traces. On the sides are grooves.
Museum Inv. No. 16721

172. Grooved hammer (*Pl. 35.172*)
Trench XXII; sq. L/5; depth below surface: -1.30 m
L: 23 cm; W: 16.5 cm; Th: 7.8 cm; Weight: 4.355 kg
Rectangular shape, oval section. Groove on one edge. One end broken.
Museum Inv. No. 16720

173. Grooved hammer (*Pl. 35.173*)
Without context
L: 22.3 cm; W: 8.5 cm; Th: 9.4 cm; Weight: 2.508 kg
Oval shape, rectangular in section, with one end pointed. Grooves on three of the four lateral edges. Fragmented at both ends.
Museum Inv. No. 13936

174. Grooved hammer (*Pl. 35.174*)
Without context
L: 23.5 cm; W:9 cm; Th: 6 cm; Weight: 2.180 kg
Oval shape, triangular in section.
Museum Inv. No. 13937

175. Grooved hammer (*Pl. 35.175*)
Without context
L: 14.6 cm; W: 10.7 cm; Th: 10.2cm; Weight: 2.889 kg
Rectangular. One end broken.
Museum Inv. No. 13939

176. Grooved hammer (*Pl. 35.176*)
Without context
L: 1.42 cm; W: 7.3 cm; Th: 6.8 cm; Weight: 1.344 kg
Trapezoidal shape. One end broken.
Museum Inv. No. 13940

177. Grooved hammer (*Pl. 36.177*)
Without context
L: 22.9 cm; W: 12.4 cm; Th: 7.8 cm; Weight: 2.968 kg
Trapezoidal shape, triangular in section. Three straight facets. All three edges have grooves.
Museum Inv. No. 13944

178. Grooved hammer (*Pl. 36.178*)
Without context
L: 27 cm; W: 14.5 cm; Th: 13.8 cm; Weight: 6.210 kg
Triangular shape, trapezoidal in section. Four straight facets.
Museum Inv. No. 13945

179. Grooved hammer (*Pl. 36.179*)
Without context
L: 31.4 cm; W: 10.3 cm; Th: 8.5 cm; Weight: 4.958 kg
Rectangular shape, rectangular in section. One end flat.
Museum Inv. No. 14834

180. Grooved hammer (*Pl. 37.180*)
Without context
L: 21.4 cm; W: 12.5 cm; Th: 7.8 cm; Weight: 5.520 kg
Rectangular shape, trapezoidal in section.
Museum Inv. No. 15328

181. Grooved hammer (*Pl. 37.181*)
Without context
L: 27.3 cm; W: 15.1 cm; Th: 8.8 cm; Weight: 5.250 kg
Rectangular shape, oval in section. Grooves on two edges. One surface broken.
Museum Inv. No. 15329

182. Grooved hammer (*Pl. 37.182*)
Without context
L: 1.46 cm; W: 2.07 cm; Th: 6.8 cm; Weight: 1.713 kg
Oval shape, triangular in section at one end, rectangular at the other.
Museum Inv. No. 13938

183. Grooved hammer (*Pl. 37.183*)
Without context
L: 19.3 cm; W: 8.6 cm; Th: 6.4 cm; Weight: 1.517 kg
Rectangular shape, triangular in section. One end rounded and broken.
Museum Inv. No. 13941

184. Grooved hammer (*Pl. 37.184*)
Without context
L: 21 cm; W: 10.8 cm; Th: 6.2 cm; Weight: 2.174 kg
Oval. On both ends traces of use.
Museum Inv. No. 13943

185. Grooved hammer (*Pl. 37.185*)
Without context
L: 17.5 cm; W: 9.8 cm; Th: 8 cm; Weight: 1.754 kg
Rectangular shape, rectangular in section. One end has use traces. Groove on one edge.
Museum Inv. No. 13948

186. Grooved hammer (*Pl. 38.186*)
Without context
L: 27.5 cm; W: 8.8 cm; Th: 11.3 cm; Weight: 4.897 kg
Rectangular shape, rectangular in section. Grooves on three edges. Active end battered.
Museum Inv. No. 16434

187. Hammer (*Pl. 38.187*)
Trench XV
L: 10.4 cm; W: 10 cm; Th: 6.1 cm; Weight:1.553 kg
Rectangular shape, rectangular with rounded edges in section. One of the surfaces smooth. One broken end, the other with traces of wear.
Museum Inv. No. 15589

188. Hammer (*Pl. 38.188*)
Without context
L: 7.5 cm; W: 9.1 cm; Th: 4.8 cm; Weight: 0.532 kg
Oval at one end, the other broken. Two flat facets.
Museum Inv. No. 13942

189. Hammer (*Pl. 38.189*)
Without context
L: 16.7 cm; W: 9 cm; Th: 8 cm; Weight: 1.911 kg
Rectangular shape, rectangular in section. One end rounded.
Museum Inv. No. 13947

190. Whetstone (?) (*Pl. 38.190*)
Trench I; sq. A–B/2; depth below surface: -0.80 m
L: 7.8 cm; W: 4.5 cm; Th: 2.2 cm; Weight: 0.032 kg
Fragment of sandstone, elongated rectangular shape. On one surface traces of wear.
Museum Inv. No. 14961

191. Whetstone (?) (*Pl. 38.191*)
Trench XV; sq. A/; depth below surface: -0.30 m
L: 8 cm; W: 3.6 cm; Th: 0.8 cm; Weight: 0.054 kg
Rectangular shape, rectangular in section. One of the surfaces smooth. Both ends broken.
Museum Inv. No. 15590

192. Whetstone (?) (*Pl. 38.192*)
Trench XV; sq. A/1; depth below surface: -0.30 m
L: 18.3 cm; W: 4.5 cm; Th: 2.5 cm; Weight: 0.420 kg
Oval shape, oval in section. one of the surfaces polished. One end broken.
Museum Inv. No. 15591

Iron

One iron object was found in Trench III, a unique find so far. Its date is unknown and no obvious parallels are yet available.

193. Pick (*Pl. 38.193*)
Trench III; sq. β/2'; depth below surface: 1.65 m
Abs. Ht.: 287.85 m
L: 20 cm; W: 6.5 cm; Th: 10.5 cm; edge L: 4.5 cm; W: 2.5 cm
Iron
Pick with shaft-hole, straight edge and sharp blade in a V shape. The blade is triangular, in the upper part with symmetrical conical protrusions. Part of a wooden handle is preserved in the hole. Completely preserved though battered.
Preservation state: good
Museum Inv. No. 17953

Appendix 2a
Conservation of archaeological wood from Băile Figa

GELU ŢICULEANU

Archaeological wood, like any organic material, when deposited in the soil suffers extensive damage due to physical, chemical and biological agents. Successful archaeological wood conservation depends on a knowledge of wood structure and types, as well as the environment in which it was preserved.

The soil at Băile Figa in which the wood structures and objects were found consists mainly of salty mud. The high content of salt (sodium chloride) and water in the mud ensured good conditions for wood preservation. The wet conditions ensured the elasticity of the wood fibre, so that it did not crack. A high salt concentration has a dual role: it promotes both the preservation and the deterioration of wood. The salt blocks the penetration of microbes into the wood. On the other hand, salt within the wood can cause damage when the wood dries out; the salt crystallizes, exerting pressure on the cell walls. In this case damage is irreversible.

After the wooden objects were removed from the original environment, they were protected from direct sunlight, along with the soil adhering to them, and humidity was kept stable,. They were transported in suitable containers, following the original site conditions, until they arrived at the Laboratory for Conservation and Restoration of the Museum of Bukovina, where expert wood conservators took charge of them.

In order for these pieces of wood to be preserved for exhibition and scientific study, they had to pass through several complex and lengthy stages of preservation:

- Applying prevention and control treatment against biologically harmful agents;
- Measuring the salinity of the wood;
- Clearing debris from the soil as well as other impurities that affected the wood over time;
- Desalination;
- Impregnation with polyethylene glycol (PEG);
- Removal of surplus impregnation material;
- Controlled drying.

Biocidation

To preserve archaeological wood, special tanks were built in the Suceava Laboratory, with electric mixers and standard thermostats. The first measure was the application of biocidal agents on the wood to prevent (or fight) a possible biological attack (*Fig. 1a*). The agent used was a Preventol R80 3–4% solution dissolved in distilled water and absolute alcohol. Treatment was performed through brushing, spraying and insertion.

Preventol R80 was also added to the PEG solution, in order for it to penetrate the wood at the same time as the PEG. Biocidal agents need to be applied throughout the entire conservation process and after the wood has dried; monitoring is required at all stages.

a

b

Fig. 1. a: excessive humidity and micro-climate fluctuations favour and maintain biological attack on archaeological wood; b: residue deposits of various pollutants on archaeological wood.

Cleaning

Removal of soil impurities, iron oxides and other foreign elements found in the wood fibres was performed through repeated bathing in distilled water solutions in which cleaning agents and biocides were introduced; for these procedures, a soft bristle brush and other utensils were used.

a

b

Fig. 2. a: crystallization of salts in the wood fibres; b: in the desalination solution.

After several rinses (for each item separately), spatulas, brushes and hair brushes were used in order to remove adhering deposits (*Fig. 1b*), as well as cleaning agents of the C2000 type at 30–40% solution in distilled water, in order to increase the solubility of soil deposits. These operations are repeated every time the solution for desalination is changed.

Desalination

Determination of salinity. The first solution measurements were made after three weeks, to establish the solution's physical parameters (salinity, electrical conductivity, pH, TDS). Measurements were repeated each time before the solution in the tanks was changed, more frequently within the first six months, followed by check-ups every 30 days. Wood desalination ended when it was considered that the measurement of the solution showed that its properties were equivalent to those of drinking water.

Desalination is performed in specially constructed tubs depending on the size of the artifact and the particular wood species (*Figs 2a–b*). Soft deciduous species are separated from hard deciduous ones because wood density differs, thus requiring different desalination procedures. As well as length, wood pieces were selected according to volume, those of smaller volume (dowels, small shovels, rods) were grouped separately, since they will desalinate in less time than those of larger volume.

Impregnation

In order to perform impregnation (dimensional and mechanical stabilization), the salt must be removed from the wood structure and replaced with a substance that is compatible with the wood fibres, so that the wood can retain its elasticity and hardness. This may take 8 to12 months (or more).

Many substances that can be used for archaeological wood impregnation have been tested; these are: polyethylene glycol (PEG), sucrose, pine resin, synthetic resins (epoxy) etc. Impregnation with PEG was the first viable method used for treating archaeological wood from water-saturated environments.

PEG is a polymer mixture with the formula $H_2OCH\ (CH_2OH_2)\ CH_2OH$. Polyethylene glycols with an average molecular weight between 300 and 600 are liquid, those with an average molecular weight between 1000–1500 are semi-liquid, have the consistency of Vaseline, and those with the average molecular weight between 3000 and 6000 are solid, and have the consistency of wax. In the present case, we used polyethylene glycols with an average molecular weight of 1500 and 4000, which gave good results over time (*Figs 3a–b*). They have a melting point between 53–55°C, are soluble in alcohol (ethanol, methanol, isopropanol) as well as in hot water, and after impregnation they are the least hygroscopic. PEG is soluble both in water and in alcohol. Usually water is used for large artifacts because it is cheaper and does not pose a fire hazard.

Impregnation with PEG is a relatively simple method, which removes excess water and inflates the fibres of the treated wood. After cleaning the wood surface, it is placed in a ventilated container or tank containing a solution of PEG and a solvent (distilled water or ethyl alcohol). The container temperature will gradually increase during a period of time ranging from several days to 2–3 weeks, until it reaches 53–55 °C. During this time the proportion of PEG in the solution is increased by gradually adding it, while the water evaporates, until it reaches a concentration of 70–100%, depending on the nature of the wood. The PEG gently penetrates the wood fibres, replacing water.

All the wood artifacts from Băile Figa are now stored at the Laboratory for Conservation and Restoration of the Museum of Bukovina in Suceava, this being the only laboratory specializing in conservation and restoration of archaeological wood in Romania.

a

b

Fig. 3. Intermediate phases during impregnation with PEG.

Appendix 2b
Report on the conservation of eleven small finds, April 2010

NYSSA MILDWATERS – STEVEN J. ALLEN

York Archaeological Trust Conservation Laboratory Report Number 2009/30

Description

In November 2008 a small assemblage of waterlogged wooden artefacts were brought to the Conservation Laboratory at York Archaeological Trust for stabilisation. The objects came from two locations in Romania: Caila and Figa. The assemblage consisted of eleven objects all of which were in a relatively good condition upon arrival. Although virtually all of the finds had some degree of orange-brown iron corrosion present across their surfaces, Figa 1440 and Figa 1436 had a noticeably greater amount present. Portions of the surface of Figa 1440 also had a hard gritty black material covering the surface, but below the iron staining. Figa 1182 and Figa 1436 both appear to have portions missing; both had relatively clean break edges but they and several of the other object show small amounts of surface damage and splintering, generally around the objects' edges. Figa 1440 also had a small but slightly flexible crack at the tapered end. The longest of the three fragments from Trench I also appeared to have patches of the wood's original surface bark surface present, but it appeared degraded.

Methodology

Salt and Corrosion Removal

The objects were removed from their packaging and carefully washed under cold running water to remove any remaining soil or other debris from their outer surfaces. As the material was known to have been recovered from a salty burial environment the chloride levels of each object or collection of objects was tested. The results of initial chloride and pH testing were as follows:

Table 1. pH and Cl content of wooden objects.

	pH	Cl (ppm)
Figa 1440	6.4	3600
Figa 1436	6.2	2400
Figa 1182	7.2	2800
Figa Trench I	6.5	2900
Figa Trench I	4.6	3800
Caila 1	6.7	1600
Caila 2	7.6	2100
Caila 3	6.4	1800

As the above table shows all of the wood had very high concentrations of chloride ions present. As the existence of such high levels of chloride ions within objects is undesirable due to the potential for future deterioration, the finds were placed in a bath of running water in order to wash out as many of the chloride ions as possible. The chloride concentration of the water was tested at intervals in order to establish how far the levels had dropped. Once the chloride concentration reached an acceptable level (in this case approximately 29.65 ppm) the wood was removed from the water bath.

Following the wood's desalination, the objects were placed in a bath of 5% Ethylenediaminetetra Acetic Acid (EDTA) in order to remove as much of the surface corrosion and iron staining presence as possible prior to freeze drying. Upon removal from the bath all the objects were thoroughly rinsed with water to remove any remaining chemical residue. The EDTA successfully removed virtually all the iron staining and corrosion products present from the surface of the wood.

Condition Assessment

In order to establish the appropriate polyethylene glycol (PEG) regime for the wood prior to freeze drying, it was necessary to establish both the density of the wood as well as to identify the species of each object. Species identification was undertaken by wood specialist Steve Allen. Once species identification was completed, each of the objects was weighed both in air and again whilst immersed in water. The weights were then entered into the PEGcon programme, created by Clifford Cook and the Canadian Conservation Institute and Parks Canada, which calculated the actual density and moisture content of the wood and compared these with the normal density of undeteriorated wood of the same species to produce an ideal PEG regime. Whilst weighing the objects each object was also pin tested to establish the hardness of the wood. Although the finds were weighed and tested prior to the start of any treatment commencing, it was felt that as the finds were undergoing several stages of treatment each object should be weighed both in water and air after both desalination and EDTA treatment, in case any change in weight occurred. The results of the three weighing and the PEGcon PEG calculations are shown in *Tables 2–4*.

Table 2. Initial weighing.

Find/Timber No.	Species	Weight in water (g)	Weight in air (g)	Normal density (g/cm³)	Actual density (g/cm³)	Moisture content (max)	Pin test (A-C)	PEGcon PEG calculations PEG 400	PEG 4000
Figa 1440	*Fraxinus excelsior L.*	107.0	667.0	0.54	0.573	107.8	C	21.3	0
Figa 1436	*Quercus spp.*	120.0	622.0	0.56	0.717	72.8	C	31.6	0
Figa 1182	*Quercus spp.*	116.0	594.0	0.56	0.728	70.7	C	32.5	0
Figa Trench 1	*Quercus spp.*	15.0	98.0	0.56	0.542	117.8	C	19.5	2.5
Figa Trench 1	*Quercus spp.*	12.0	84.0	0.56	0.5	133.3	C	17.3	7.8
	Salix spp.(flat)	5.0	45.0	0.36	0.375	200.0	C	16.8	0
	Salix spp.(pole)	9.0	140.0	0.36	0.206	418.5	C	8.0	13.9
Caila 1	*Salix spp. (short)*	9.0	56.5	0.36	0.568	109.3	C	30.6	0
	Salix spp. (long)	6.0	49.0	0.36	0.419	172.2	C	19.5	0
Caila 2	*Sambucus nigra L.*	9.0	64.0	0.38	0.491	137.0	C	24.5	0
Caila 3	*Sambucus nigra L.*	9.0	92.0	0.38	0.325	240.7	C	13.9	5.8

Table 3. Following desalination.

Find/ Timber No.	Species	Weight in Water (g)	Weight in Air (g)	Normal density (g/cm³)	Actual density (g/cm³)	Moisture content (max)	Pin test (A-C)	PEGcon PEG calculations PEG 400	PEG 4000
Figa 1440	*Fraxinus excelsior L.*	106.0	669.0	0.54	0.565	110.4	C	20.8	0
Figa 1436	*Quercus spp.*	119.0	629.0	0.56	0.700	76.2	C	30.2	0
Figa 1182	*Quercus spp.*	114.0	600.0	0.56	0.704	75.4	C	30.5	0
Figa Trench 1	*Quercus spp.*	15.0	101.0	0.56	0.523	124.4	C	18.5	5.0
Figa Trench 1	*Quercus spp.*	12.0	81.0	0.56	0.522	125.0	C	18.4	5.1
	Salix spp. (flat)	5.0	46.5	0.36	0.361	210.0	C	15.9	0
	Salix spp. (pole)	14.0	143.0	0.36	0.326	240.5	C	14.0	3.6
Caila 1	*Salix spp. (short)*	10.0	58.0	0.36	0.625	93.3	C	35.9	0
	Salix spp. (long)	8.0	51.0	0.36	0.558	112.5	C	29.8	0
Caila 2	*Sambucus nigra L.*	9.0	65.0	0.38	0.482	140.7	C	23.8	0
Caila 3	*Sambucus nigra L.*	11.0	95.0	0.38	0.393	187.9	C	17.8	0

Table 4. Following EDTA.

Find/Timber No.	Species	Weight in Water (g)	Weight in Air (g)	Normal density (g /cm³)	Actual density (g/ cm³)	Moisture content (max)	Pin test (A-C)	PEGcon PEG calculations PEG 400	PEG 4000
Figa 1440	*Fraxinus excelsior L.*	105.0	664.0	0.54	0.563	110.8	C	20.7	0
Figa 1436.	*Quercus spp.*	119.0	621.0	0.56	0.711	73.9	C	31.1	0
Figa 1182	*Quercus spp*	113.0	591.0	0.56	0.709	74.3	C	30.9	0
Figa Trench 1	*Quercus spp.*	14.0	97.5	0.56	0.503	132.1	C	17.4	7.4
Figa Trench 1	*Quercus spp.*	12.0	83.5	0.56	0.504	131.9	C	17.5	7.3
	Salix spp. (flat)	5.0	45.5	0.36	0.37	203.3	C	16.5	0
	Salix spp. (pole)	13.5	139.0	0.36	0.323	243.2	C	13.8	3.9
Caila 1	*Salix spp. (short)*	9.0	56.5	0.36	0.568	109.3	C	30.6	0
	Salix spp. (long)	6.5	49.0	0.36	0.459	151.3	C	22.2	0
Caila 2	*Sambucus nigra L.*	9.0	64.5	0.38	0.486	138.9	C	24.1	0
Caila 3	*Sambucus nigra L.*	11.5	92.5	0.38	0.426	168.1	C	19.9	0

Fraxinus excelsior L. (ash)
Quercus spp. (oak, exact species not determinable
Salix spp. (willows, exact species not determinable)
Sambucus nigra L. (elder)

Treatment

Taking into consideration the PEG regime suggested by the PEGcon programme as well as past experience, it was decided to put all the objects through a two-stage PEG treatment with a final concentration of 15% PEG 400 and 10% PEG 4000. The finds were placed initially in a 5% solution of PEG 400, which was increased by 5% every two weeks until the target concentration was reached. An initial 0.2% of solution of biocide (Linkcide KMC, a broad spectrum biocide) was added at the start of treatment to inhibit any mould and bacterial growth.

Treatment commenced on 12 June 2009 and was completed by the end of October. The wood was left to soak in its final solution for a further 13 weeks until freeze-drying space became available.

The 2m Birchover freeze dryer at Walmgate was used to dry the wood. The wood was loaded into the specimen chamber and frozen to -30°C before freeze-drying commenced. This took place over 5 weeks (Birchover Freeze Dryer Run no.80) commencing 25 January 2010. Monitoring of the conditions inside the equipment was undertaken through recording of chamber and condenser temperature and vacuum.

Following completion of freeze- drying, surface finishing was undertaken. Each object was dry-brushed, then lightly swabbed over with warm water and tissue paper, then dried using a hot air blower to remove excess PEG deposits from the surfaces. None of the artefacts had any detached fragments and no further remedial work was required.

Each artefact was placed inside a cut length of layflat tubing, perforated to avoid any condensation whilst in storage. A length of Correx board was cut to support each piece, and Jiffy® foam padding was employed to cushion the artefacts. The objects were then heat-sealed into their tubing and new labels attached to replace those used during stabilisation.

The assemblage was then packed into a purpose-made Correx box lined with Plastazote® and Jiffy® foam padding to cushion the objects.

Appendix 3
Dendrochronological investigations at Băile Figa, Romania, 2006–7

TOMASZ WAŻNY

Fifty-three chainsawed sections were collected from the site of Băile Figa as part of an excavation project led by Valerii Kavruk and Anthony Harding. Samples 101–107 were collected in October 2006 while samples with numbers between 1 and 95 were taken in May 2007. All samples are *Quercus* sp.

Method

This study was carried out using classical dendrochronological methods (i.e. BAILLIE 1982; HILLAM 1998; WAZNY 2001). The tree-ring widths were measured on a cross-section of the samples. The surface was cleaned using sharp knives and razor blades. The annual rings were measured by means of a LINTAB measuring stage. The accuracy of the measurements was 0.01 mm. The tree-ring series were compared and synchronized with European master chronologies. Cross-matching was undertaken using the programs CATRAS v. 4.35 (ANIOL 1989), TSAP (RINN 1996), DENDRO for WINDOWS (TYERS 2004), aand CORINA (Ken Harris, Cornell), and the results were supported by visual comparison.

Results

The tree-ring series from Figa do not cross-date with any existing European oak chronologies. The samples were crossed against the site/master chronologies from Germany, Switzerland, Poland, and the Aegean with no significant scores found.

The timbers were divided into three groups according to internal cross-dating results (*Tables 1–3, Figs 1–3*). There is no correlation between these groups, which suggests that they represent different periods. A fourth group consists of undated samples (*Table 4*).

Samples 1–8 represent a solid wooden wall that diverts from the stream at a slight angle. All planks were split radially. The wooden wall was supported and sealed by clay. The same type of construction was made 250 m upstream (samples 29–36), but the quality of construction was poorer.

Table 1. Group I.

Sample	Provenience	Marker	Rings total	Rings sapwood	Bark	Relative dates
1	radial plank, excavation	017	245	18	Y	**36–280**
3	radial plank, excavation	017	194	17	Y	**87–270**
4	radial plank, excavation	017	197 +1	18	Y, EW	**84–280**
5	radial plank, excavation	017	147 +1	17	Y, EW	**134–280**
6	radial plank, excavation	017	216 +1	9	N	**57–272**
8	radial plank, excavation	017	102	0	N	**65–166**
17	single radial plank, excavation, upper part of the trench	005	171 +1	0	N	**46–216**
23		007	123	0	N	**124–256**
29	radial plank	009	170 +1	0	N	**86–255**
31	radial plank	009	73 +1	0	N	**134–206**
33	radial plank	009	225	0	N	**45–269**
34	radial plank	009	148	0	N	**114–261**
35	radial plank	009	175	0	N	**39–213**
45		012	124	0	N	**140–263**
46	radial plank	012	203	0	N	**2–204**
47	radial plank	012	212	0	N	**51–262**
70	radial plank	009	73	0	N	**159–231**
95	Second sample nr 35, but taken from different plank belonging to the same construction	009	173	0	N	**88–260**
101	2006, previous No 59 (d12)		174	0	N	**92–265**
105	2006, previous No 273		171	0	N	**86–256**

Table 2. Group II.

Sample	Provenience	Marker	Rings total	Rings sapwood	Bark	Relative dates
37	loose piece of wood below the dam, quartered timber, 16x7 cm	009	57 +1	9	N	**1034–1090**
38	quartered timber, 13x5 cm	010	94 +1	14	Y	**1001–1094**
39	quartered timber, 16x6 cm, dam	011	72	16	Y	**1023–1094**
40	quartered timber, 10x10 cm, dam	011	68	14	Y	**1027–1094**
41	quartered timber, 10x7 cm, dam	011	56	12	Y	**1040–1095**
42	quartered timber, dam	011	69	16	Y	**1026–1094**
102	2006, previous No 300		90	16	Y	**1005–1094**

Table 3. Group III.

Sample	Provenience	Marker	Rings total	Rings sapwood	Bark	Relative Dates
62	round, tributary	013	90	15	Y, EW	**19–108**
63	tributary	013	102 +1	10	N	**1–102**
64	Round and hollowed (pipe?), tributary	013	74	10	Y	**34–107**

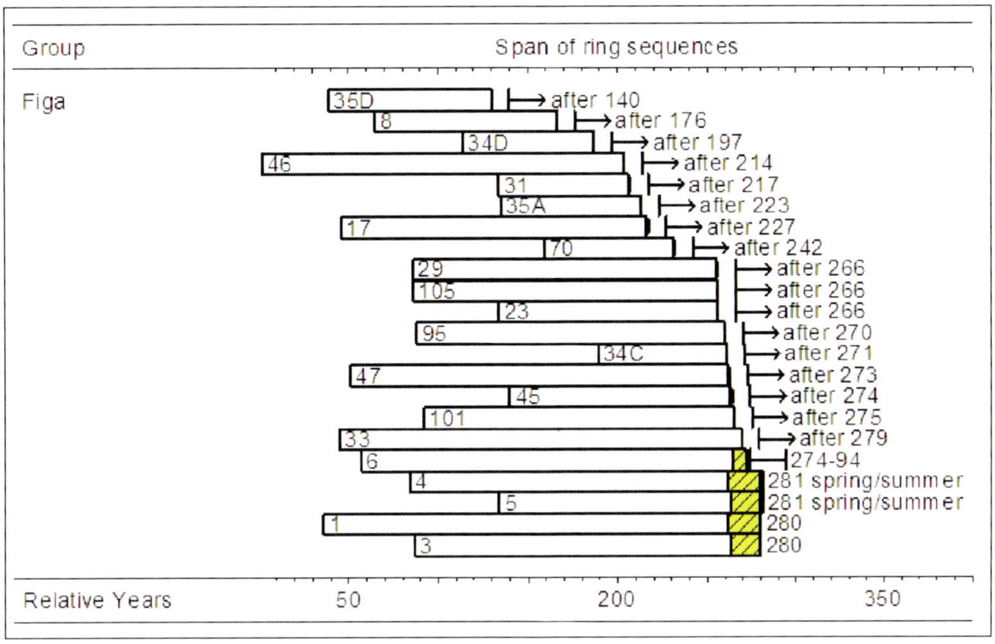

Fig. 1. Relative dating of samples of Group I.

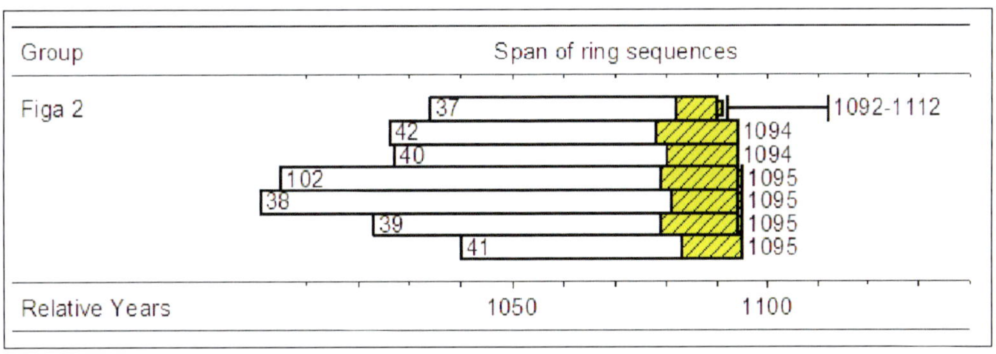

Fig. 2. Relative dating of samples of Group II.

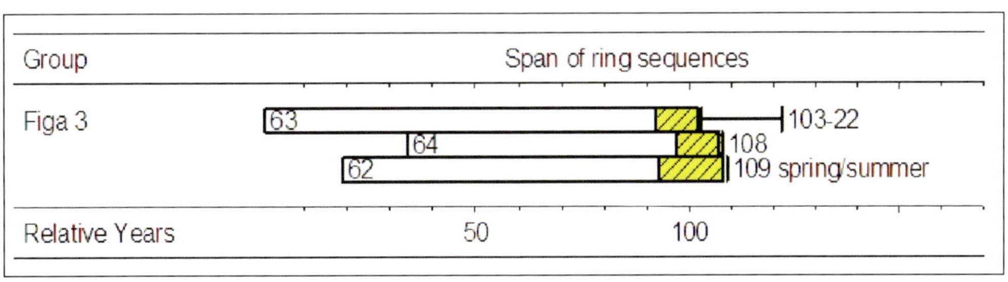

Fig. 3. Relative dating of samples of Group III.

Table 4. Group IV (undated samples).

Sample	Provenience	Marker	Rings total	Rings sapwood	Bark	Remarks
22		006	53	2	N	
25	loose fragments	008	56 +1	0	N	
26	loose fragments	008	30 +1	8	N	
27	loose fragments	008	31 +1	9	N	
30		009	30	0	N	
32		009	97	20	Y	
43	Half-split timber, diam. 11cm, dam	011	18	7	Y	
44	Section, diam. 16 cm, dam	011	51 +1	16	N	
48			52	21	Y	
49			97 +1	0	N	
53		015	70 +1	0	N	
54		015	62 +1)	8	Y, EW	
55		015	36 +1	0	N	
56		015	122 +1	0	N	
57			30 +1	9	N	
58	tributary	014	58	10	Y	
59	tributary	014	22	0	N	
60	tributary	014	27 (+1)	0	N	
61	wooden peg, tributary	014	56 +1	0	N	
103			122	15	Y	2006, original no 411
104			105	0	N	2006, original no 273
106			120	0	N	2006, original no d11?
107			119	0	N	

All samples in Group IV are different trees and do not crossdate with each other. They all have less than 100 rings except sample 56 (n=122).

Conclusions

The first dendrochronological examinations in this region of Romania yielded a 280-year oak chronology based on 20 timbers and two shorter chronologies – 95 and 107 years long. Selected decade-long samples from the chronologies were sent to the radiocarbon laboratory for wiggle-matching. New material from this site would contribute to the extension of our chronologies and would increase the chances to achieve precise absolute dating.

Appendix 4a
Wood species identified at Băile Figa, Beclean, Bistrița-Năsăud county

IONEL POPA – ANCA SEMENIUC

Introduction

Wood species identification was carried out as part of a collaboration with the Museum of the Eastern Carpathians, Sfîntu Gheorghe, on the site of Băile Figa. The goal was to establish the wood species composition for the wooden constructions preserved *in situ*, and for the worked wood pieces preserved *ex situ*. The results reveal the use of local wood species with a high proportion of oak (*Quercus* spp.), amounting to over 95% in the case of wood installations and 55% on worked wood. Other species identified in this site are elm (*Ulmus* spp.), poplar (*Populus* spp.), elder (*Sambucus nigra*), alder (*Alnus glutinosa*), hazel (*Corylus avelana*), and some others.

Knowing the composition structure of wood species identified in a particular archaeological site gives very useful information, both historically and environmentally. Such information show the existence of preferences for certain tree species due to workability and durability, use of local resources or those from other areas (an indication of trade in timber), or qualitative aspects of full or partial use of trees and their parts.

Materials and methods

Sampling and wood species identification were carried out in 2008 and 2010 at the request of the Museum of the Eastern Carpathians. Analyses were done on two sets of samples: the wooden elements and installations preserved on site (July 2008); and wooden objects undergoing conservation in the Museum of Bukovina, Suceava (December 2010).

Given the restrictions imposed by the heritage character of the pieces analysed, information was assembled from external appearance, bark characteristics, macroscopic structure and micro-section analysis of the wood structure. For identification of the wood species, determination keys and standard microscopic structures were used (SCHOCH *et al.* 2004; SCHWEINGRUBER 1990).

The samples were cut into relatively regular shapes with an fine-toothed electric saw; the transverse surface was subsequently prepared for macroscopic analysis using a rotary microtome. Sections for microscopic analysis were obtained by slicing with the microtome, the resulting slices being 10µ thick. To highlight the cell structure, the micro-sections were coloured with a 1:1 mixture of safranin and atrablue (SCHWEINGRUBER 1990). Fragile samples were first fixed in paraffin blocks to provide stability during analysis.

Results

On-site installations Due to the large number of wooden structure elements identified and numbered in survey and excavation (667 items), and the predominance of *Quercus* spp. (oak), the total number of pieces of wood was divided in two sets. The first consists of 250 structural elements, from which samples were taken from each piece of wood identifiable by number. The second consists of the remaining structural elements, about 20% of which were identified from the external characteristics of bark and wood as being *Quercus* spp., and of other pieces for which an identification as oak was uncertain. Thus from the total number of 667 elements, 281 samples were taken (42%). In the case of the first set, samples were taken from 160 items (64%), and from the second set 121 samples (29%).

The following results were obtained:

- a total of 248 samples (88%) belong to the genus *Quercus* spp. (*Fig. 1*). Within this genus differences between species cannot be ascertained only from the preserved wood, but judging by the current composition of the nearby forest, one can say that the species is *Quercus petraea* (sessile oak);
- 5 samples were identified as poplar (*Populus* spp.);
- 3 samples were identified as beech (*Fagus sylvatica*);
- 2 samples are hazel (*Corylus avelana*) and 1 *Clematis* spp. The wattle fences are mostly made of these two species;
- 21 samples could not be identified from cross-sectional analysis because of the advanced state of decay of the wood. They certainly do not belong to the genus *Quercus* spp., and given the current composition of the surrounding forest, they are either birch, hornbeam, alder, or poplar;
- In the second set of structural elements, which were not sampled, the wood species identified in the field from external characteristics is *Quercus* spp. in over 90% of cases;

In summary, the type of wood used for the *in situ* structures was predominantly *Quercus* spp. (95%). It was observed that no differentiation was made in usage of the various parts of the tree: both the trunk and the branches were used.

***Objects* ex situ:** 105 pieces were analysed, of which the species was determined for 103; the remaining two could not be determined because they were too crumbly. Macroscopic analysis enabled species identification for those species with circular pore distribution (*Quercus* spp. and *Ulmus* spp.). Differentiation between oak and elm was possible using microsections observed under the microscope (*Fig. 1*), in the case of the elm the vessel layout being tangential and in the case of elder wavy and radial (*Fig. 2*).

The pegs used in the wooden troughs were a special case. The smaller ones, with no traces of working, were easy to identify, and the wood was established as elder (*Sambucus nigra*). For the large worked pieces, microscopic sections were made, identification being based on microscopic wood structure. There is a diffuse distribution of pores and 2–4 rows of flattened vascular tracheids at the boundary between the annual rings, key elements for determining elder (*Sambucus* spp.) (*Fig. 3A*).

For the two wooden troughs, which are in the process of conservation, Trough 2 (inventory no.14959) is made of oak and Trough 3 (inventory no. 14960) of black alder (*Alnus glutinosa*), identification in the second case being achieved by a microscopic analysis of the wood structure (key elements in identifying the genus being the diffuse distribution of pores and aggregate rays) (*Fig. 3B*).

In summary, by comparison with the species used in wooden structures, there is a greater wood diversity:

- 56 wood samples are of *Quercus* spp. (55%). In view of the current composition of the nearby forest it is likely that the species used is sessile oak (*Quercus petraea*);
- 3 samples belong to the genus *Populus* spp., the species beings probably white poplar (*Populus alba*) or trembling poplar (*Populus tremula*);
- 1 piece was identified as beech (*Fagus sylvatica*);
- 1 piece was identified as black alder (*Alnus glutinosa*) (Trough 3, inventory no. 14960);
- 3 samples are hazel (*Corylus avelana*) and one willow (*Salix* spp.);
- 31 pegs analysed are of black elder (*Sambucus nigra*);
- 7 pieces are elm (*Ulmus* spp.), most probably Field elm (*Ulmus minor*).

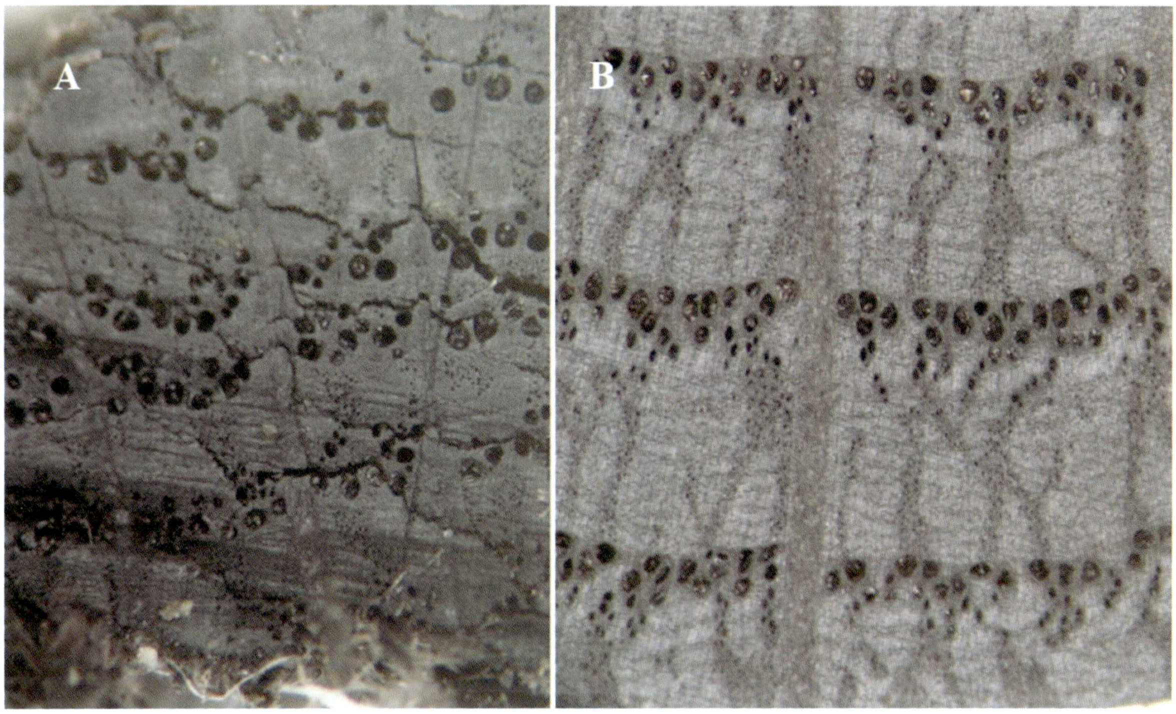

Fig. 1. Macroscopic structure of Quercus *spp. (A: sample 75; B: sample 624).*

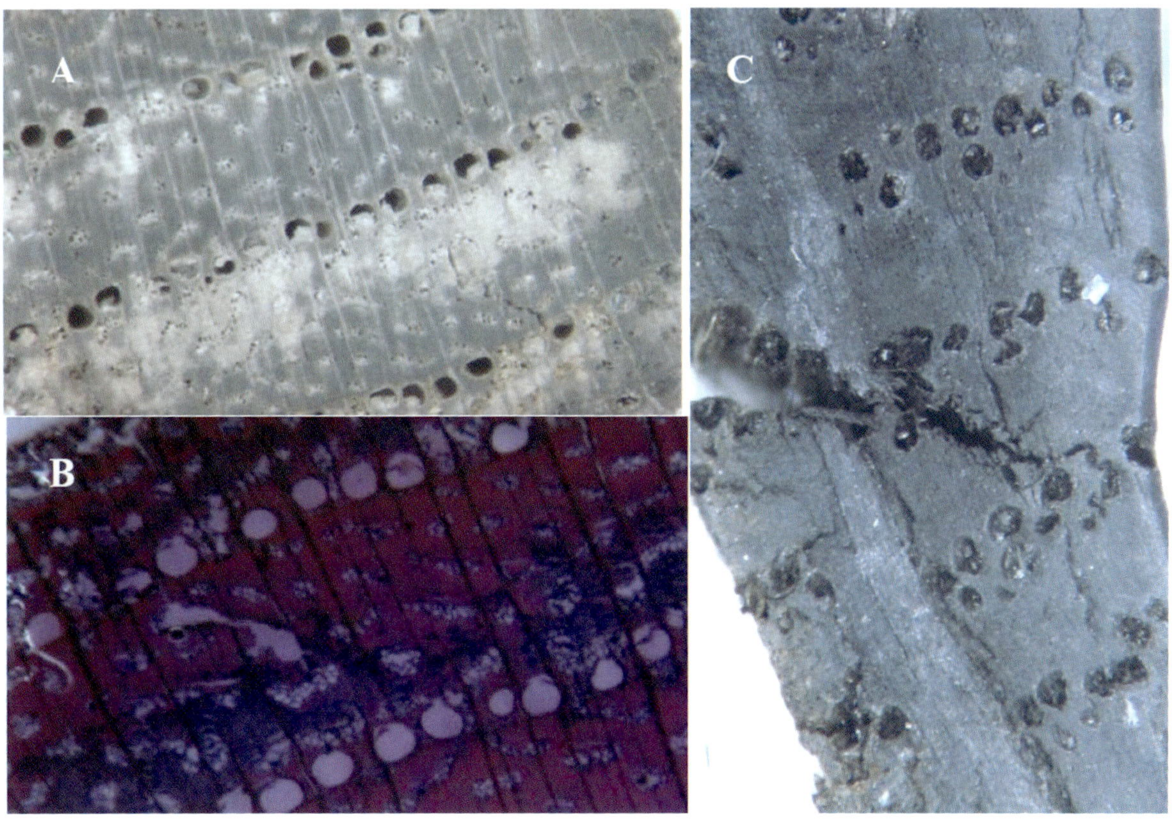

Fig. 2. Structural differences between elm and oak (A: macrosection of elm from sample 15580_b; B: microsection of elm from sample 15580_c; C: macrosection of oak from sample 16723).

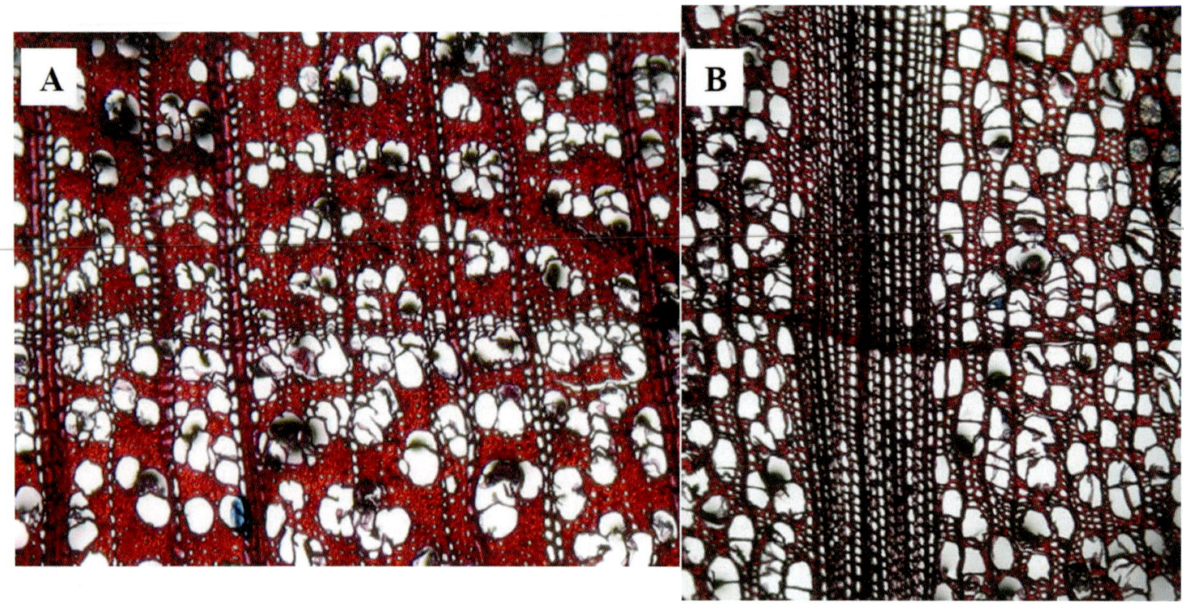

Fig. 3. Microscopic structure of elder (A) sample 14968 and alder (B) – Trough 3 (inv. no. 14960).

Conclusion

In summary, the spectrum of the identified species is similar to that in the local present-day forest ecosystem, a typical oak hills area. In addition, species specific to meadow areas are present (poplar, alder, willow). One can thus say that the Băile Figa structures were built using local wood species, most parts of the trees (branches and trunk) being used in the process.

Appendix 4b
Identification of timber samples

ROWENA GALE

Introduction

This report presents the species identification of nine samples of waterlogged wood from a trough structure associated with a Bronze Age saltern at Băile Figa, and two samples from Sânpaul.

Methods

The wood was structurally degraded and some fragments were extremely hard and possibly partially mineralized. The wood was prepared using standard methods (GALE – CUTLER 2001). Thin sections of wood were mounted on microscope slides and examined using transmitted light on a Nikon Labophot-2 microscope at magnifications up to x400. The anatomical structures were matched to reference slides. When possible, the maturity of the wood was assessed (i.e., heartwood/ sapwood).

Băile Figa 2007

Results

The taxa identified and comments on individual samples are shown on *Table 1*. The taxa identified included:

Table 1.

Sample	Trench/sector	Description	Taxon	Comments
1	Trench I, E sector	Troaca/ trough	Oak *(Quercus* sp.)	Deciduous oak, heartwood from largewood.
2	Trench I, W sector	Binding under trough	Willow *(Salix* sp.) or poplar *(Populus* sp.)	Twisted stem, diameter 15mm, very degraded and slightly mineralized.
3	Trench I, E sector	Items in fill of trough	Birch *(Betula* sp.)	Base of lateral shoot, diameter 1O mm, plus a slither of supporting stem. Bark *in situ,* ?oblique tool mark on distal end of shoot.
			Birch *(Betula* sp.)	Sliver of bark and wood
			Willow *(Salix* sp.) or poplar *(Populus* sp.)	Thin strip from outer surface of stem
			Unidentified ?monocotyledon	Length of narrow culm/ stem or leaf, very degraded
4	Trench I, W sector	Troaca/ trough	Alder *(Alnus glutinosa)*	Several fragments, probably largewood
5	Trench I, E sector	Binding round trough	Willow *(Salix* sp.) or poplar *(Populus* sp.)	Fragmented binding but probably similar in character to sample 2
6	Trench I, E sector	Binding round posts	Willow *(Salix* sp.) or poplar *(Populus* sp.)	(Non-twisted) stem, diameter 7mm, bark *in situ*

Sample	Trench/sector	Description	Taxon	Comments
7	Trench I, E sector	Post A	Oak *(Quercus* sp.)	Wood plus loose bark. Sapwood, probably fairly juvenile. Single fragment isolated for dating
8	Trench I, E sector	Peg from trough base	Elder *(Sambucus* sp.)	Roundwood diameter 20 mm. The hollow centre is probably the result of natural decay of the thin-walled pith cells in the central axis rather than deliberate boring to make a tube (although elder tubes have traditionally been used to blow/flame the embers of fire)
9	Trench III	Timber from plank wall ?no. 58	Oak *(Quercus* sp.)	Thin, short slivers of heartwood

Betulaceae. *Alnus* sp., European alder and *Betula* sp., birch
Caprifoliaceae. *Sambucus* sp., elder
Fagaceae. *Quercus* sp., oak (deciduous)
Salicaceae. *Salix* sp., willow, or *Populus* sp., poplar. In most respects these taxa are anatomically similar, particularly in juvenile growth, but in this instance the structure suggests willow as the more likely.

With the exception of samples 1 and 9, the samples include juvenile wood and provide excellent material for ^{14}C dating. Samples 1 and 9 consist of oak heartwood probably from substantial trunks or largewood and, although these could be used for dating, the results would be less accurate since the wood may already be of some considerable age.

Sânpaul 2005

Sample 3

The sample contained several pieces of juvenile oak *(Quercus* sp.) wood, probably all from fairly narrow roundwood. A fragment of roundwood containing 12 growth rings was isolated for dating.

Sample 4

This sample consisted of 4 narrow slivers of oak *(Quercus* sp.). Each piece included heartwood.

Bibliography

Alexandrowicz, Stefan Witold, Garlicki, Aleksander and Rutkowski, Jacek (1981). Podstawowe jednostki litostratygraficzne miocenu zapadliska przedkarpackiego. *Kwartalnik Geologiczny* 26, 470–71.

Alexianu, Marius and Weller, Olivier (2007). Recherches ethnoarchéologiques sur le sel: les enquêtes de 2004 et les premiers résultats obtenus dans la zone de Poiana Slatinei à Lunca (dép. Neamţ, Roumanie). In: Dan Monah, Gheorghe Dumitroaia, Olivier Weller and John Chapman (eds), *L'exploitation du sel à travers le temps*. Piatra Neamţ: Editura Constantin Matasă. 299–318.

Alexianu, Marius, Weller, Olivier and Brigand, Robin (2007). *Izvoarele de apă sărată din Moldova subcarpatică. Cercetări etnoarheologice*. Iaşi: Casa Editorială Demiurg.

Alth, A. (1868). *Pogląd na źródła słone i naftowe tudzież na warzelnie soli kuchennej w Galicji i Bukowinie*. Kraków: Sprawozdanie Komisji Fizjograficznej.

Andreyeva-Grigorovich, Aida S., Oszczypko, Nestor, Savitskaya, Natalia A., Ślączka, Andrzej and Trofimovich, Natalia A. (2003). Correlation of Late Badenian salts of the Wieliczka, Bochnia and Kalush areas (Polish and Ukrainian Carpathian Foredeep). *Annales Societatis Geologorum Poloniae* 73, 67–89.

Andriţoiu, Ioan and Vasiliev, Valentin (1993). Quelques considérations concernant la culture Noua en Transylvanie. *Cultură şi Civilizaţie Dunărea de Jos* 10, 101–23.

Andronic, Mugur (1989). Cacica – un nou punct neolitic de exploatare a sării. *Studii şi cercetări de istorie veche şi arheologie* 40/2, 171–77.

Andronic, Mugur, Buzea, Dan, Cavruc, Valeriu, Harding, Anthony, Munteanu, Roxana, Matei, Gheorghe and Renţa, Elena (2006). Repertoriul selectiv al siturilor arheologice cu vestigii ale exploatării sării din România. In: Valeriu Cavruc and Andrea Chiricescu (eds), *Sarea, Timpul şi Omul*. Sfîntu Gheorghe: Editura Angustia. 55–86.

Aniol, Roland W. (1989). *Computer aided tree ring analysis system. User's Manual*. Schleswig: Privately published stencil.

Anthony, David W. (1990). Migration in archaeology: the baby and the bathwater. *American Anthropologist* 92/4, 895–914.

Anthony, David W. (1997). Prehistoric migration as social process. In: John Chapman and Helena Hamerow (eds), *Migrations and invasions in archaeological explanation*. Oxford BAR International Series 664. 21–32.

Armstrong, Kayt (2010). *Archaeological geophysical prospection in peatland environments*. Unpublished PhD dissertation, Bournemouth University.

Bąbel, Maciej (2004). Badenian evaporite basin of the northern Carpathian Foredeep as a drawdown salina basin, *Acta Geologica Polonica* 54 (3), 313–337.

Bader, Tiberiu (1979). Die Suciu de Sus-Kultur in Nordwestrumänien. *Praehistorische Zeitschrift* 54, 3–31.

Bàldi, Katalin, Benkovics, László and Sztanó, Orsolya (2002). Badenian (Middle Miocene) basin development in SW Hungary: subsidence history based on quantitative paleobathymetry of foraminifera. *International Journal of Earth Sciences (Geol. Rundschau)* 91, 490–504.

Balintoni, Ioan and Petrescu, Iustinian (2002). A hypothesis on the Transylvanian halite genesis. *Studia Universitatis Babes-Bolyai, Geologia, Special Issue* 1, 51–61.

Bánffy, Eszter (forthcoming 2013). *The early Neolithic of the Duna-Tisza Interfluve, Southern Hungary*. BAR Archaeolingua Central European Series 7. Oxford: Archaeopress.

Biró, Katalin (2004). What for? Possible trade items in the archaeological record. In: Dorin Nicola, Roxana Munteanu, Constantin Preoteasa and Gheorghe Dumitroaia (eds), *Arheologia pre- şi protoistorică a sării (Piatra Neamţ, 21–24 octombrie, 2004)*. Piatra Neamţ: Editura Constantin Matasă. 106.

Blakely, Richard J. (1995). *Potential Theory in Gravity and Magnetic Applications*. Cambridge: Cambridge University Press.

Bodi, George (2007). Experiments with salt crystallization. In: Dan Monah, Gheorghe Dumitroaia, Olivier Weller and John Chapman (eds), *L'exploitation du sel à travers le temps*. Piatra Neamţ: Editura Constantin Matasă. 89–97.

Bonte, Pierre and Izard, Michele (2007). *Dicţionar de etnologie şi antropologie*. Bucureşti: Polirom.

Boroffka, Nikolaus (1994). *Die Wietenberg-Kultur. Ein Beitrag zur Erforschung der Bronzezeit in Südosteuropa*. Universitätsforschungen zur prähistorischen Archäologie 19. Bonn: Habelt.

Boroffka, Nikolaus (2006). Resursele minerale din România şi stadiul actual al cercetărilor privind mineritul preistoric. *Apulum* 43/1, 71–94.

Brunning, Richard (2007). *Structural Wood in Prehistoric England and Wales*. Unpublished PhD thesis, University of Exeter.

Bukowski, Krzysztof (2011). *Badeńska sedymentacja salinarna na obszarze między Rybnikiem a Dębicą w świetle badań geochemicznych, izotopowych i radiometrycznych*. Rozprawy, Monografie Akademii Górniczo-Hutniczej w Krakowie 236.

Bukowski, Krzysztof, Czapowski, Grzegorz, Karoli, Stanislav and Bąbel, Maciej (2007). Sedimentology and geochemistry of the Middle Miocene (Badenian) salt-bearing succession from East Slovakian Basin (Zbudza Formation). In: B.Charlotte Schreiber, Stefano Lugli and Maciej Bąbel (eds), *Evaporites Through Space and Time*. Geological Society Special Publication, 285. London: Geological Society of London. 247–64.

Bukowski, Zbygniew (1963). O możliwości wykorzystywania solanek w okresie halsztackim na terenie Wielkopolski i Kujaw. *Archeologia Polski* 8/1, 246–73.

Bukowski, Zbygniew (1985). Salt production in Poland in prehistoric times. *Archaeologia Polona* 24, 27–71.

Bulleid, Arthur and Gray, Harold St George (1911). *The Glastonbury Lake Village, vol. 1*. Glastonbury: Glastonbury Antiquarian Society.

Butkovič, Štefan (1978). *Dejiny t'ažby soli v Solivare*. Košice: Východoslovenské vydavatel'stvo pre Muzeum SRR v Prešove / Krajské stredisko štátnej pamiatkovej staristlivosti a ochrany prírody v Prešove.

Buzea, Dan (2010). Experimentul ‚Troaca'. *Angustia* 245–56.

Cavruc, Valeriu (1997). The final stage of the Early Bronze Age in South-Eastern Transylvania (in the light of new excavations at Zoltan). *Thraco–Dacica* 18, 97–133.

Cavruc, Valeriu (2005). The Ciomortan Group in the light of new researches, *Marmatia* 8/1, 82–91.

Cavruc, Valeriu (2008). The present stage of the researches regarding prehistoric salt production in the Carpatho-Danubian region. *Angustia* 12, 79–90.

Cavruc, Valeriu (2010). Consideraţii privind schimbul cu sare în mileniile VI–II î. Hr. în spaţiul carpato-dunărean. *Tyrageţia* 4 (19), 7–34.

Cavruc, Valeriu and Chiricescu, Andrea (eds) (2006). *Sarea, Timpul şi Omul*. Sfîntu Gheorghe: Editura Angustia.

Chapman, John and Gaydarska, Bisserka (2003). The provision of salt to Tripolye mega-sites. In: O.G. Korvin-Piotrovskiy (ed), *Tripil'ski poselennya-giganti: Materiali mizhnarodnoy konferentsiy / Tripolian Settlements-Giants, the international symposium materials*. Kiiv: Korvin Press. 203–11.

Chintăuan, Ioan (2002). *Apele Minerale şi Staţiunile din judeţul Bistriţa-Năsăud*. Cluj-Napoca: Editura Supergraph.

Chintăuan, Ioan (2005). Pan used for salt extraction from brines. *Studii şi Cercetări, Geologie-Geografie* 10, 75–78.

Chintăuan, Ioan and Rusu, Ioan (1988). Consideraţii cu privire la utilizarea sării şi a apelor sărate din nord-estul Transilvaniei (judeţul Bistriţa-Năsăud). *File de Istorie* 5, 238–78.

Chiricescu, Andrea (2006). Repertoriul selectiv al fântânilor de apă sărată din estul Transilvaniei. In: Valeriu Cavruc and Andrea Chiricescu (eds), *Sarea, Timpul şi Omul*. Sfîntu Gheorghe: Editura Angustia. 164–71.

Chiricescu, Andrea (2013). *Civilizaţia tradiţională a sării în sud-estul Transilvaniei: raport de cercetare*. Sfîntu Gheorghe: Angustia.

Ciobanu, Doina (2003). Exploatarea sării, element de viaţă cotidiană în spaţiul românesc. In: *Sesiunea Naţională de Etnografie*. Slatina: 40–59.

Ciobanu, Doina (2006). Exploatarea sării în spaţiul Carpato-Dunărean în perioada post-romană şi cea a marilor migraţii. In: Valeriu Cavruc and Andrea Chiricescu (eds), *Sarea, Timpul şi Omul*. Sfîntu Gheorghe: Editura Angustia. 87–91.

Ciugudean, Horia (1996). *Epoca timpurie a bronzului în centrul şi sud-vestul Transilvaniei*. Bucureşti: Institutul Român de Tracologie. Bibliotheca Thracologica 13.

Ciugudean, Horia (2003). Noi contribuţii privind bronzul timpuriu din Transilvania. *Apulum* 40, 89–122.

Czapowski, Grzegorz & Bukowski, Krzysztof (2009). Złoża soli w Polsce – stan aktualny i perspektywy zagospodarowania, *Przegląd Geologiczny* 57, 798–811.

Czapowski, Grzegorz and Bukowski, Krzysztof (2010). Geology and resources of salt deposits in Poland: the state of the art. *Geological Quarterly* 54, 509–18.

Dănilă, Ştefan and Marinescu, George (1974). Unelte, arme şi obiecte de piatră şlefuită descoperite pe territoriul judeţului Bistriţa-Năsăud. *File de Istorie* 3, 11–55.

Danu, Mihaela, Gauthier, Emilie and Weller, Olivier (2010). Human impact and vegetation history on salt spring exploitation (Halabutoaia-Tolici, Petricani, Neamt, Romania). *International Journal of Conservation Science* 1/3, 167–73.

Deák, Andrea (2008). Mineritul tradiţional al sării – o ocupaţie demult dispărută? *Angustia* 12, 237–50.

De Leeuw, Arjan, Bukowski, Krzysztof, Krijgsman, Wout and Kuiper, Klaudia F. (2010). Age of the Badenian salinity crisis; impact of Miocene climate variability on the circum-Mediterranean region. *Geology* 38, 715–18.

Dicker, Józef (1924). Górnictwo na Rusi halickiej w XV i pierwszej połowie XVI wieku. *Rocznik Przemyski* 5, 1–28.

Dobák, Imre (2010). Environment and security in Western Ukraine – short overview. *Hadmérnök* 5(1), 110–18.

Drăgănescu, Liviu (2006). Sarea gemă din extra- şi intracarpaticul României. In: Valeriu Cavruc and Andrea Chiricescu (eds), *Sarea, Timpul şi Omul*. Sfîntu Gheorghe: Editura Angustia. 13–16.

Duchoň, Marek (2007). *Potulky po Sol'nej Bani*. Prešov: Ateliér JANA.

Đurman, Alexander (1997). Tin in south-eastern Europe? *Opuscula Archeologica* 21, 7–14.

Earwood, Caroline (1993). *Domestic wooden artefacts in Britain and Ireland from Neolithic to Viking times*. Exeter: University of Exeter Press.

Fries-Knoblach, Janine (2001). Gerätschaften, Verfahren und Bedeutung der eisenzeitlichen Salzsiederei in Mitteleuropa. *Leipziger Forschungen zur Ur- und Frühgeschichtlichen Archäologie* 2, 1–46.

Furmánek, Václav and Marková, Klára (2008). *Včelnice: archív dávnej minulosti* Archeologické Pamätníky Slovenska, 9. Nitra: Slovenská akadémia vied.

Gale, Rowena and Cutler, David F. (2001). *Plants in Archaeology*. Otley/London: Westbury Publishing and Royal Botanic Gardens, Kew.

Garlicki, Aleksander (1979). Sedymentacja soli mioceńskich w Polsce. *Prace Geologiczne Komisji Nauk Geologicznych PAN* 119, 1–66.

Gaydarska, Bisserka and Chapman, John (2007). Salt research in Bulgaria. In: Dan Monah, Gheorghe Dumitroaia, Olivier Weller and John Chapman (eds), *L'exploitation du sel à travers le temps*. Piatra-Neamţ: Editura Constantin Matasă. 147–60.

Gedl, Marek (1996). Młodsza epoka brązu w południowej karpackiej części województwa przemyskiego. *Rocznik Przemyski* 32, 29–49.

Georgescu, O., Frunzescu, D., Drăgănescu, L. and Stamate, I. (1994). Considérations génétiques concernant les séquences évaporitiques de Slănic Prahova. In: E. Nicorici, I. Bedelean, N. Mészáros and I. Petrescu (eds), *The Miocene from the Transylvanian Basin, Romania*. Cluj-Napoca: Editura Carpatica. 217–24.

Gläser, Roland (1991). Bemerkungen zur absoluten Datierung des Beginns der westlichen Linienbandkeramik. *Banatica* 11, 53–64.

Gligor, Mihai (2009). *Aşezarea neolitică şi eneolitică de la Alba Iulia – Lumea nouă în lumina noilor cercetări*. Cluj-Napoca: Editura Mega.

Gogâltan, Florin (1996). About the Early Bronze age in the Romanian Banat. In N. Tasić (ed), *The Yugoslav Danube Basin and the Neighbouring Regions in the 2nd Millenium B.C.*, 43–67. Belgrade-Vršac: Balkanološki institut, Srpska akademija nauka i umetnosti. Posebna izdanja 65.

Goldman, György and Szénászky, Júlia G. (1994). Die neolitische Esztár-Gruppe in Ostungarn. *A Jósa András Múzeum Évkönyve* 36, 225–30.

Görsdorf, Jochen, Marková, Klára and Furmánek, Václav (2004). Some new 14C data to the Bronze Age in the Slovakia. *Geochronometria* 23, 79–91.

Hansen, Svend (2007). *Bilder vom Menschen der Steinzeit. Untersuchungen zur anthropomorphen Plastik der Jungsteinzeit und Kupferzeit in Südosteuropa*. Archäologie in Eurasien 20. Mainz: von Zabern.

Harding, Anthony (1998). Resources and their distribution in the European Bronze Age. In: B. Hänsel (ed), *Mensch und Umwelt in der Bronzezeit Europas*. Kiel: Oetker-Voges Verlag. 149–55.

Harding, Anthony (2000). *European Societies in the Bronze Age*. Cambridge: Cambridge University Press.

Harding, Anthony (2001). Natural resources as a factor in the evolution of Late Bronze Age groups in the Carpathian area. In: Carol Kacsó (ed), *Der nordkarpatische Raum in der Bronzezeit, Symposium Baia Mare 7–10 Oktober 1998*. Biblioteca Marmatia I. Baia Mare (Muzeul Județean Maramureș): Editura Cornelius. 119–24.

Harding, Anthony (2009). Producing salt in wooden troughs: the technology of Bronze Age salt production in Transylvania. In: G. Bodi (ed), *In medias res praehistoriae. Miscellanea in honorem annos LXV peragentis Professoris Dan Monah oblata*. Iași: Editura Universității "Alexandru Ioan Cuza". 195–204.

Harding, Anthony (2011). Evidence for salt production rediscovered in the Hungarian Central Mining Museum. *The Antiquaries Journal* 91, 27–49.

Harding, Anthony and Kavruk, Valerii (2010). A prehistoric salt production site at Băile Figa, Romania. *Eurasia Antiqua* 16, 131–67.

Haustein, Michael, Gillis, Carole and Pernicka, Ernst (2010). Tin isotopy – a new method for solving old questions. *Archaeometry* 52, 816–32.

Horiuchi, Akiko, Ochiai, Nobuo, Kurozumi, Hitomi and Miyata, Yoshiki (2011). Detection of chloride from pottery as a marker for salt: A new analytical method validated using simulated salt-making pottery and applied to Japanese ceramics. *Journal of Archaeological Science* 38, 2949–56.

Horváth, Istvan (1998). *Istoricul mineritului de la Slatina Praid*. Praid: Unknown.

Hryniv, Sofia P., Dolishniy, B.V., Khmelevska, Oksana.V., Poberezhskyy, Andrey.V. & Vovnyuk, Sergyi.V. (2007). Evaporites of Ukraine: a review, *Geological Society, London, Special Publications* 285, 309–334.

Ichim, Dorinel (2006). Exploatarea sării în Moldova medievală. In: Valeriu Cavruc and Andrea Chiricescu (eds), *Sarea, Timpul și Omul*. Sfîntu Gheorghe: Editura Angustia. 125–31.

Ignat, Doina (1998). *Grupul cultural neolitic Suplacul de Barcău*. Seria Bibliotheca Historica et Archaeologica Banatica, XVI. Timișoara: Editura Mirton.

Ivanova, Svetlana V. (2010). Prirodn'ie resurs'i i ekonomika drevnich obshchestv. *Stratum Plus* 2, 49–97.

Jodłowski, Antoni (1968). Problem wykorzystywania solanek na podkarpaciu polskim w neolicie. *Acta Archaeologica Carpathica* 10, 173–81.

Jodłowski, Antoni (1971). *Eksploatacja sóli na terenie Małopolski w pradziejach i we wczesnym średniowieczu*. Studia i materiały do dziejów żup sólnych w Polsce, 4. Wieliczka: Muzeum Żup Krakowskich.

Kacsó, Carol (2003). Noi descoperiri Suciu de Sus și Lăpuș în nordul Transilvaniei – Neue Suciu de Sus- und Lăpuș-Funde im Norden Siebenbürgens. *Marmatia* 7, 105–81.

Kacsó, Carol (2008). Contribuții la cunoașterea ceramicii de tip Suciu de Sus. *Revista Bistriței* 22, 9–27.

Kaczanowska, Małgorzata (ed) (2006). *Dziedzictwo cywilizacji naddunajskich w Małopolsce na przełomie epoki kamienia i miedzi*. Kraków: Muzeum Archeologiczne w Krakowie.

Kadrow, Sławomir (2003). Charakteristyka technologiczna ceramiki kultury łużyckiej. In: S. Kadrow (ed), *Kraków-Bieżanów, stanowisko 27 i Kraków-Rząka, stanowisko 1, osada kultury łużyckiej*. Via Archaeologica. Źródła z badań wykopaliskowych na trasie autostrady A4 w Małopolsce. Kraków: Zespół do Badań Autostrad. 205–20.

Kalicz, Nándor (1994). Wenden des Spätneolithikums im oberen Theißgebiet. *A Nyíregyházi Jósa András Múzeum Évkönyve* 36, 263–90.

Kalicz, Nándor (2011). *Méhtelek. The First Excavated Site of the Méhtelek Group of the Early Neolithic Körös Culture in the Carpathian Basin*. BAR/Archaeolingua, Central European Series 6. Oxford: Archaeopress.

Kalicz, Nándor (2012). Méhtelek-Nádas. The first excavated site of the Méhtelek fazies of the Early Neolithic Körös culture in the Carpathian Basin. In: Alexandra Anders and Zsuzsanna Siklósi (eds), *The First Neolithic Sites in Central/South-East European Transect. Volume III. The Körös Culture in Eastern Hungary*. BAR International Series 2334. 113–23.

Kalicz, Nándor and Raczky, Pál (1984). Preliminary report on the 1977–1982 excavations at the Neolithic and Bronze Age tell settlement at Berettyóújfalu-Herpály. Part I: Neolithic. *Acta Archaeologica (Budapest)* 36, 85–136.

Kityk, Vasilij I. , Bokun, Aleksandr N., Panov, Genadij M., Slivko, Yelena P. and Shaidetskaya, Vasilina S. (1983). *Galogennye formatsii Ukrainy*. Kiev: Zakarpatskii progib. Naukova dumka.

Kojumdgieva, Emilia and Popov, N. (1988). Litostratigrafiya na neogenskite sedimenti v Severozapadna Bulgariya. *Paleontology, Stratigraphy and Lithology* 25, 3–26.

Kónya, Peter and Michnovič, Imrich (eds) (1997). *Dejiny obce Sol'*. Prešov: L.I.M. Prešov pre Obecný úrad Sol'.

Kopaka, Katerina and Chaniotakis, Nikos (2003). Just taste additive? Bronze Age salt from Zakros, Crete. *Oxford Journal of Archaeology* 22/1, 53–66.

Krézsek, Csaba and Bally, Albert W. (2006). The Transylvanian Basin (Romania) and its relation to the Carpathian fold and thrust belt: Insights in gravitational salt tectonics. *Marine and Petroleum Geology* 23, 405–42.

Krushelnitska, Larisa (1993). Novi pam'yatki kultury Gava-Goligradi. In: Larisa Krushelnitska (ed), *Pam'yatki Galshtatskogo Periodu v Medzhirichchi Visli, Dnistra i Prip'yati*. Kiev: Naukova Dumka. 56–122.

Krushelnitska, Larisa (2002a). O nouă aşezare din epoca bronzului timpuriu în Ucraina subcarpatică. *Angustia* 7, 99–114.

Krushelnitska, Larisa (2002b). Pamyatka solevarinnya rubezhu rannoy i serednoy bronzy v Pivnichno-Schnidnich Karpatakh. *Zapiski Naukove Tovaristvo im. T. Shevchenka* 244, 140–55.

Kunysz, Antoni (1964–65 [1968]). Rola źródeł słonych w rozwoju osadnictwa wczesnośredniowiecznego w górnym dorzeczu Sanu. *Rocznik Województwa Rzeszowskiego* 5, 11–22.

Lazarovici, Gheorghe (1998). Once again about the ceramics from Gornea-Vodneac, of the Early Bronze in Banat. In: Christian Schuster (ed), *Die Kulturen der Bronzezeit in dem Gebiet des Eisernen Tores. Kolloquium in Drobeta Turnu Severin*. Jugoslawish-rumänische Kommission für die Erforschung der Region des Eisernen Tores, 2. Bucharest: Rumänisches Institut für Thrakologie. 47–70.

Lazarovici, Gheorghe and Lazarovici, Cornelia-Magda (2011). Some salt sources in Transylvania and their connections with the archaeological sites in the area. In: Marius Alexianu, Olivier Weller and Roxana-Gabriela Curca (eds), *Archaeology and anthropology of salt. Proceedings of the International Colloquium, 1–5 October 2008. Al. I. Cuza University (Iaşi, Romania)*. British Archaeological Reports, International Series 2198. Oxford: Archaeopress. 89–110.

Lazarovici, Gheorghe and Maxim, Zoia (1995). *Gura Baciului*. Cluj-Napoca: Muzeul Național de Istorie a Transilvaniei.

Leidinger, Wendelin (1983). Frühe Salzgewinnung in Werl, Kreis Soest, Westfalen. *Archäologisches Korrespondenzblatt* 13, 269–74.

Leidinger, Wendelin (1996). Salzgewinnung an den Solquellen der Saline Werl. In: Rüdiger Just and Uwe Meissner (eds), *Das Leben in der Saline – Arbeiter und Unternehmer*. Schriften und Quellen zur Kulturgeschichte des Salzes 3 (also separately published as a pamphlet, 27 pp.). Halle/Saale: Technisches Halloren- und Salinemuseum. 189–215.

Lenneis, Eva and Lüning, Jens (2001). *Die altbandkeramischen Siedlungen von Neckenmarkt und Strögen*. Universitätsforschungen zur prähistorischen Archäologie 82. Bonn: Habelt.

Lenneis, Eva, Stadler, Peter and Windl, H. (1996). Neue 14C-Daten zum Frühneolithikum in Österreich. *Préhistoire Européenne* 8, 97–116.

Loke, Meng Heng and Barker, R.D. (1996). Rapid least-squares inversion of apparent resistivity pseudosections by a quasi-Newton method. *Geophysical Prospecting* 44, 131–52.

Lüning, Jens (1988). Frühe Bauern in Mitteleuropa im 6. und 5. Jahrtausend v. Chr. *Jahrbuch des Römisch-Germanischen Zentralmuseums Mainz* 35, 27–93.

Lüning, Jens (2005). Bandkeramische Hofplätze und absolute Chronologie in der Bandkeramik. In: Jens Lüning, Chr Frirdich and Andreas Zimmermann (eds), *Die Bandkeramik im 21. Jahrhundert. Symposium Brauweiler 2002*. Rahden/Westf.: Marie Leidorf. 49–74.

Majchrzycka, Anna (2009). Warzelnictwo soli w pradziejach i we wczesnym średniowieczu w Wielkopolsce i na Kujawach. Stan badań. Możliwości poznawcze. Uniwersytet im. Adama Mickiewicza w Poznaniu, Wydział Historyczny. Unpublished Masters dissertation.

Marc, Dorel (2006). Sisteme de transport şi de comercializare tradiţională a sării. In: Valeriu Cavruc and Andrea Chiricescu (eds), *Sarea, Timpul şi Omul*. Sfîntu Gheorghe: Editura Angustia. 152–57.

Marinescu, George (1980). Cercetări şi descoperiri arheologice în judeţul Bistriţa-Năsăud (VII). *Marisia* 10, 31–51.

Marinescu, George (1984). Cercetări şi descoperiri arheologice în judeţul Bistriţa-Năsăud (VIII). *Marisia* 13–14, 23–32.

Marinescu, George (1988). Noi descoperiri de epoca bronzului la Bistriţa. *File de Istorie* 5, 21–37.

Marinescu, George (1993). Cercetări şi descoperiri arheologice de epoca bronzului în Nord-Estul Transilvaniei. *Revista Bistriţei* 7, 5–27.

Marinescu, George (1994). Cercetări şi descoperiri arheologice aparţinând bronzului timpuriu (Cultura Coţofeni) în Nord-Estul Transilvaniei. *Revista Bistriţei* 8, 9–19.

Marinescu, George (1995). Cercetări şi descoperiri arheologice de epoca bronzului în Nord-Estul Transilvaniei (II). *Revista Bistriţei* 9, 49–69.

Marinescu, George and Dănilă, Ştefan (1974). Obiecte de bronz descoperite pe teritoriul judeţului Bistriţa-Năsăud. *File de Istorie* 3, 65–88.

Marinescu, George and Dănilă, Ştefan (1976). Cercetări şi descoperiri arheologice în judeţul Bistriţa-Năsăud. *File de Istorie* 4, 24–56.

Maślankiewicz, Kazimierz (1985). *Z dziejów górnictwa solnego w Polsce*. Warszawa: Wydawnictwo Naukowo-Techniczne.

Mateiciucová, Inna (2004). Mesolithic traditions and the origin of the Linear Pottery culture (LBK). In: Alena Lukes and Marek Zvelebil (eds), *LBK dialogues. Studies in the formation of the Linear Pottery culture*. Oxford: BAR International Series 1304. 91–107.

Mateiciucová, Inna (2008). *Talking Stones: the chipped stone industry in Lower Austria and Moravia and the beginnings of the Neolithic in Central Europe (LBK), 5700–4900 BC*. Dissertationes Archaeologicae Brunenses/Pragensesque 4. Brno: Muni Press.

Maxim, Ioan Al. (1971). Un depozit de unelte dacice pentru exploatarea sării. *Acta Musei Napocensis* 8, 457–63.

Maxim, Zoia (1999). *Neo-Eneoliticul din Transilvania*. Biblioteca Musei Napocensis 19. Cluj-Napoca: Muzeul Național de Istorie Transilvaniei.

McGrail, Sean (1978). *Logboats of England and Wales with comparative material from European and other countries*. Oxford: British Archaeological Reports British series No. 51 and National Maritime Museum, Geeenwich, Archaeological series No. 2.

Medeleț, Florin (1995). Über das Salz in Dakien, *Archäologie Österreichs* 6/2, 53–57.

Mircea, Magda and Alexianu, Marius (2007). A re-evaluation of salt resources for the Cucuteni-Tripolye area. In: D. Monah, G. Dumitroaia, O. Weller and J. Chapman (eds), *L'exploitation du sel à travers le temps*. Piatra Neamț: Editura Constantin Matasă. 199–219.

Moga, Iulian (2008). Symbolic functions of salt and related similes in oriental curses and blessings. *Arheologia Moldovei* 29, 95–130.

Moga, Iulian (2009). Salt extraction and imagery in the Ancient Near East. *Journal for Interdisciplinary Research on Religion and Science* 4, 175–213.

Monah, Dan (1991). L'exploitation du sel dans les Carpates orientales et ses rapports avec la culture de Cucuteni-Tripolje. In: V. Chirica and D. Monah (eds), *Le paléolithique et le néolithique de la Roumanie en contexte Européen*. Iaşi: Bibl. Arch Iassiensis 4. 387–400.

Monah, Dan (2002). L'exploitation préhistorique du sel dans les Carpathes orientales. In: Olivier Weller (ed), *Archéologie du sel: techniques et sociétés dans la pré- et protohistoire européenne / Salzarchäologie. Techniken und Gesellschaft in der Vor- und Frühgeschichte Europas*. Rahden/ Westf.: Verlag Marie Leidorf. 135–46.

Monah, Dan (2007 [2008]). Sarea în preistoria României. *Arheologia Moldovei* 30, 87–116.

Monah, Dan (2008). Arheologia preistorică a sării în România. Scurt istoric. L'archéologie préhistorique du sel en Roumanie. Bref historique. In: Dan Monah, Gheorghe Dumitroia and D. Garvăn (eds), *Sarea, de la prezent la trecut*. Piatra Neamț: Constantin Mătasă. 13–39.

Monah, Dan (2012). L'approvisionnement en sel des tribus chalcolithiques sédentaires et des tribus des steppes du Nord de la Mer Noire. In: Vassil Nikolov and Krum Bacvarov (eds), *Salz und Gold: die Rolle des Salzes im prähistorischen Europa. Akten der internationaler Fachtagung (Humboldt-Kolleg) in Provadia, Bulgarien 30 September – 4 October 2010 / Salt and Gold: The Role of Salt in Prehistoric Europe Proceedings of the International Symposium (Humboldt-Kolleg) in Provadia, Bulgaria 30 September – 4 October 2010*. Provadia – Veliko Tarnovo: Verlag Faber. 127–42.

Montagnari Kokelj, Emanuela (2007). Salt and the Trieste karst (north-eastern Italy). In: Dan Monah, Gheorghe Dumitroaia, Olivier Weller and John Chapman (eds), *L'exploitation du sel à travers le temps*. Piatra-Neamț: Editura Constantin Matasă. 161–89.

Motzoi-Chicideanu, Ion (2000). Observații asupra cimitirului din epoca bronzului de la Pietroasa Mică. *Romanian Journal of Archaeology (Archaeological Professional Association – Romania)* http://apar.archaeology.ro/imc_artrja.htm#2.

Mowat, Robert J.C. (1996). *The logboats of Scotland*. Oxford: Oxbow Monograph 68.

Munteanu, Roxana, Garvăn, Daniel, Nicola, Dorin, Preoteasa, Constantin and Dumitroaia, Gheorghe (2007). Cucuieţi-Slatina Veche (Romania). Prehistoric exploitation of a salt resource. In: Dan Monah, Gheorghe Dumitroaia, Olivier Weller and John Chapman (eds), *L'exploitation du sel à travers le temps*. Piatra Neamţ: Editura Constantin Matasă. 57–70.

Nikolov, Vasil (ed) (2008). *Praistoricheski solodobiven tsentr Provadiya-Solintsata. Razkopki 2005–2007 g.* Sofia: Bl'garska Akademiya na Naukite / Natsionalen Arkheologicheski Institut i Muzej.

Nikolov, Vasil (ed) (2009). *Provadiya-Solintsata. Arkheologicheski razkopki i izsledvaniya prez 2008 g. Predvaritelen otchet*. Sofia: Publisher not stated.

Pany, Doris and Teschler-Nicola, Maria (2007). Working in a salt mine: everyday life for the Hallstatt females? *Lunula* 15, 89–97.

Paul, Iuliu (1992). *Cultura Petreşti*. Bucureşti: Editura Museion.

Pavlů, Ivan and Vokolek, Vít (1992). Early Linear Pottery culture in the East Bohemian region. *Památky archeologické* 83, 41–87.

Peryt, Tadeusz Marek (2006). The beginning, development and termination of the Middle Miocene Badenian salinity crisis in Central Paratethys. *Sedimentary Geology* 188–189, 379–96.

Petre-Govora, Gheorghe (1988). Descoperiri arheologice din Oltenia privind epoca timpurie a bronzului. *Thraco-Dacica* 9, 137–47.

Petre-Govora, Gheorghe (1995). *O preistorie a nord-estului Olteniei*. Râmnicu-Vâlcea: Petras.

Petryczenko, O.I., Panow, G.M., Peryt, T.M., Srebrodolski, B.I., Pobereżski, A.W. and Kowalewicz, W.M. (1994). Zarys geologii mioceńskich formacji ewaporatowych ukraińskiej części zapadliska przedkarpackiego. *Przegląd Geologiczny* 42, 9, 734–37.

Piso, Ioan (2004–2005 [2007]). Un nouveau *conductor salinarum* en Dacie. *Acta Musei Napocensis* 41–42/1, 179–82.

Popa, Cristian Ioan and Totoianu, Radu (2010). *Aspecte ale epocii bronzului în Transilvania (între vechile şi noile cercetări)*. Bibliotheca Mvsei Sabesiensis 1. Sebeş: Muzeul Municipal „Ioan Raica" Sebeş.

Popov, Sergej V., Rögl, Fred, Rozanov, Alexei Y., Steininger, Fritz F., Shcherba, I.G. and Kovac, Michal (2004). Lithological-Paleogeographic maps of Paratethys 10 maps Late Eocene to Pliocene. *Courier Forschungsinstitut Senckenberg* 1–46.

Preisig, Eduard (1877). Geschichte des Máramaroser Bergbaues. *Oesterreichische Zeitschrift für Berg- und Hüttenwesen* 25/28–30, 301–03, 11–13, 21–23. Tafel 12.

Pusch, Georg Gottlieb (1833–36). *Geognostische Beschreibung von Polen so wie der übrigen Nordkarpathen-Ländern*. Atlas Univ. Tubingen. Stuttgart & Tubingen: J.G. Cotta.

Pydyn, Andrzej (1999). *Exchange and cultural interactions. A study of long distance trade and cross-cultural contacts in the Late Bronze Age and Early Iron Age in Central and Eastern Europe*. BAR International Series 813. Oxford: Archaeopress.

Quitta, Hans (1960). Zur Frage der ältesten Bandkeramik in Mitteleuropa. *Prähistorische Zeitschrift* 38, 1–38, 153–88.

Rajchel, Lucyna, Zuber, Andrzej, Duliński, Marek and Rajchel, Jacek (2004). Występowanie i geneza wód chlorkowych Soli. *Przegląd Geologiczny* 52, 1179–86.

Ramminger, Britta (2011). Zur wirtschaftlichen Organisation der ältestbandkeramischen Siedlung Schwanfeld am Beispiel der Felsgesteinforschung. In: J. Lüning (ed), *Schwanfeldstudien zur ältesten Bandkeramik*. Universitätsforschungen zur prähistorischen Archäologie 196. Bonn: Habelt. 119–221.

Rinn, Frank (1996). *Time Series Analysis and Presentation. Version 3.0 reference manual*. Heidelberg: Frank Rinn Engineering and Distribution.

Rögl, Fred (1998). Palaeogeographic considerations for Mediterranean and Paratethys Seaways (Oligocene to Miocene), *Annalen des Naturhistorischen Museums in Wien* 99A, 279–310.

Runnels, Curtis and van Andel, Tjeerd (1988). Trade and the origins of agriculture in the Eastern Mediterranean. *Journal of Mediterranen Archaeology* 1, 83–109.

Russu, Ioan Iosif (1956). Sclavul Atticus. Contribuţie la istoria economică a Daciei. *Studii şi Cercetări de Istorie* 7, 7–13.

Rustoiu, Aurel (2005). The salt of Maramureş and the Dacian settlements alongside of Upper Tisa, *Mousaios* 10, 353–367.

Sands, Rob (1997). *Prehistoric woodworking: the analysis of Bronze and Iron Age toolmarks*. London: Institute of Archaeology.

Sandu, Ion, Poruciuc, Adrian, Alexianu, Marius, Curcă, Roxana-Gabriela and Weller, Olivier (2010). Salt and human health: science, archaeology, ancient texts and traditional practices of eastern Romania. *Mankind Quarterly* 50, 225–56.

Schade-Lindig, Sabine and Schwitalla, Guntram (1999). Das Steinzeitdorf von Hempler. *Denkmalpflege und Kulturgeschichte (Hesse)* 1, 26–32.

Schoch, W., Heller, I., Schweingruber, Fritz Hans and Kienast, F. (2004). *Wood Anatomy of Central European Species*. Online version: www.woodanatomy.ch.

Schweingruber, Fritz Hans (1990). *Anatomy of European Woods: An Atlas for the Identification of European Trees, Shrubs and Dwarf Shrubs*. Bern: Verlag Paul Haupt.

Simon, Zsolt (2006). Mineritul de sare în Evul mediu în Transilvania şi Maramureş. In: Valeriu Cavruc and Andrea Chiricescu (eds), *Sarea, Timpul şi Omul*. Sfîntu Gheorghe: Editura Angustia. 92–96.

Šiška, Stanislav (1997). Archeologicky výskum pri sol'ných prameňoch v Soli. *Balneologický spravodajca / Balneological Bulletin* 35, 120–36.

Sófalvi, András (2005). *A Sóvidék a középkorban (The Só region in the Middle Ages)*. Székelyudvarhely: Haáz Rezső Múzeum.

Spataro, Michela (2008). Early Neolithic pottery production in Romania: Gura Baciului and Seusa La Cararea Morii (Transylvania). In: Douglass Bailey, Alasdair Whittle and Danni Hoffmann (eds), *Living well together? Settlement and materiality of South East and Central Europe*. Oxford: Oxbow. 91–100.

Staszic, Stanisław (1815 (reprint 1955)). *O ziemiorództwie Karpatów i innych gór i równin Polski*. Warszawa: Drukarna rządowa.

Stoica, Coriolan and Gherasie, Ion (1981). *Sarea şi sărurile de potasiu şi magneziu din România*. Bucharest: Editura Tehnică.

Szajnocha, Władysław (1891). Źródła mineralne Galicji, pogląd na ich rozpołożenie, skład chemiczny i powstanie. Rozprawy Wydziału Matematyczno – Przyrodniczego Akademii Umiejętności. Kraków.

Szajnocha, Władysław (1893). Kopalnie i warzelnie soli w Galicji. *Przewodnik Naukowy i Literacki* 21, Lwów.

Szathmáry, K.P. (1867). Máramarosi régiségek. *Vasárnapi Újság* 14/8, 91.

Tasić, Nenad (2000). Salt use in the Early and Middle Neolithic of the Balkan Peninsula. In: Lolita Nikolova (ed), *Technology, Style and Society. Contributions to the innovations between the Alps and the Black Sea in prehistory*. Oxford British Archaeological Reports, International Series 854. 35–40.

Telford, William M., Geldart, L.P. and Sheriff, Robert E. (1990). *Applied Geophysics, 2nd ed.* Cambridge: Cambridge University Press.

Toma, Camelia Bianca (2012). *The Geomorphosites on Salt from Transylvania Depression and Tourism and Their Touristic Value*. Unpublished dissertation, Cluj-Napoca Babeş-Bolyai University, Faculty of Geography.

Trashliev, Stoyan (1969). Structures of the Tortonian gypsum rocks in northwestern Bulgaria. *Sedimentary Geology* 3, 291–316.

Tyers, Ian G. (2004). *Dendro for Windows program guide, 3rd edition*. ARCUS Report 500B. Sheffield: ARCUS Dendrochronology Laboratory, University of Sheffield.

Ursulescu, Nicolae (1977). Exploatarea sării din saramura în neoliticul timpuriu în lumina descoperirilor de la Solca (jud. Suceava). *Studii şi cercetări de istorie veche* 28/3, 307–17.

Ursulescu, Nicolae (1984). *Evoluţia culturii Starčevo-Criş pe teritoriul Moldovei*. Suceava: Muzeul Judeţean.

Ursulescu, Nicolae (2001). Local variants of the Starčevo-Criş culture in the Carpato-Nistrean area. In: F. Draşovean (ed), *Festschrift für Gh. Lazarovici*. Timişoara: Muzeul Banatului. 59–67.

Várady, Gábor (1901). *Máramarosi emlékkönyv: hazánk ezeréves fennállása emlékére vármegyénkben rendezett ünneplések és az 1901-ik évben Máramaros-Szigeten tartott gazdasági és ipari kiállítás leírásával / összeáll.* Máramaros-Sziget [Sighetu Marmaţiei]: Berger M. Ny.

Vezető (1909). *Vezető a Magyar Királyi Földtani Intézet Múzeumában*. A Magyar Királyi Földtani Intézet népszerű kiadványai I. kötet. Budapest: Az Intézet negyven éves fennállásának emlékére. Kiadja a Magyar Királyi Földmivelésügyi Minisztérium Fennhatsóga alá Tartozó M. Kir. Földtani Intézet.

Weller, Olivier (2002a). Aux origines de la production du sel en Europe. Vestiges, fonctions et enjeux archéologiques. In: O. Weller (ed), *Archéologie du sel: techniques et sociétés dans la pré- et protohistoire européenne / Salzarchäologie. Techniken und Gesellschaft in der Vor- und Frühgeschichte Europas*. Rahden/Westf.: Verlag Marie Leidorf. 163–75.

Weller, Olivier (2002b). The earliest rock salt exploitation in Europe: a salt mountain in the Spanish Neolithic. *Antiquity* 76, 317–18.

Weller, Olivier, Dufraisse, Alexa and Pétrequin, Pierre (eds) (2008). *Sel, eau et forêt. D'hier à aujourd'hui*. Collection "Les cahiers de la MSHE Ledoux", 12. Besançon: Presses universitaires de Franche-Comté.

Weller, Olivier and Dumitroaia, Gheorghe (2005). The earliest salt production in the world: an early Neolithic exploitation in *Poiana Slatinei*-Lunca, Romania. *Antiquity* 79, http://antiquity,ac.uk/ProjGall/weller/index.html, accessed 14/08/2007.

Windakiewicz, Edward (1926). *Solnictwo, sole kamienne, potasowe i solanki, ich własności, fizjografia, górnictwo i warzelnictwo. vols I–IV*. Kraków: Skład Główny w Księgarni Jagiellońskiej.

Wollmann, Volker (1996). *Mineritul metalifer, extragerea sării şi carierele de piatră în Dacia Romană / Der Erzbergbau, die Salzgewinnung und die Steinbrüche im Römischen Dakien*. Bibliotheca

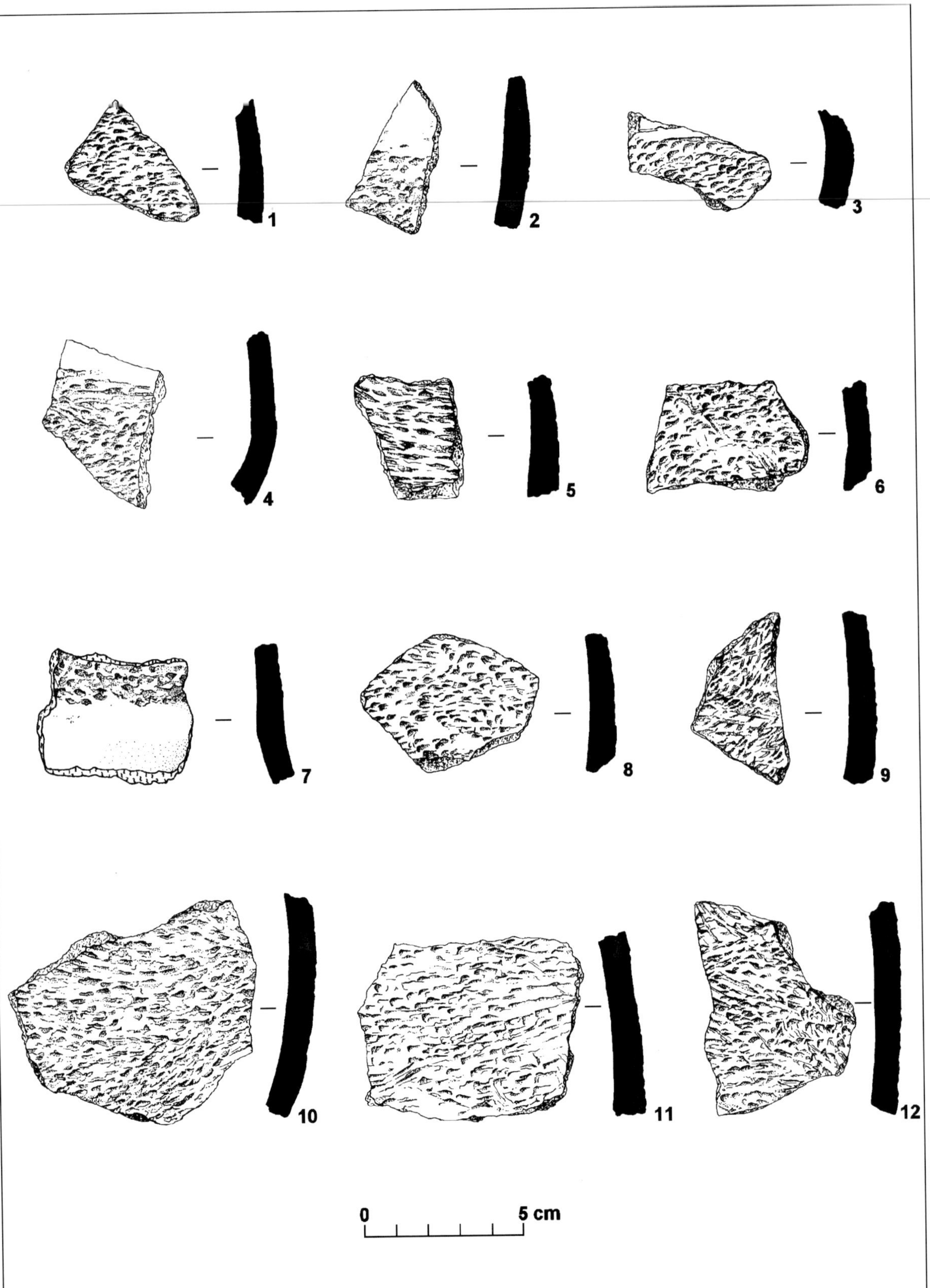

Plate 2. Trench III. 1–12: Bronze Age sherds, textile-impressed. 1–6, 8–12: pit 2a; 7: pit 2.

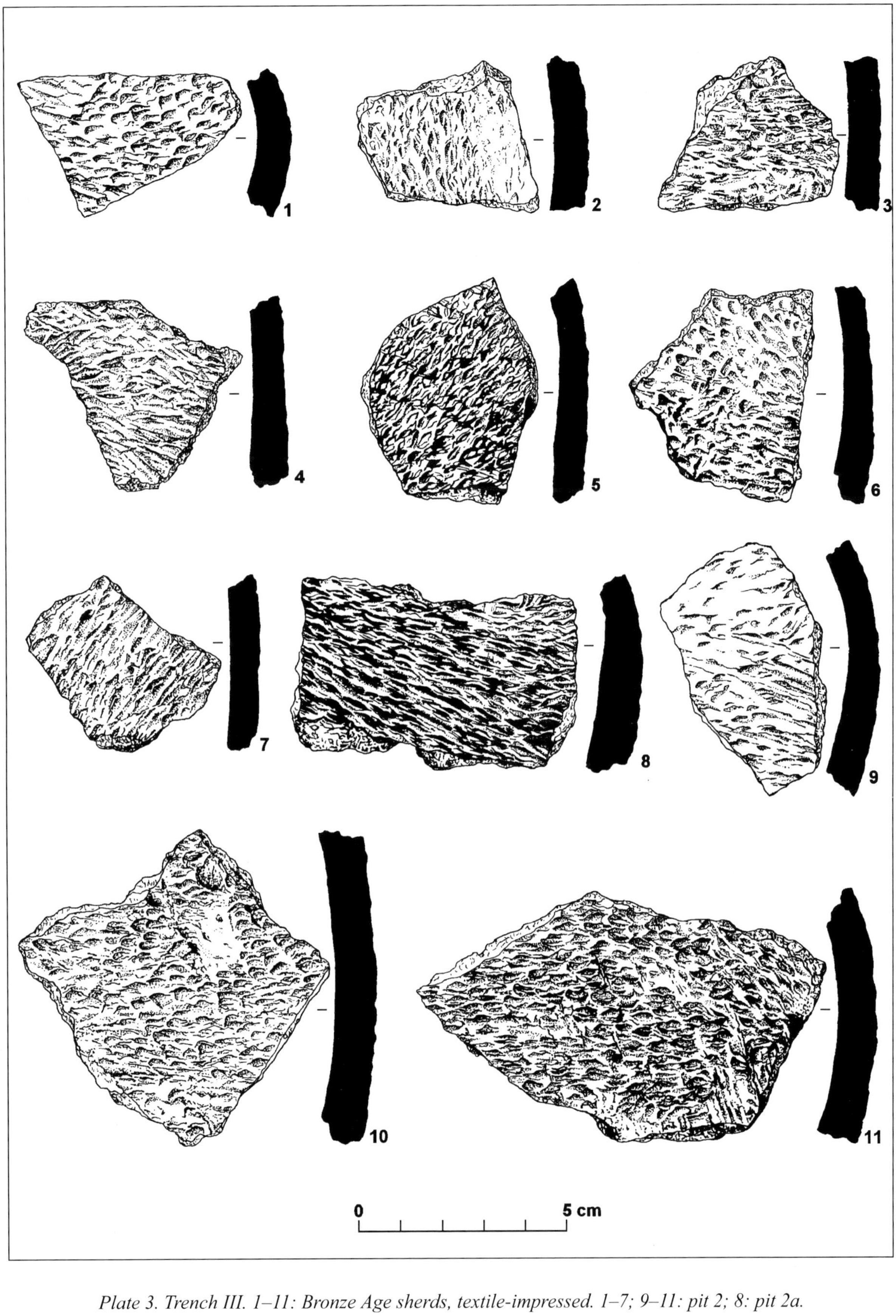

Plate 3. Trench III. 1–11: Bronze Age sherds, textile-impressed. 1–7; 9–11: pit 2; 8: pit 2a.

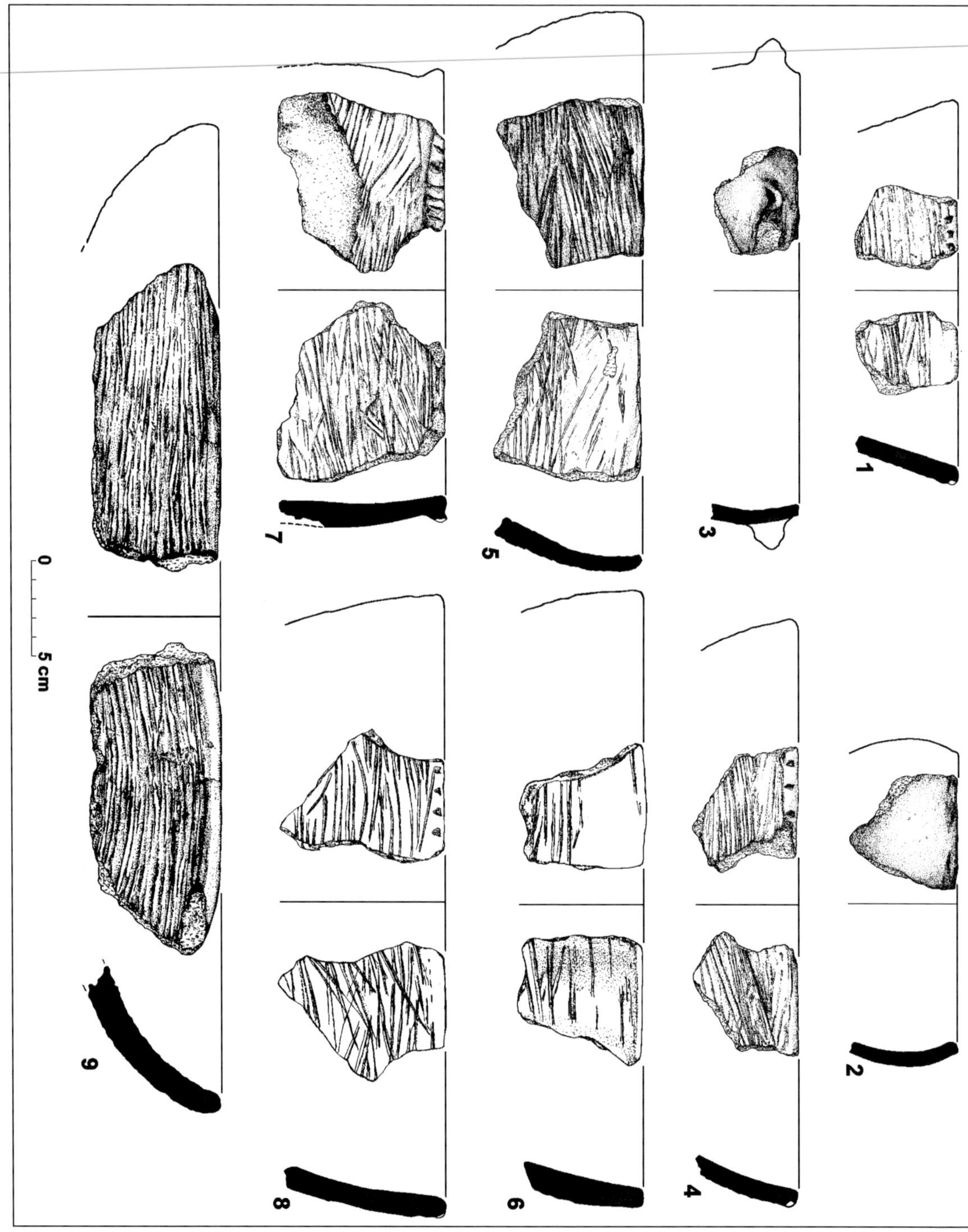

Plate 4. Trench III. 1–9: Bronze Age sherds. 1–2, 4–8 Pit 1; 3: unstratified; 9: no context.

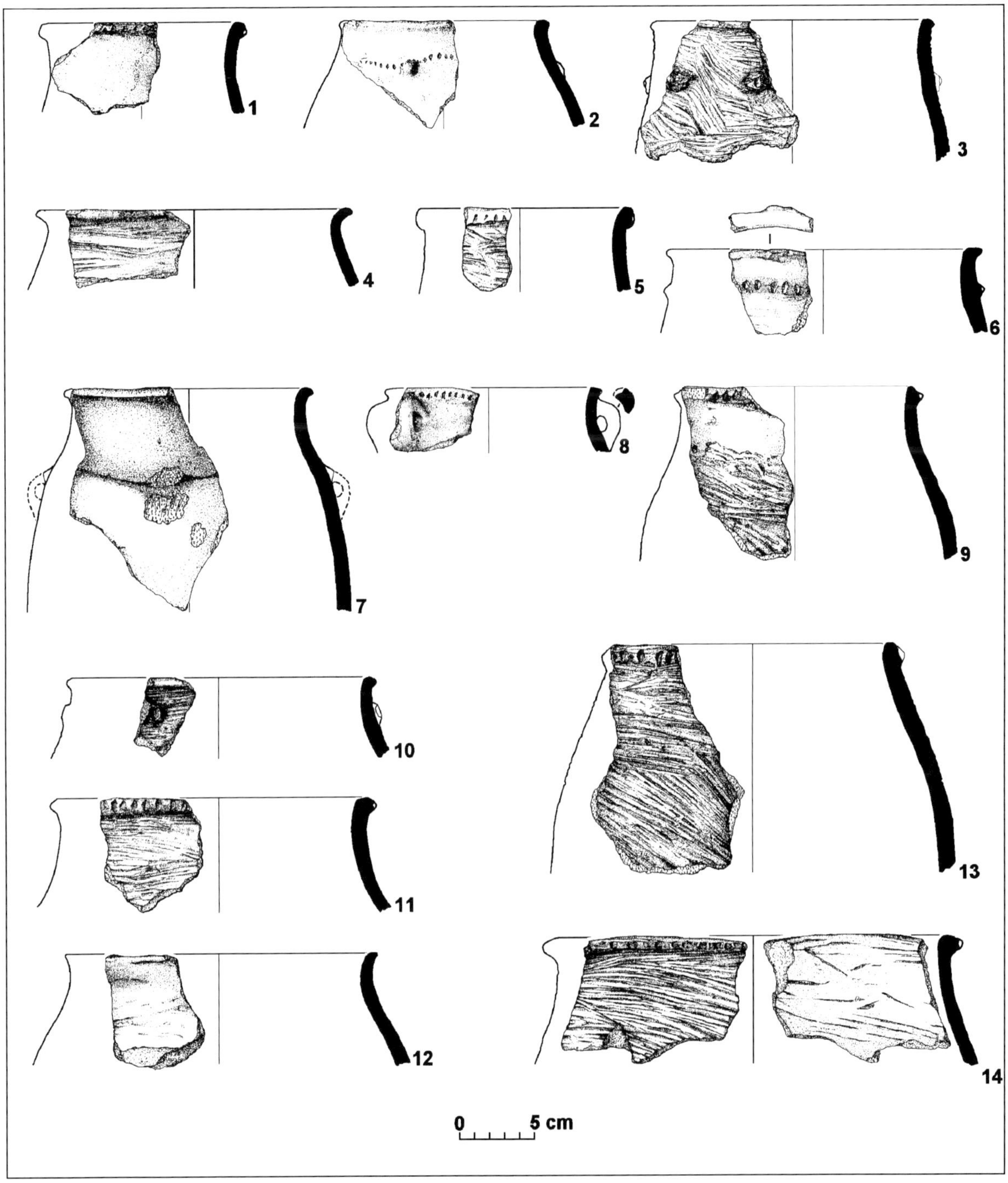

Plate 5. Trench III. 1–14: Bronze Age sherds.
1: no context; 2: unstratified; 3, 5– 6, 8, 9–12: Pit 2; 4, 7, 13–14: Pit 1.

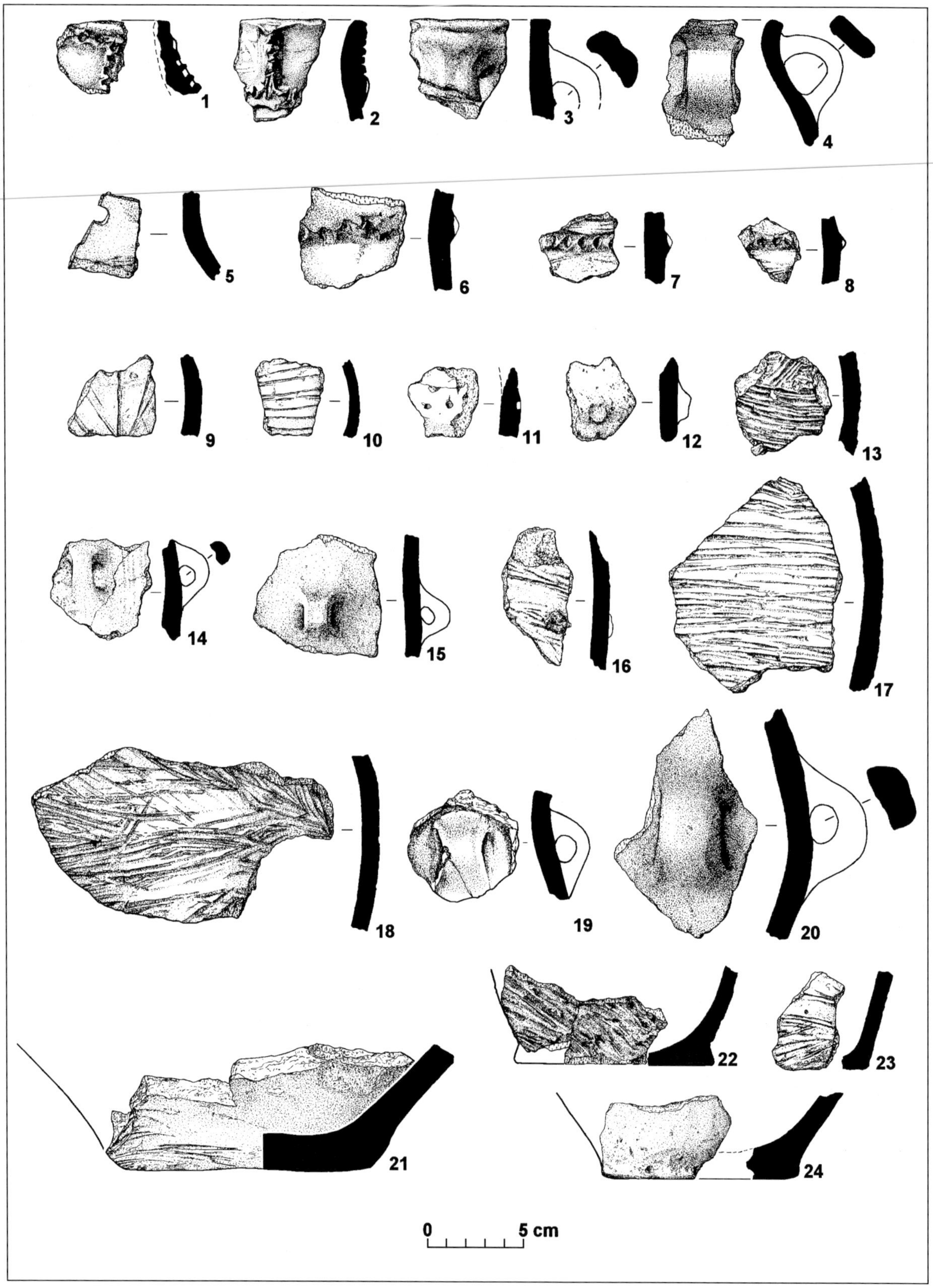

Plate 6. Trench III. 1–24 : Bronze Age sherds.
1, 3, 4, 6–11, 13–15, 17, 18, 22: Pit 2; 2, 5, 9, 19, 24: unstratified; 12, 16: Pit 1; 20, 21, 23: no context.

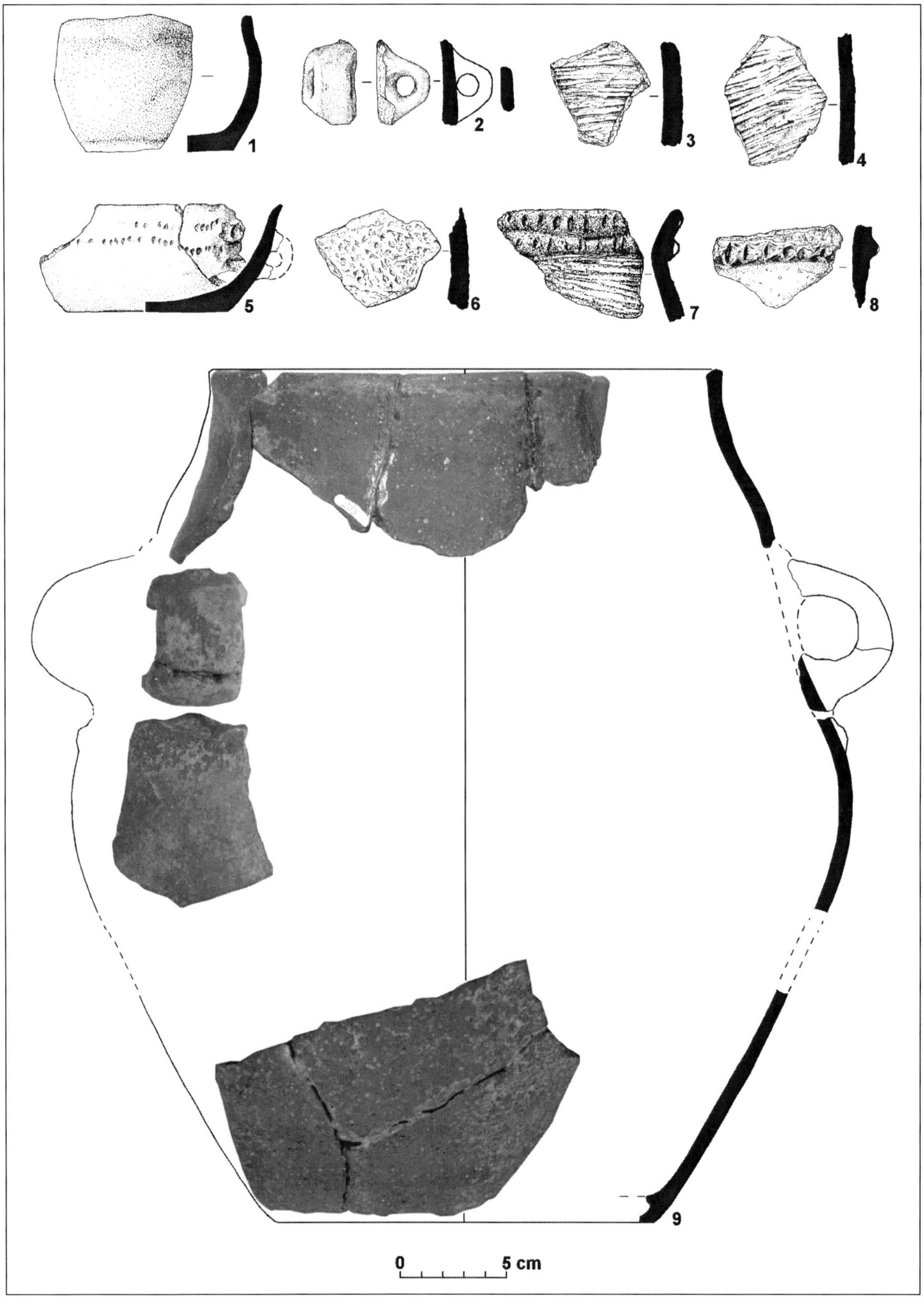

Plate 7. Trench VII. 1–9 Bronze Age sherds from Pit 3.

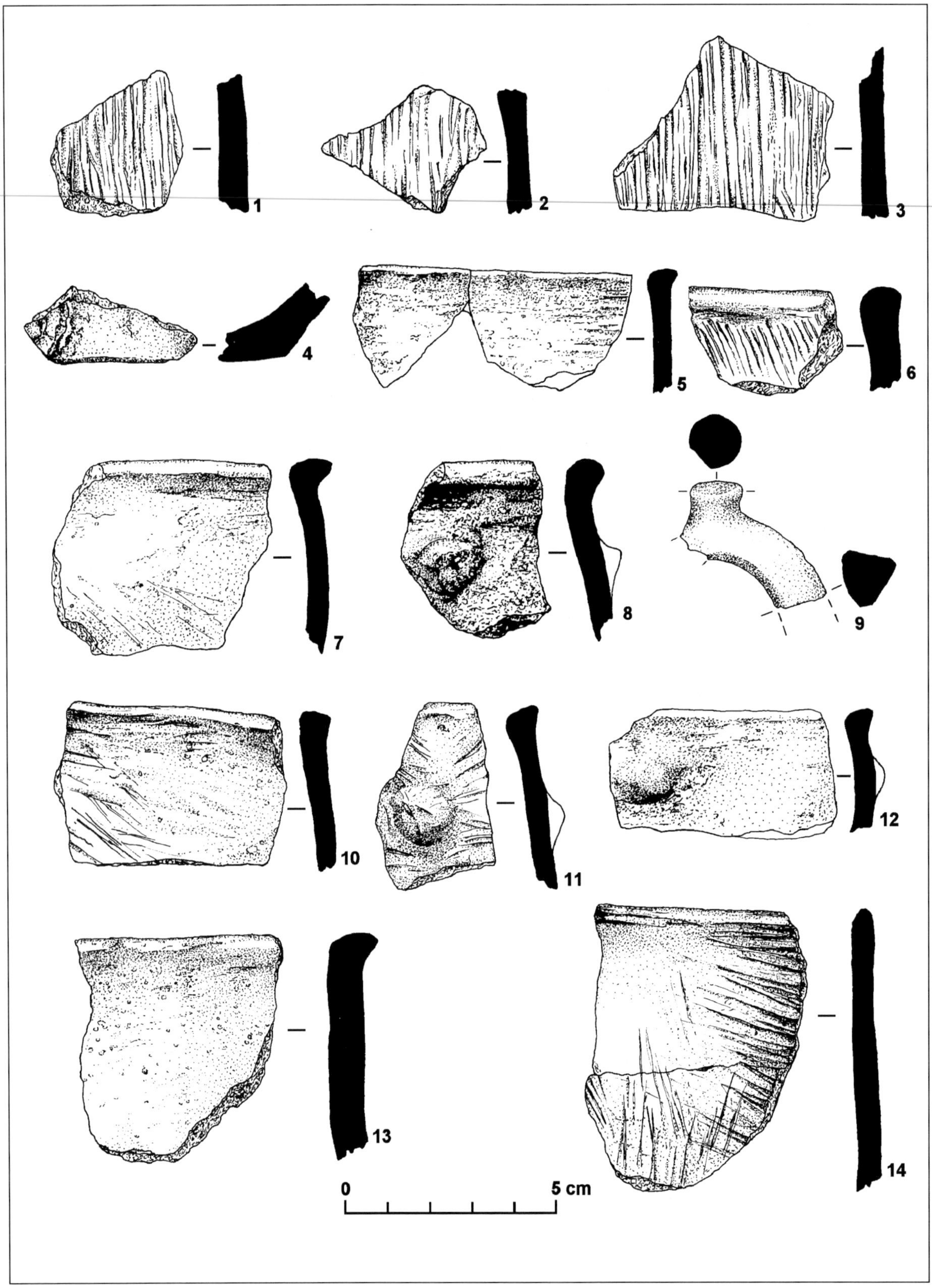

Plate 8. Bronze Age sherds from Trenches V and VI:
1–7, 10, 12–15 Trench VI; 8, 11 Trench VI extension; 9 Trench V.

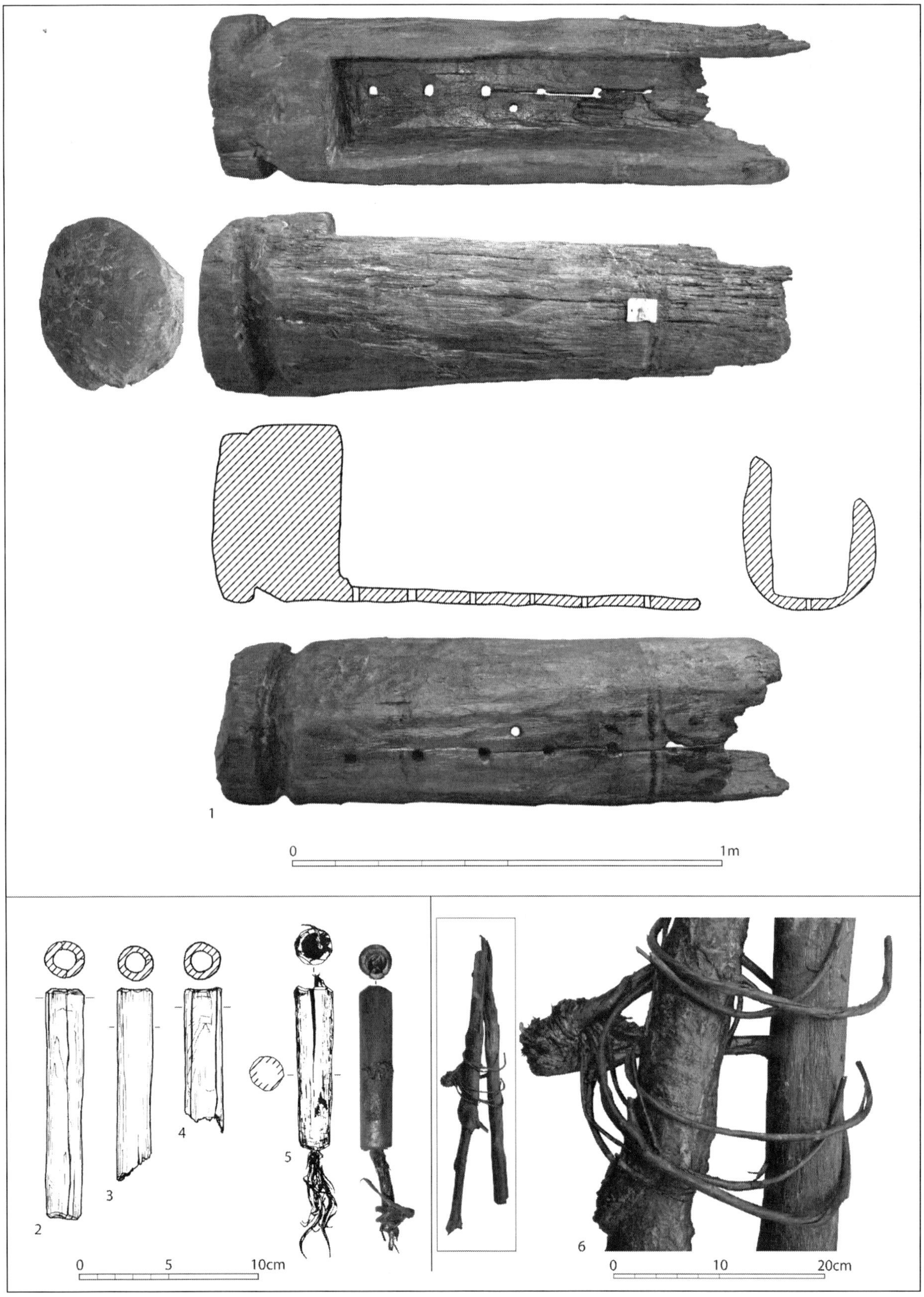

Plate 9. Trench I. – 1 trough (2). – 2–5 pegs – 6 binding. Scales 1: 1:12.5; 2–5: 1:3.

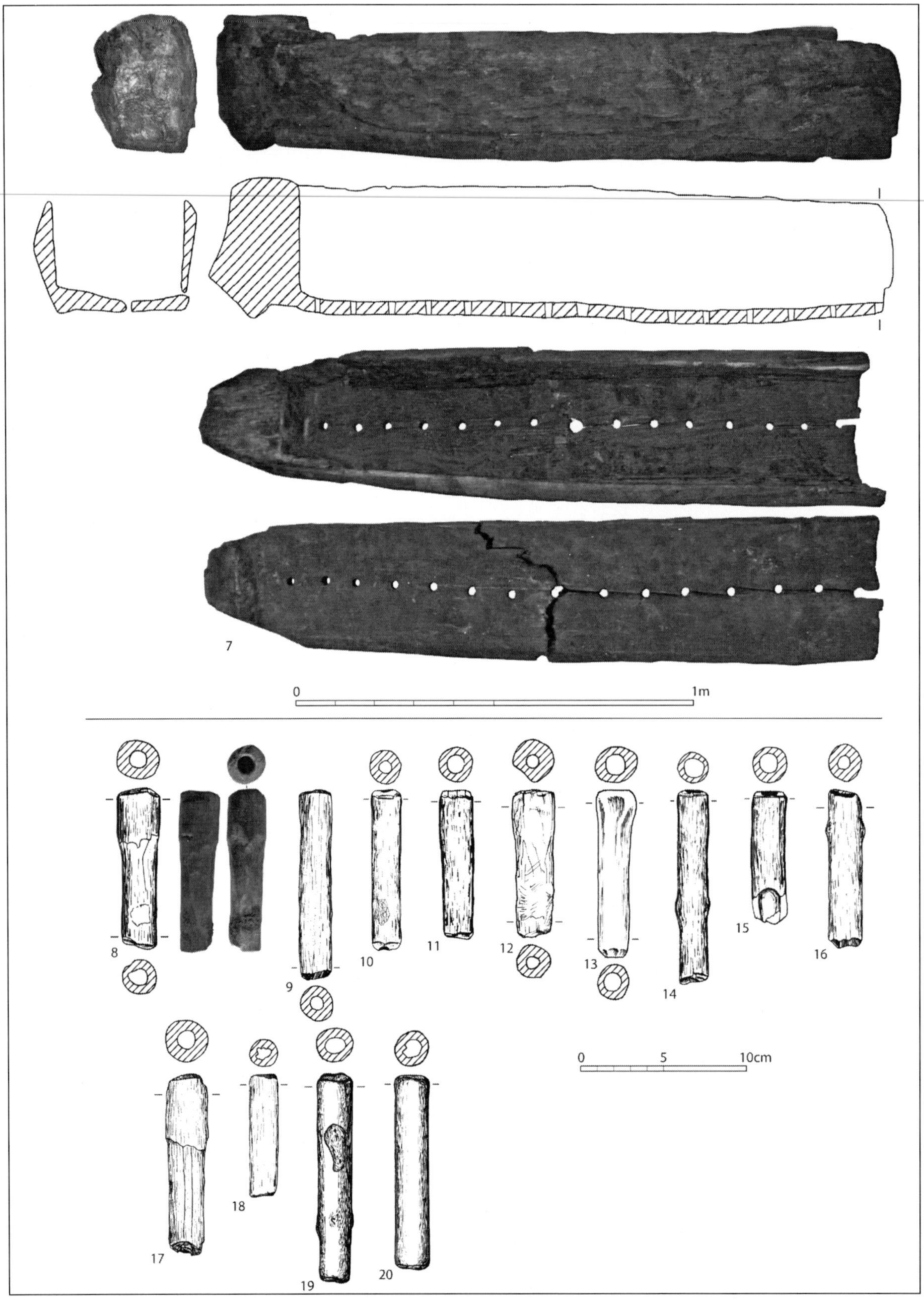

Plate 10. Trench I. – 7 trough (3). – 8–20 pegs. Scales 7: 1:12.5; 8–20: 1:3.

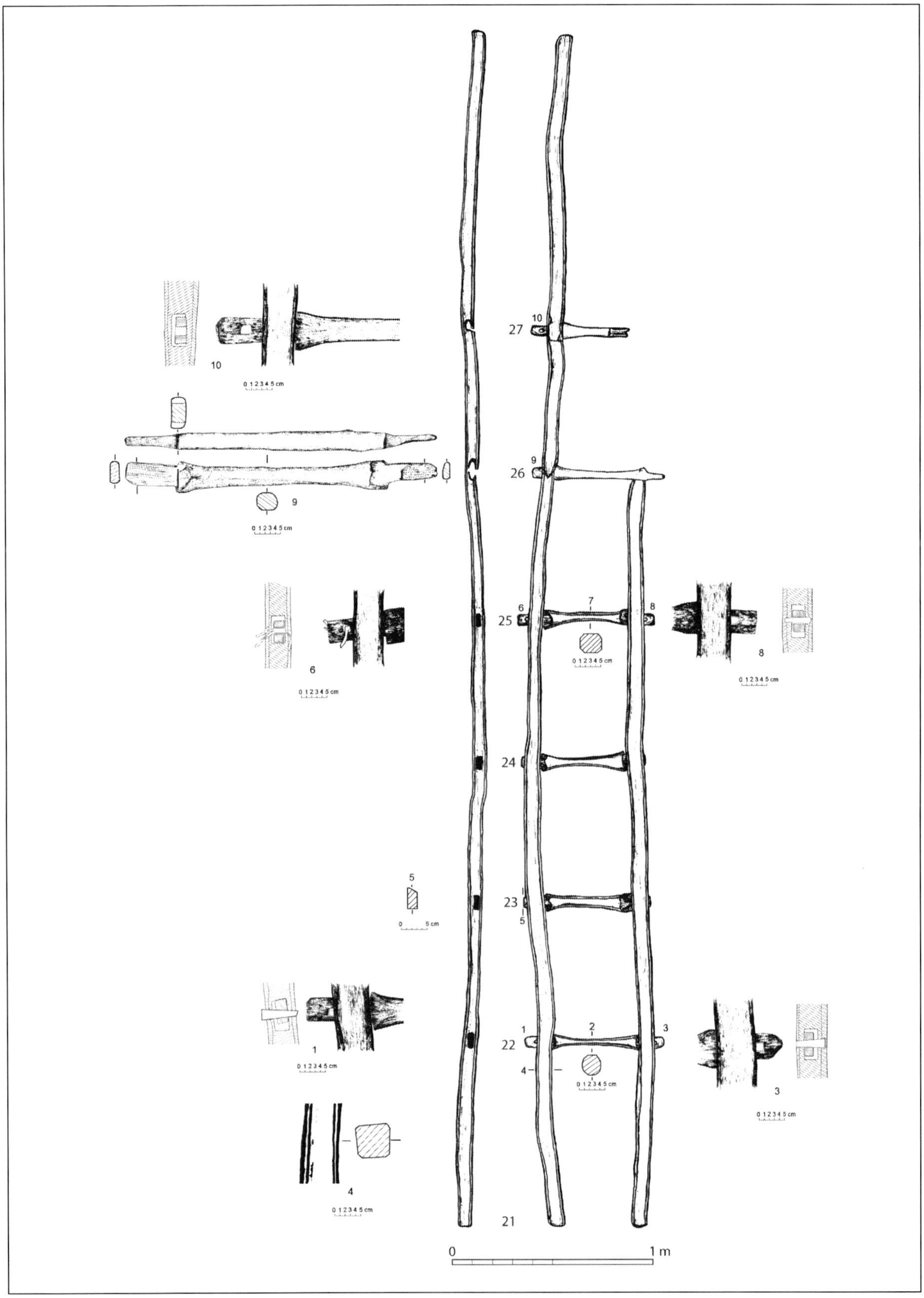

Plate 11. Trench I. – 21 ladder. – 22–27 rungs. Scales 21: 1:25; details 1:10.

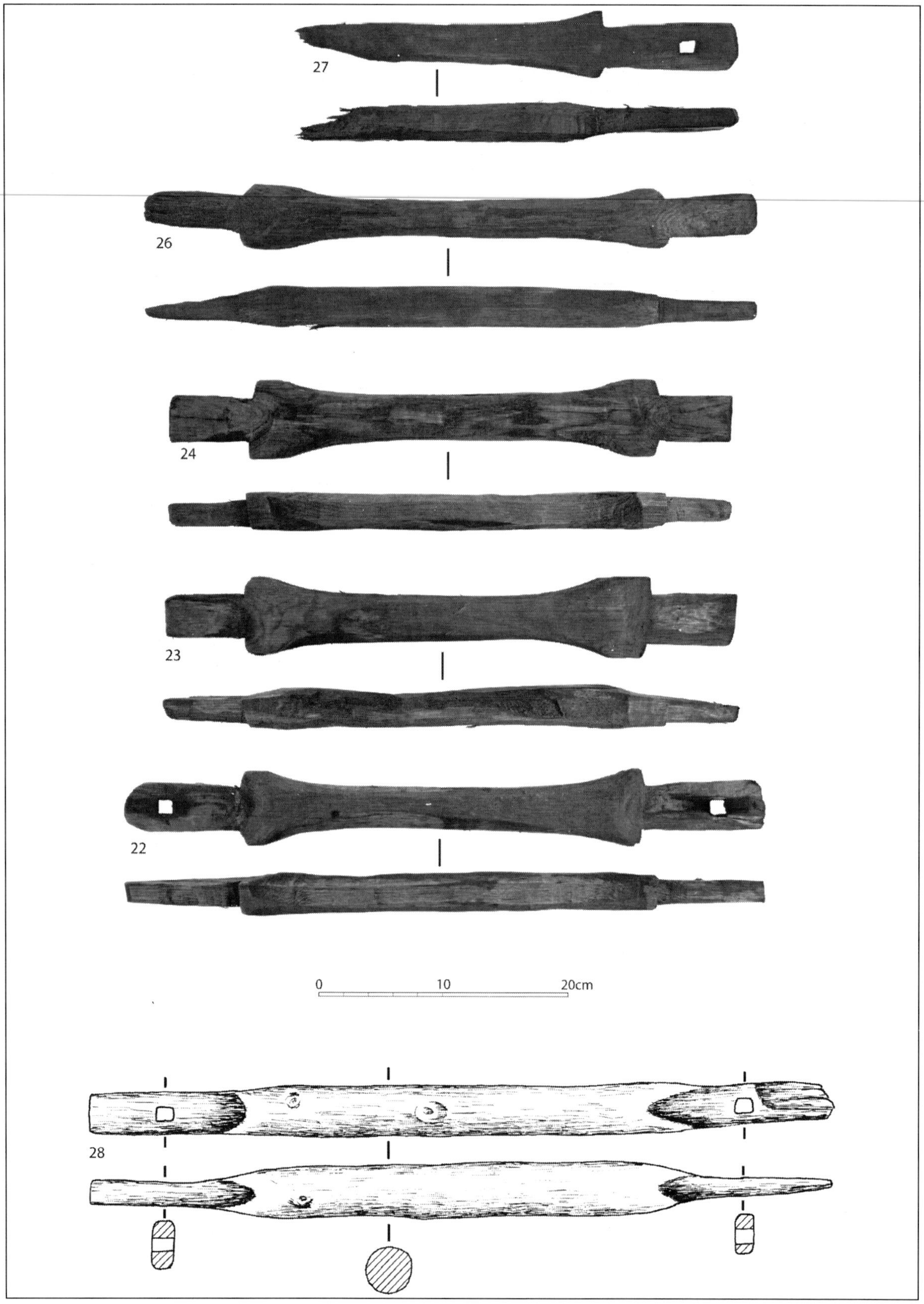

Plate 12. Trench I. – 22–24. 26–28 ladder rungs. Scale 1:4.

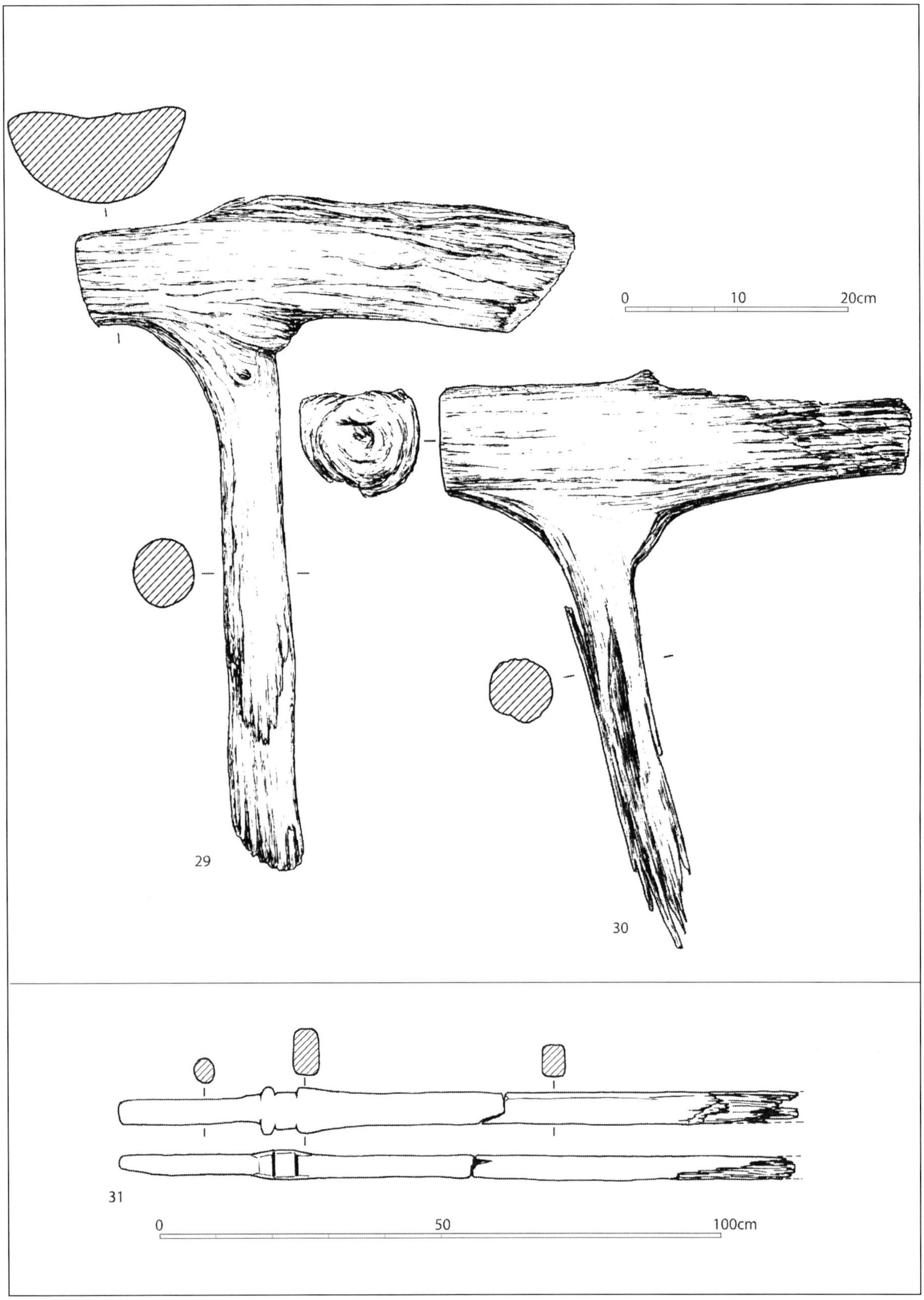

Plate 13. Trench I. – 29. 30 mallets. – 31 possible axle. Scales 29. 30: 1:5; 31: 1:10.

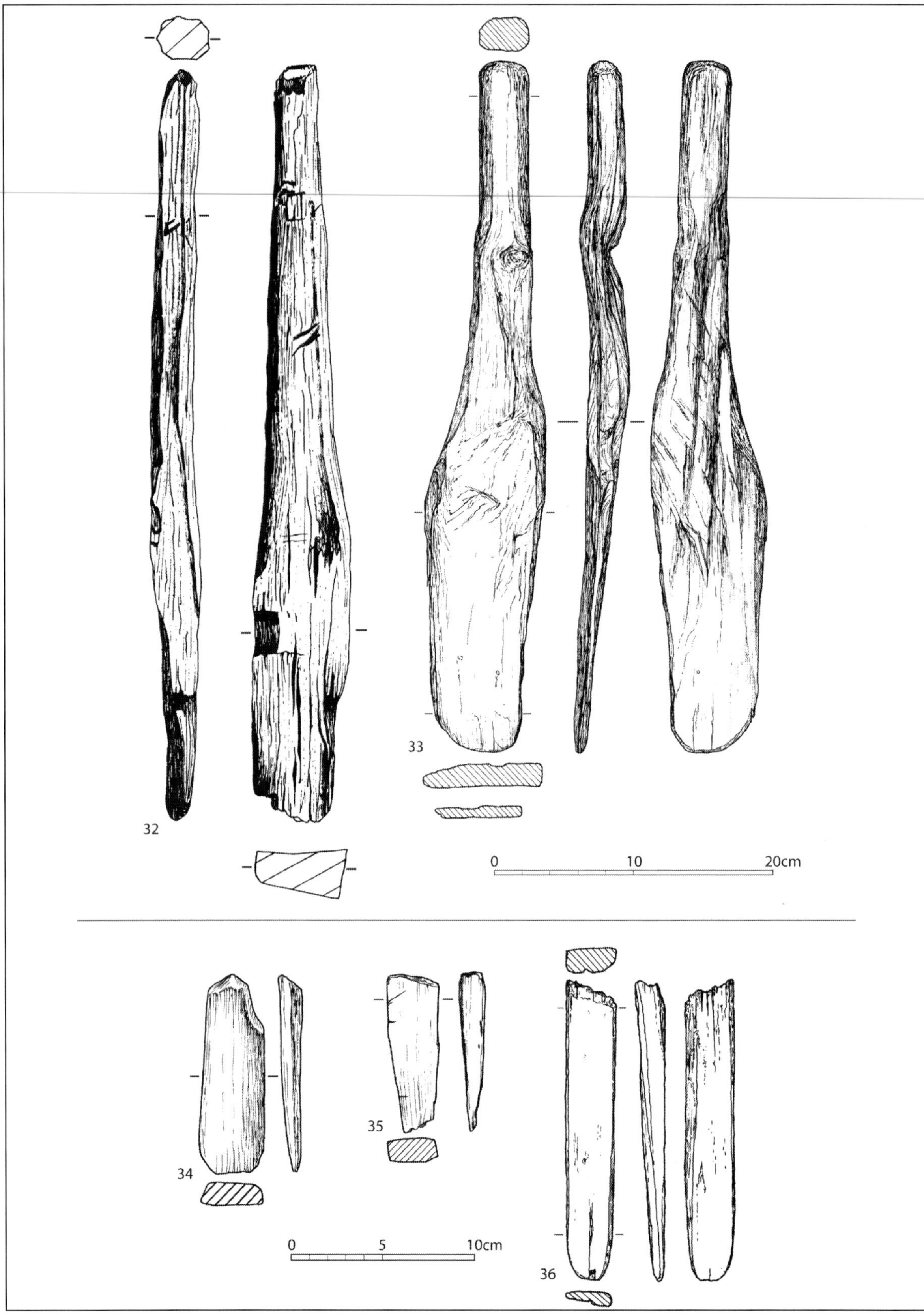

Plate 14. Trench I. – 32. 33 small shovels. – 34–36 small wedges. Scales 32. 33: 1:4; 34–36: 1:3.

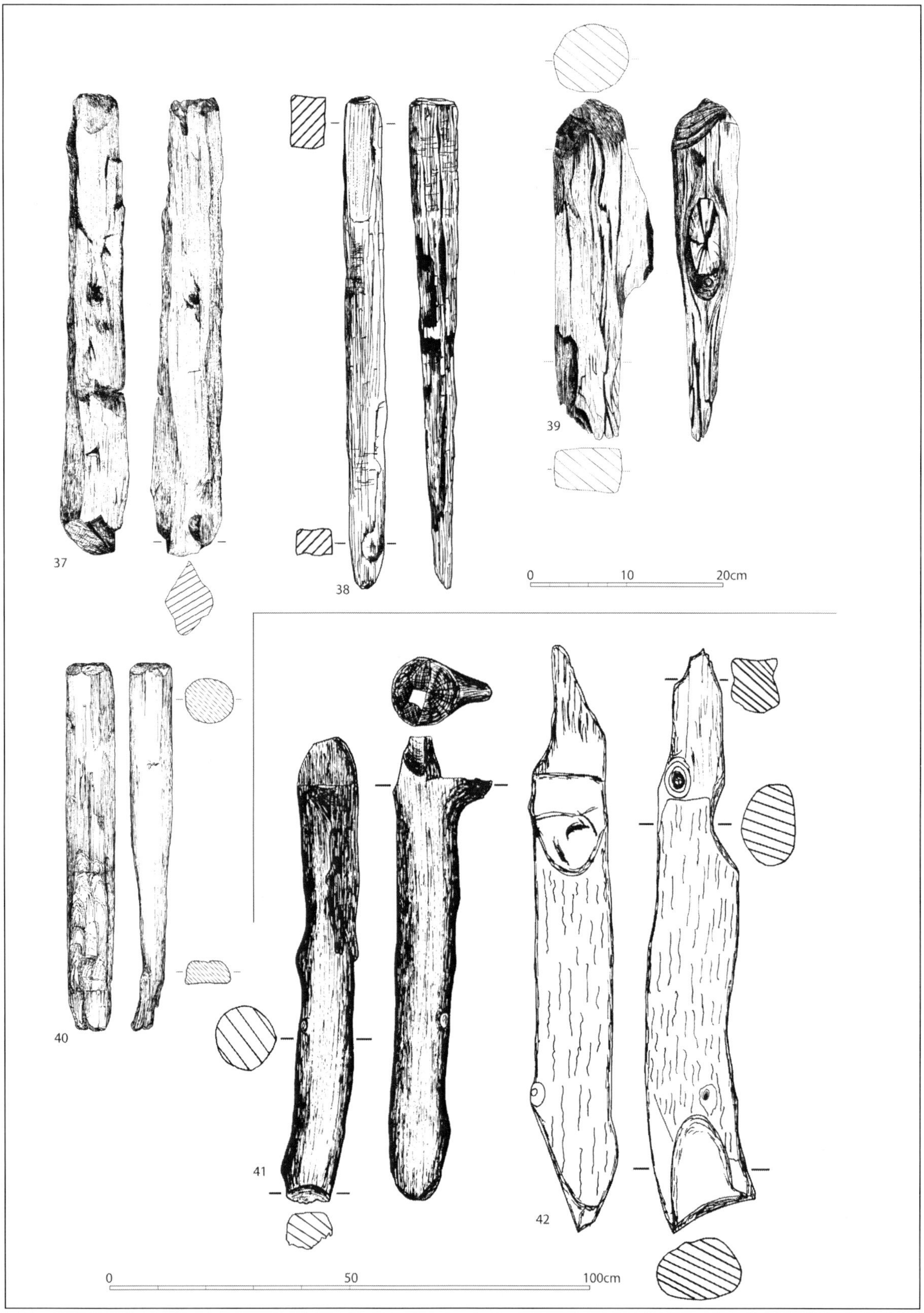

Plate 15. Trench I. – 37–40 large wedges. – 41. 42 worked pieces of large wood. 37–40: 1:5; 41. 42: 1:10.

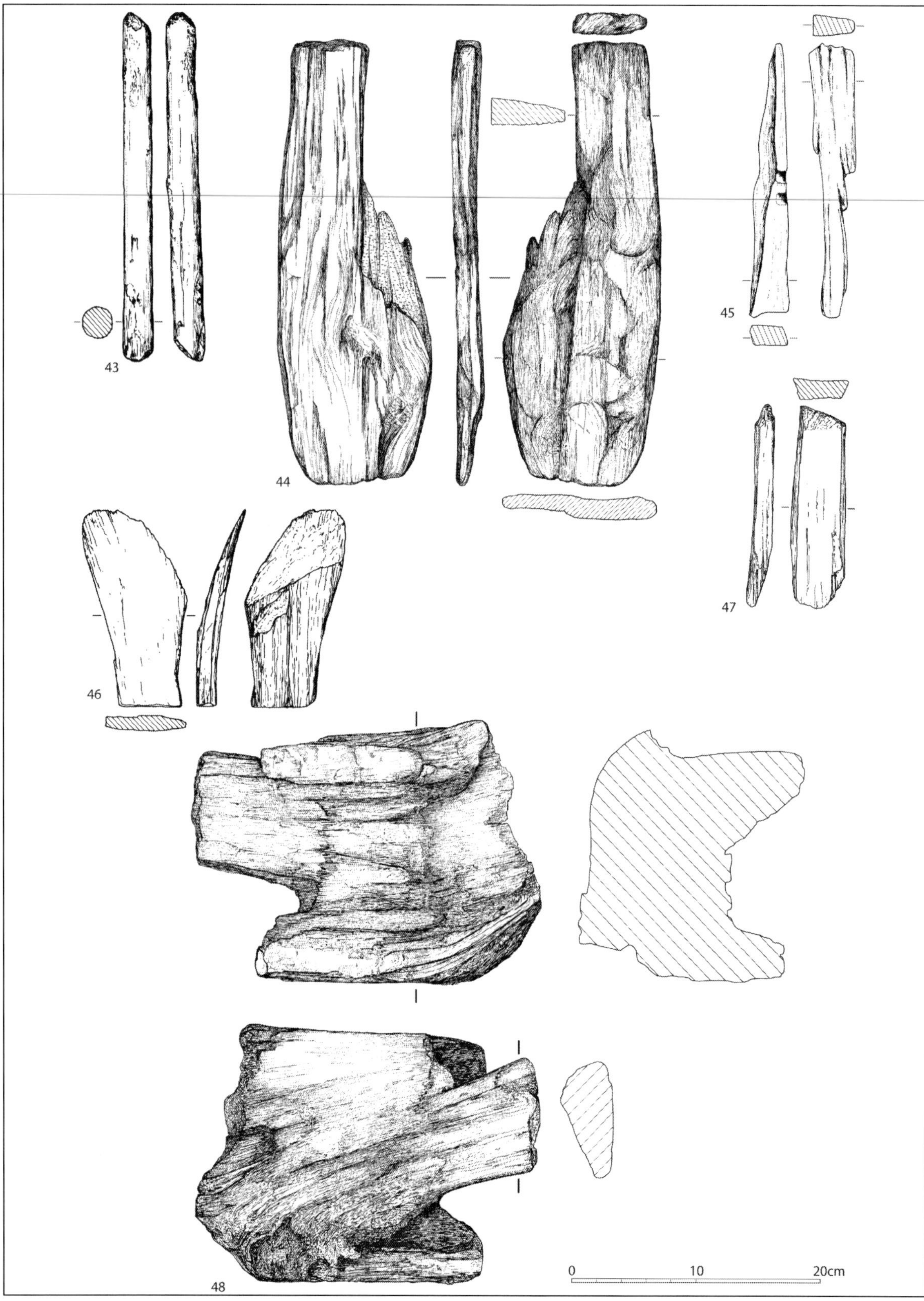

Plate 16. Trench I. – 43 worked roundwood. – 44 flat-worked piece of wood. – 45–47 Splinter. – Trench III. – 48 trough. Scale 1:4.

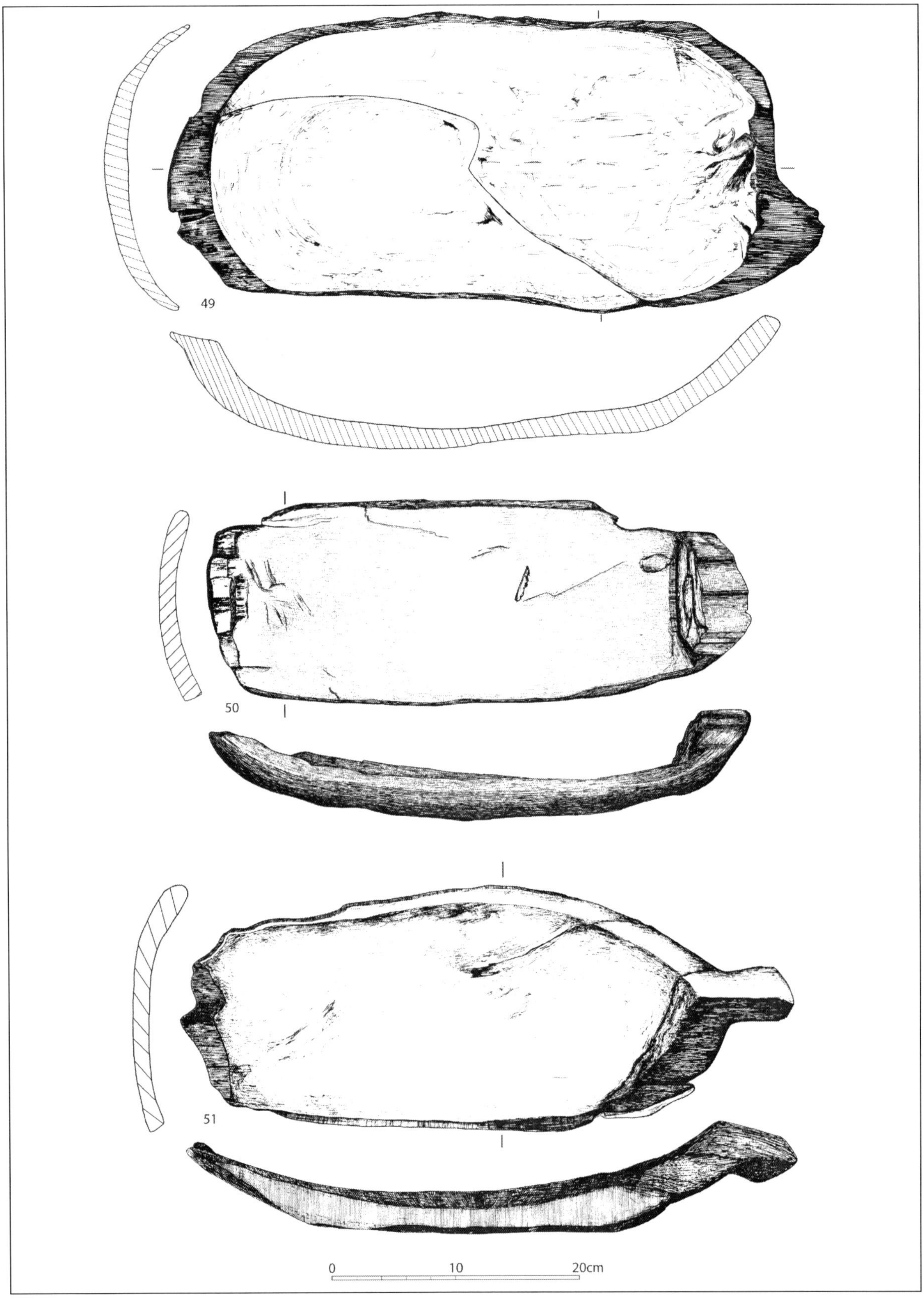

Plate 17. Trench III. – 49–51 small troughs. Scale 1:4.

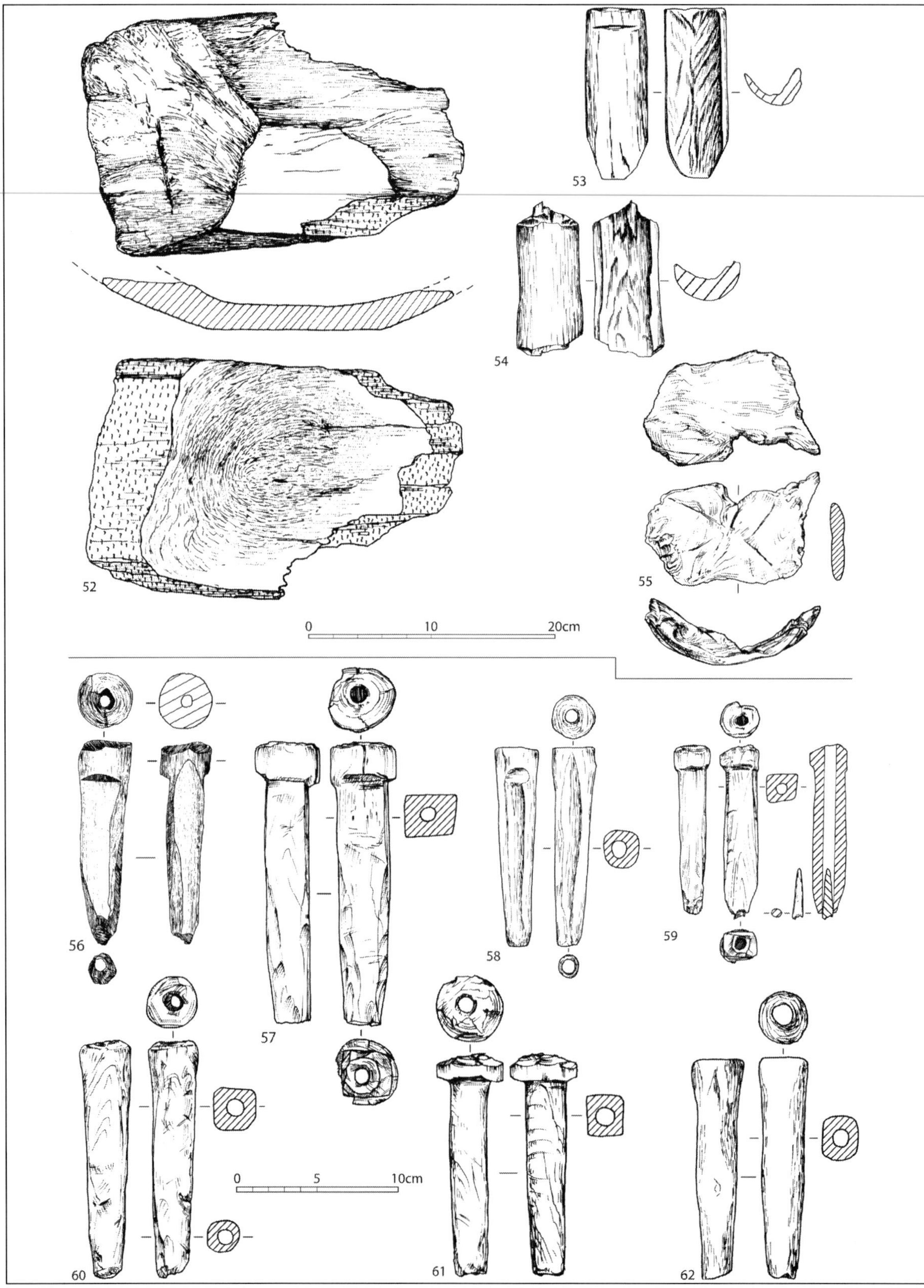

Plate 18. Trench III. – 52 piece of small trough. – 53. 54 small channelled pieces. – 55 hollowed piece of wood. – 56–62 pegs. Scales 52–55: 1:4; 56–62: 1:3.

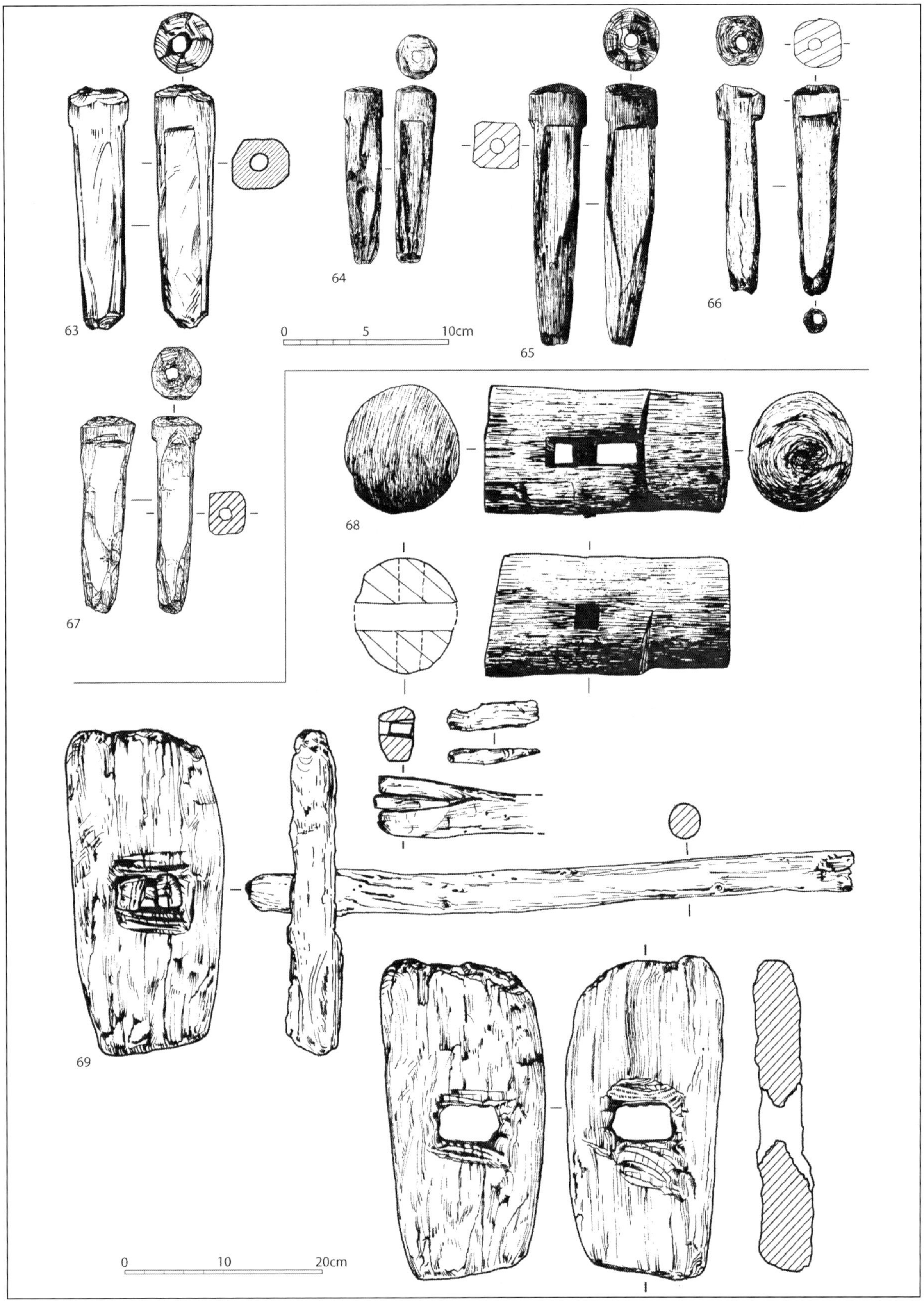

Plate 19. Trench III. – 63–67 pegs. – 68 possible wheel hub. – 69 mallets. Scales 63–67: 1:3; 68. 69: 1:5.

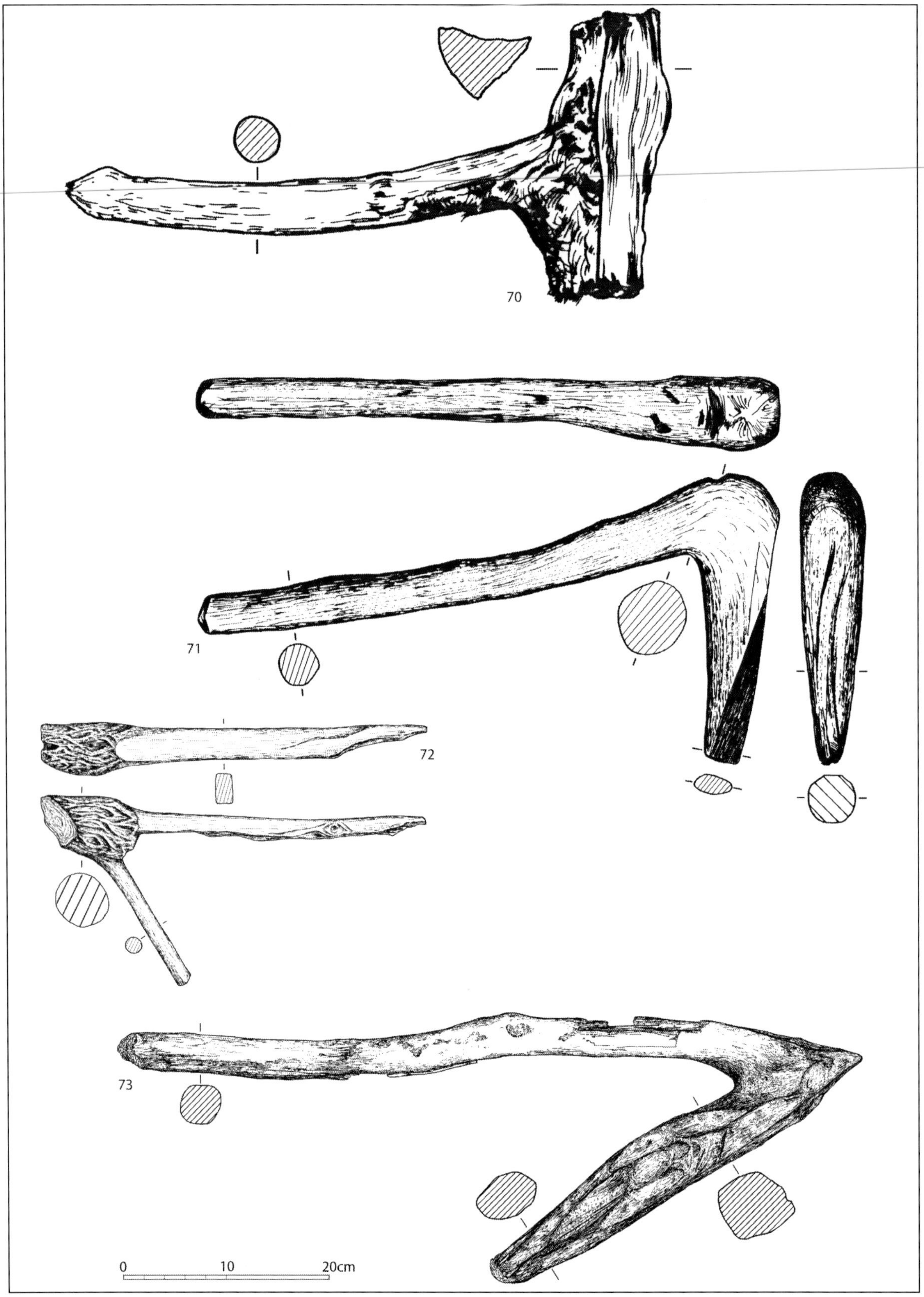

Plate 20. Trench III. – 70 mallet. – 71. 73 tool hafts. – 72 haft?. Scale 1:5.

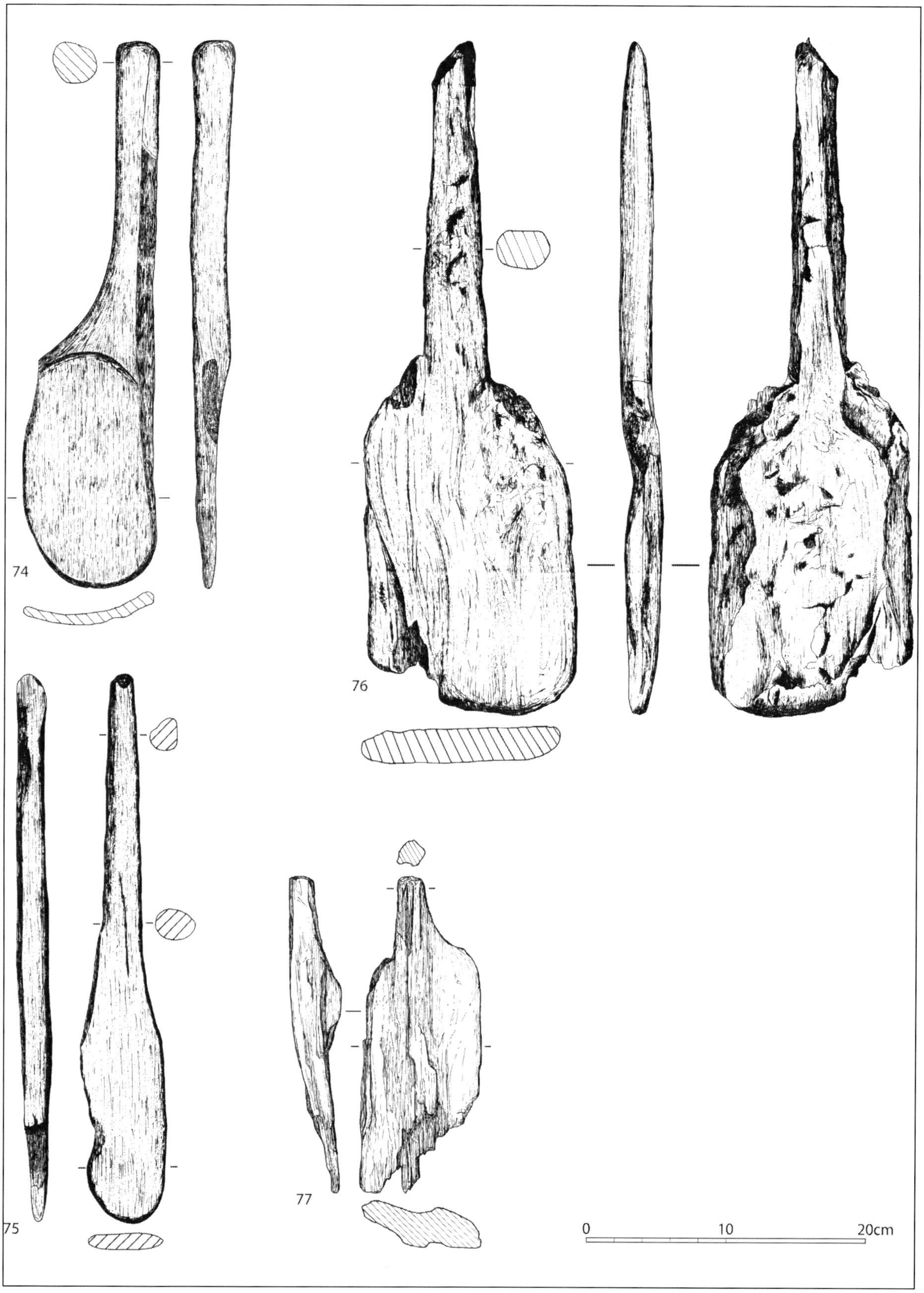

Plate 21. Trench III. – 74–77 shovels. Scale 1:5.

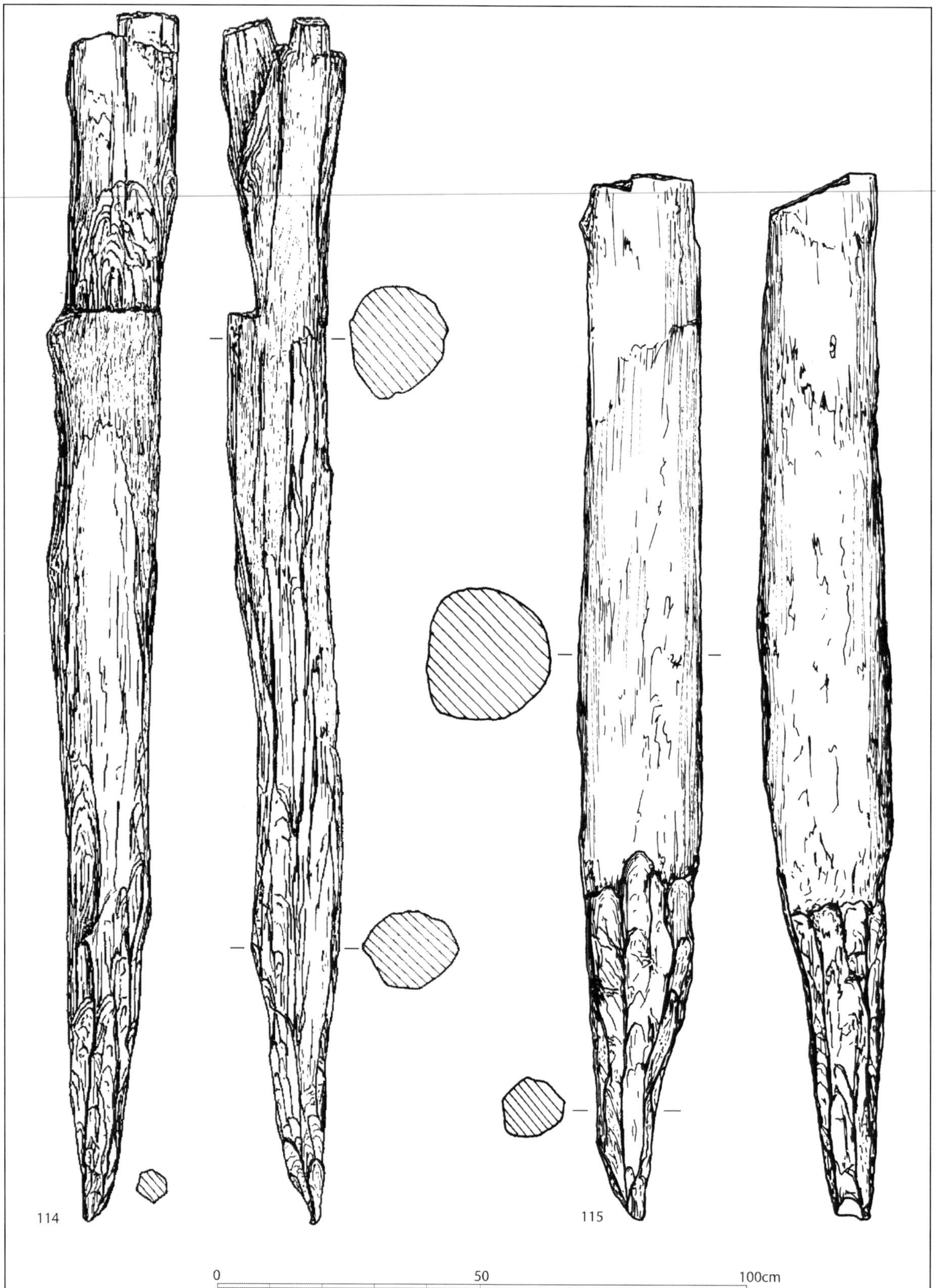

Plate 26. Trench III, pieces from wooden structures. – 114. 115 posts. Scale 1:10.

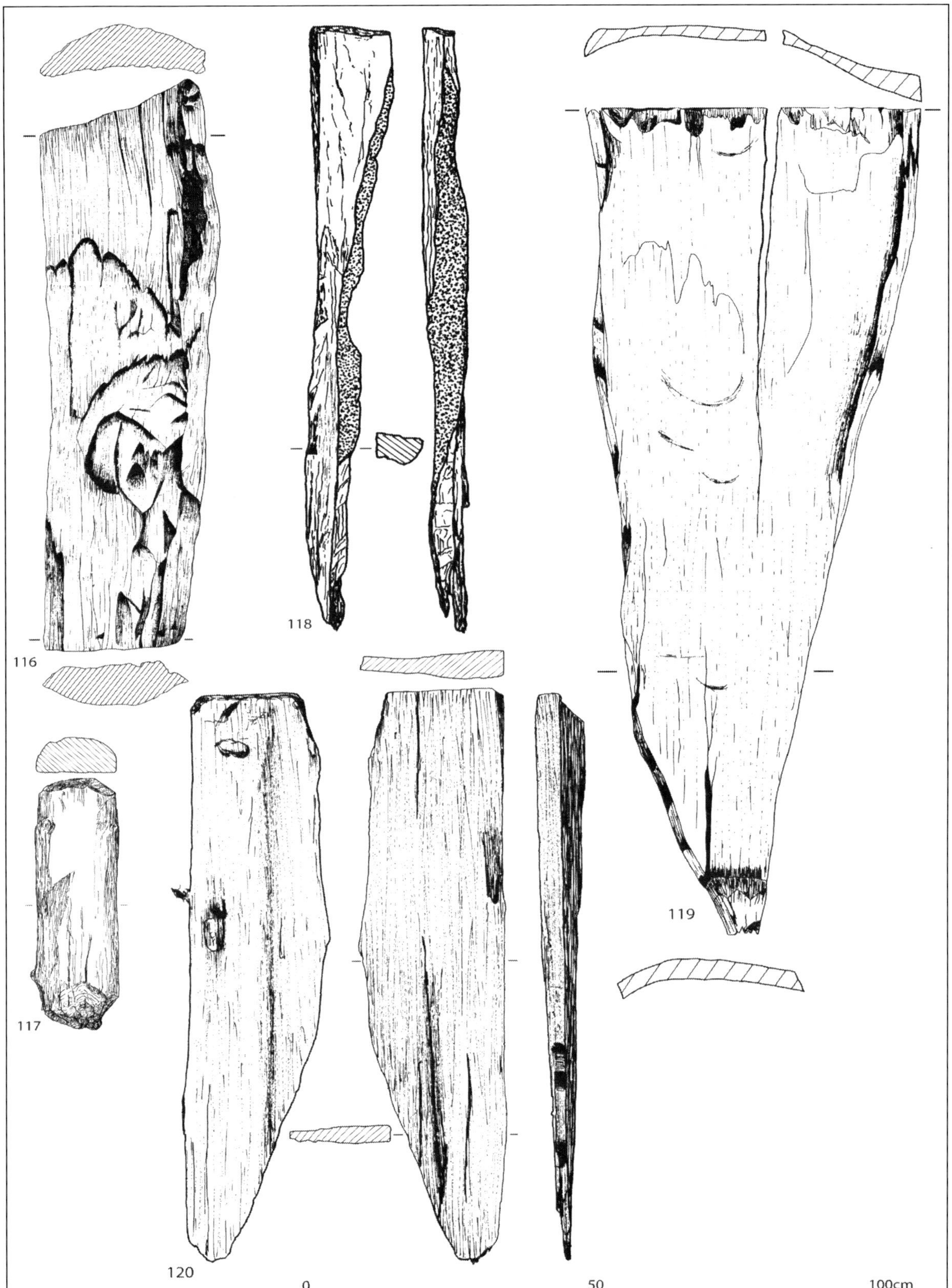

Plate 27. Trench III, pieces from wooden structures. – 116 wood piece. – 117 worked piece of wood. – 118. 120 planks. – 119 large piece of flat-worked wood. Scale 1:10.

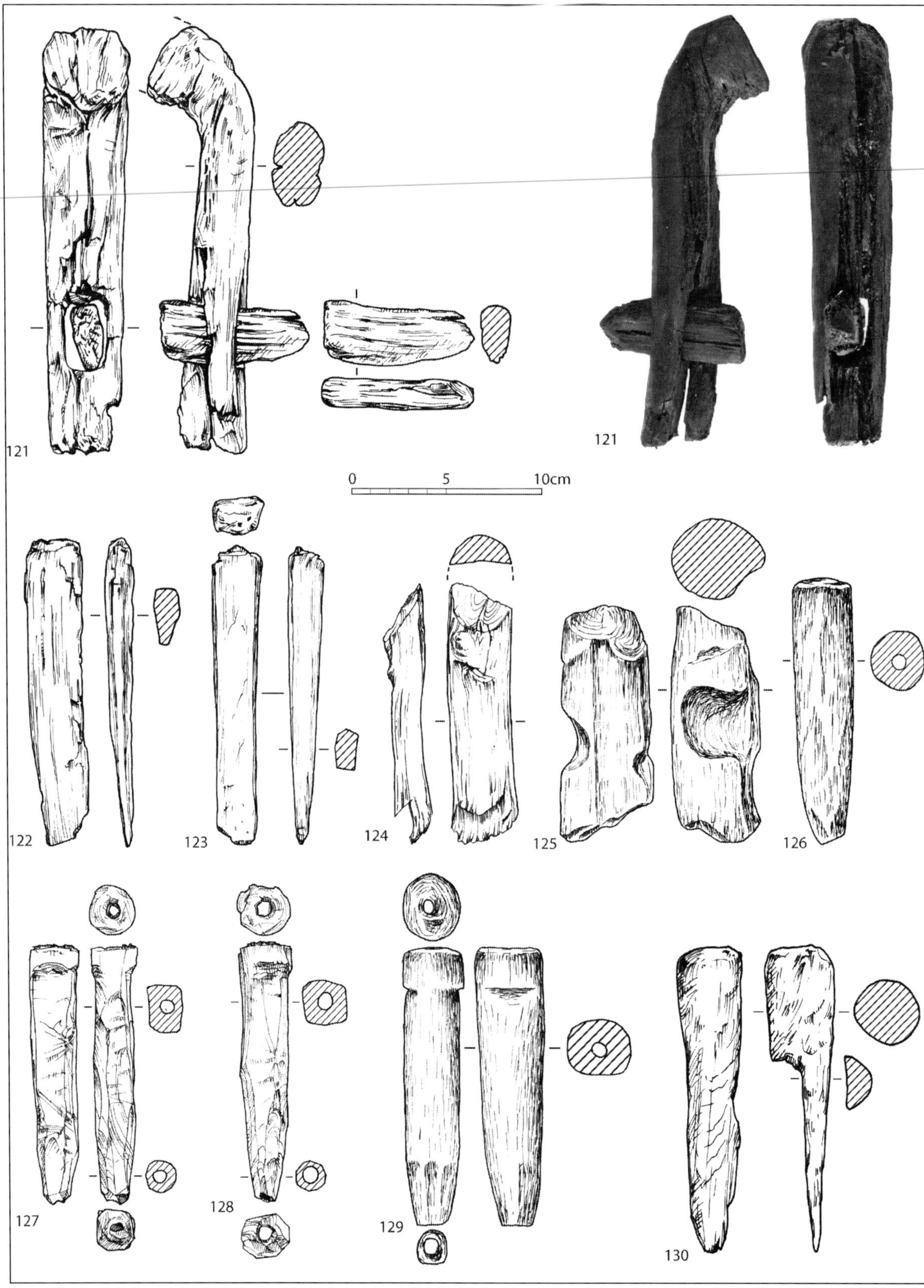

Plate 28. Trench XV. – 121 V-shaped object. – 122. 123 small wedges. – 124 small splinter. – 125 worked roundwood. – Trench XXII. – 126–129 pegs. – 130 peg stopper. Scale 1:3.

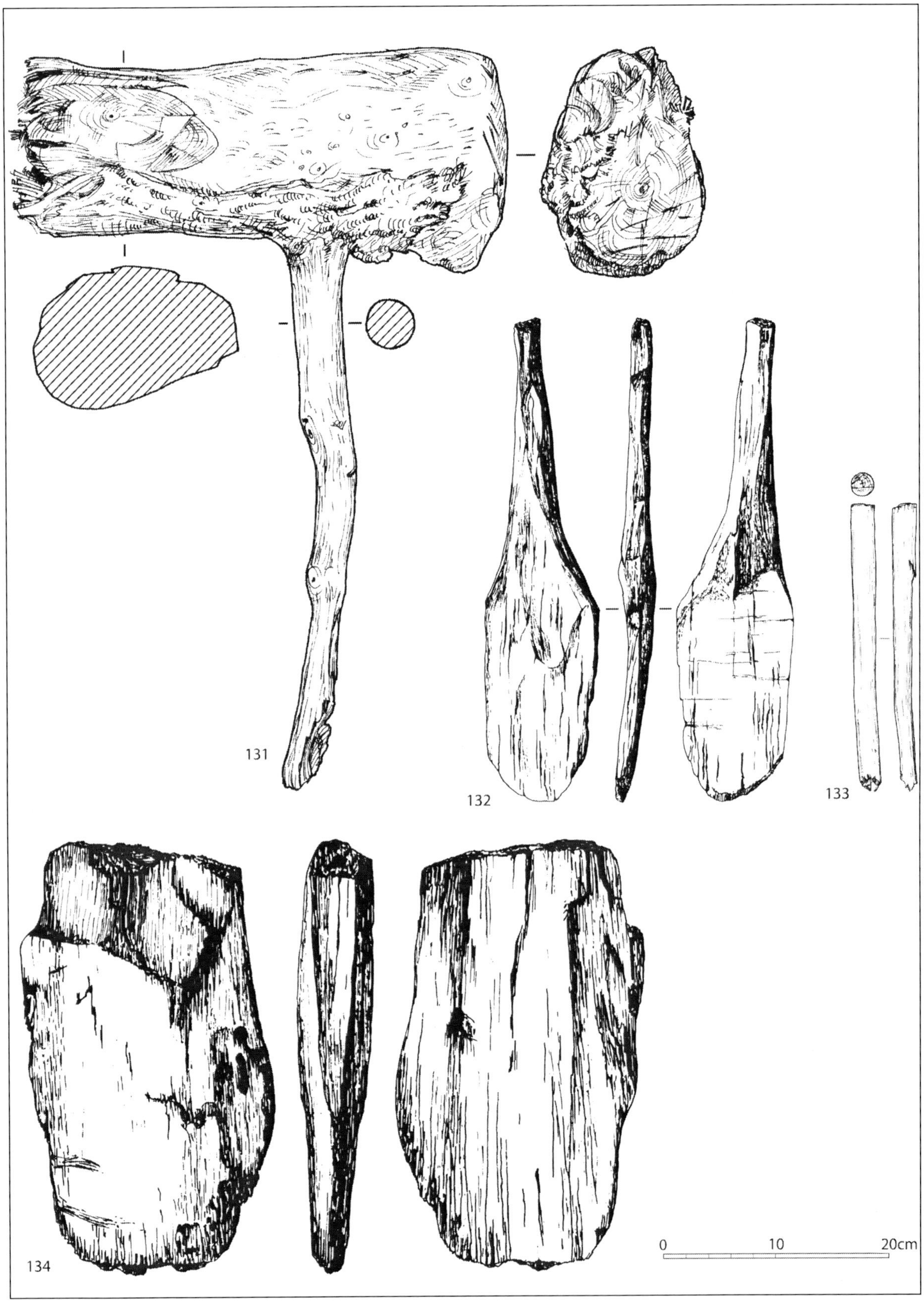

Plate 29. Trench XXII. – 131 mallet. – 132 shovel. – 133 worked roundwood. – 134 split board. Scale 1:5.

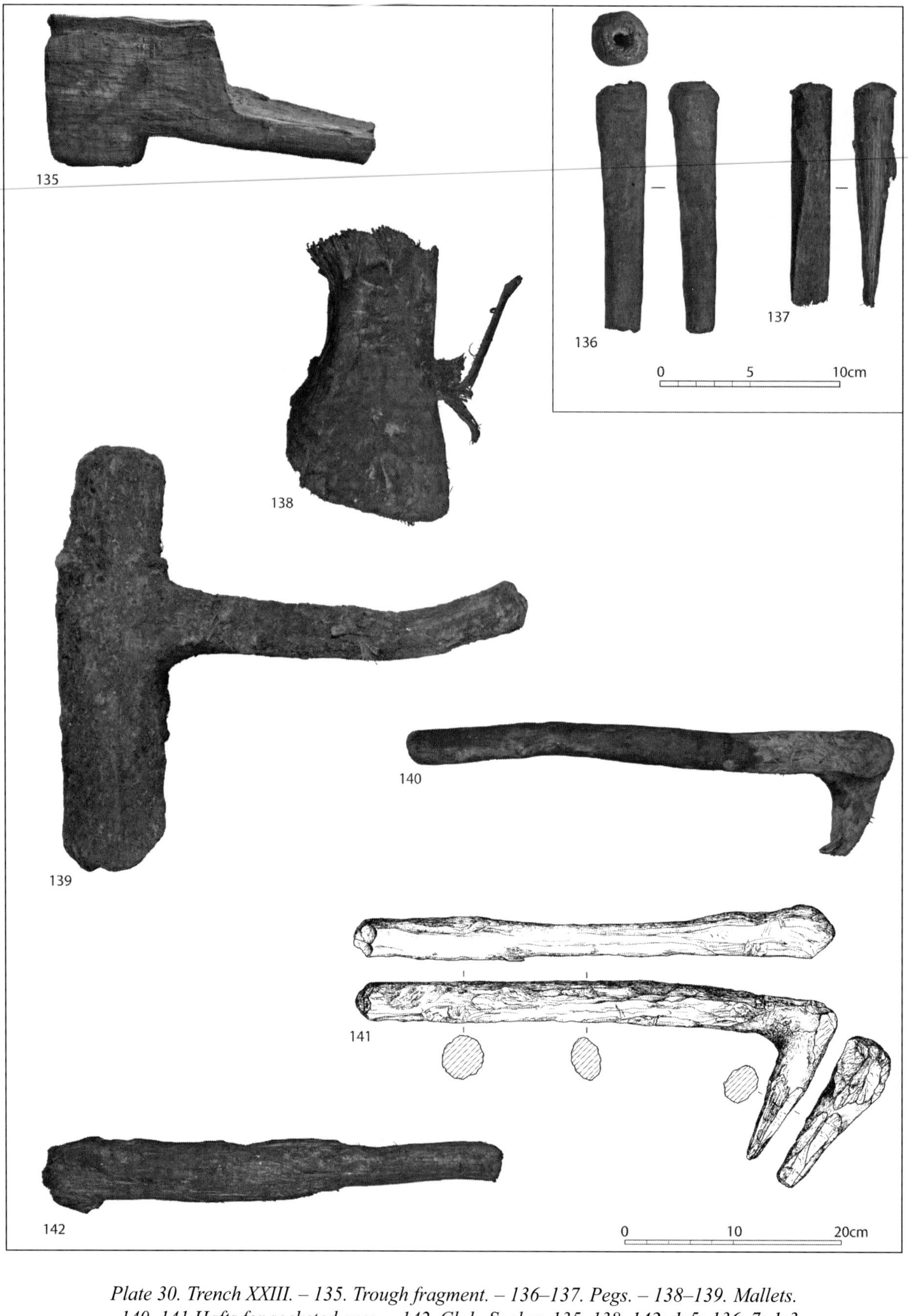

Plate 30. Trench XXIII. – 135. Trough fragment. – 136–137. Pegs. – 138–139. Mallets. – 140–141 Hafts for socketed axes. – 142. Club. Scales: 135, 138–142: 1:5; 136–7: 1:3.

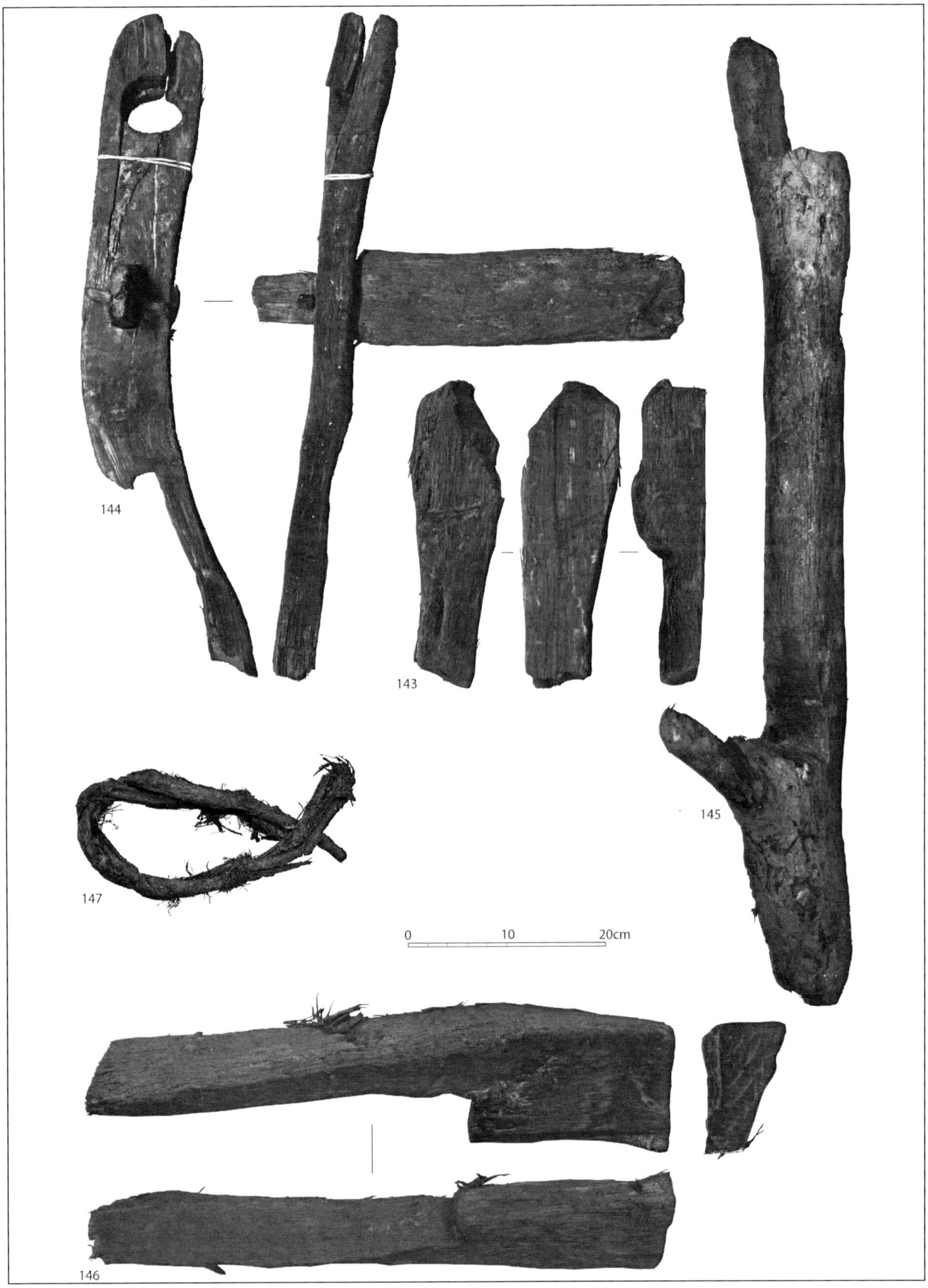

Plate 31. Trench XIII. – 143. Worked piece. – 144. Composite object. – 145. Worked object. – 146. Piece of worked plank. – 147. Binding. Scale 1:5.

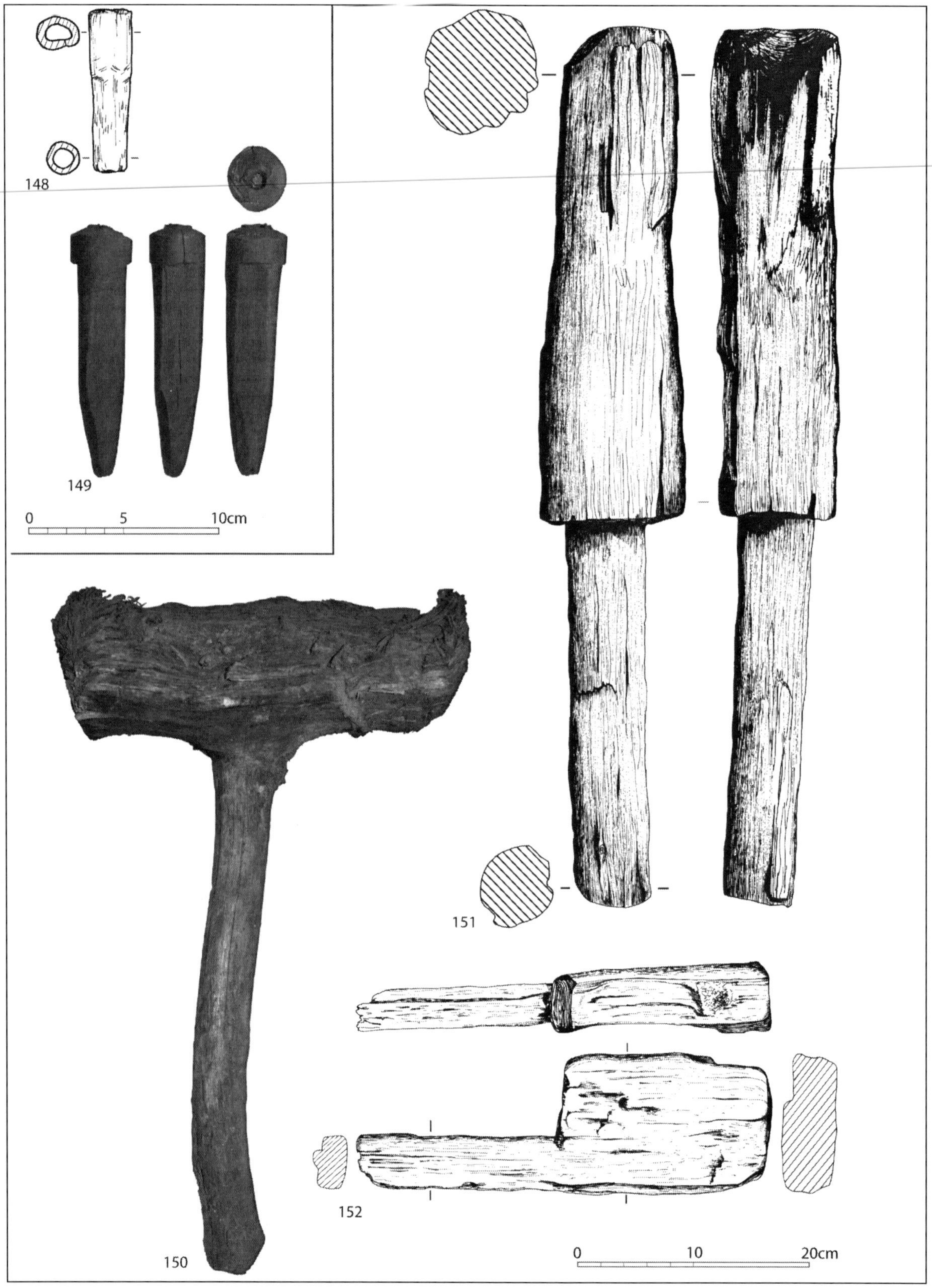

Plate 32. Objects collected in the stream. – 148–149. Pegs. – 150–151. Mallets. – 152. Split board. Scales: 148–9: 1:3; 150–152: 1:5.

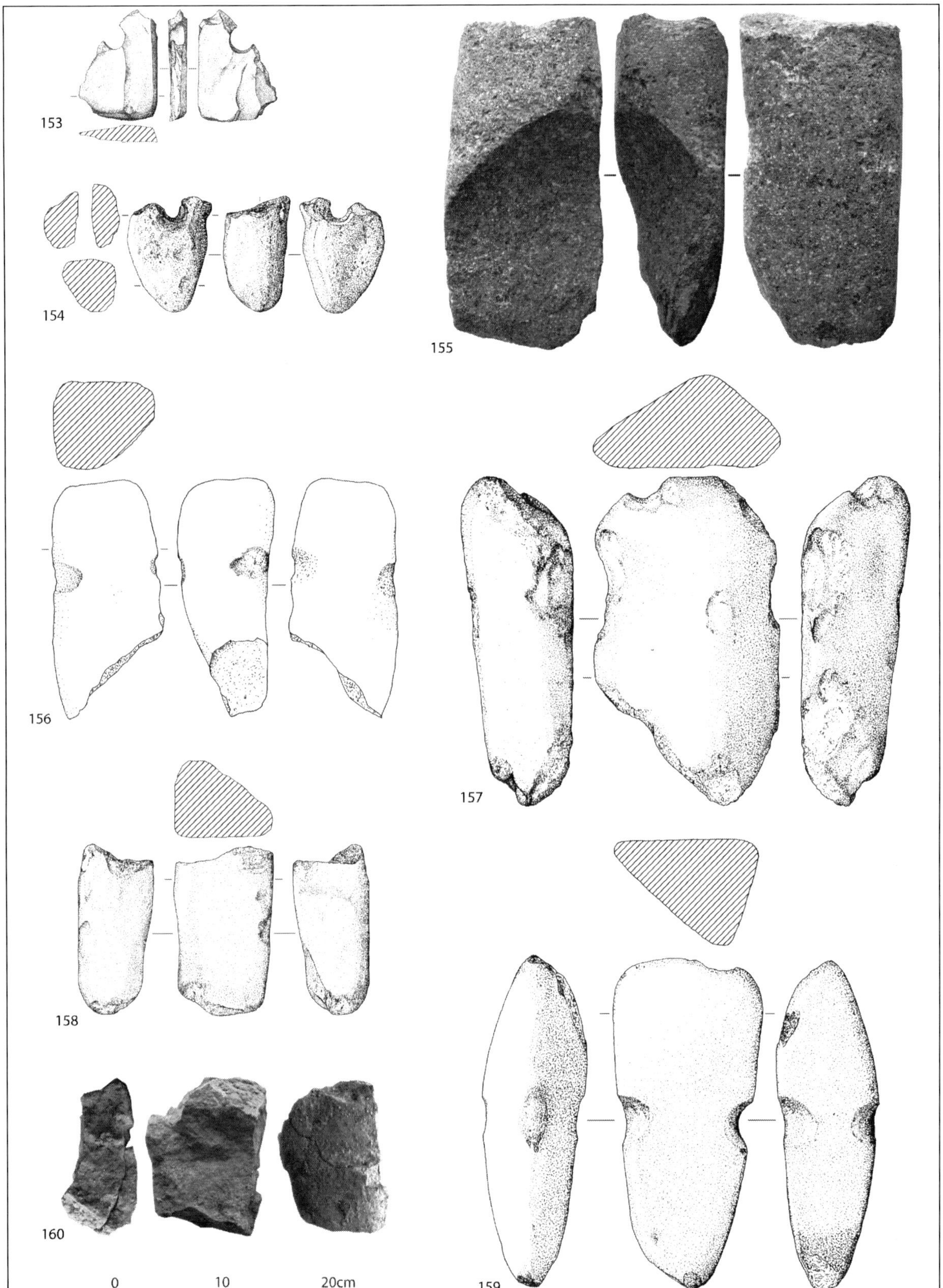

Plate 33. Trench I: 153. 155–159; Trench III: 160; without context: 154. – 153–154 hammer with shaft-hole. – 155–159 grooved hammer. – 160 painted grooved hammer. Scale 1:5.

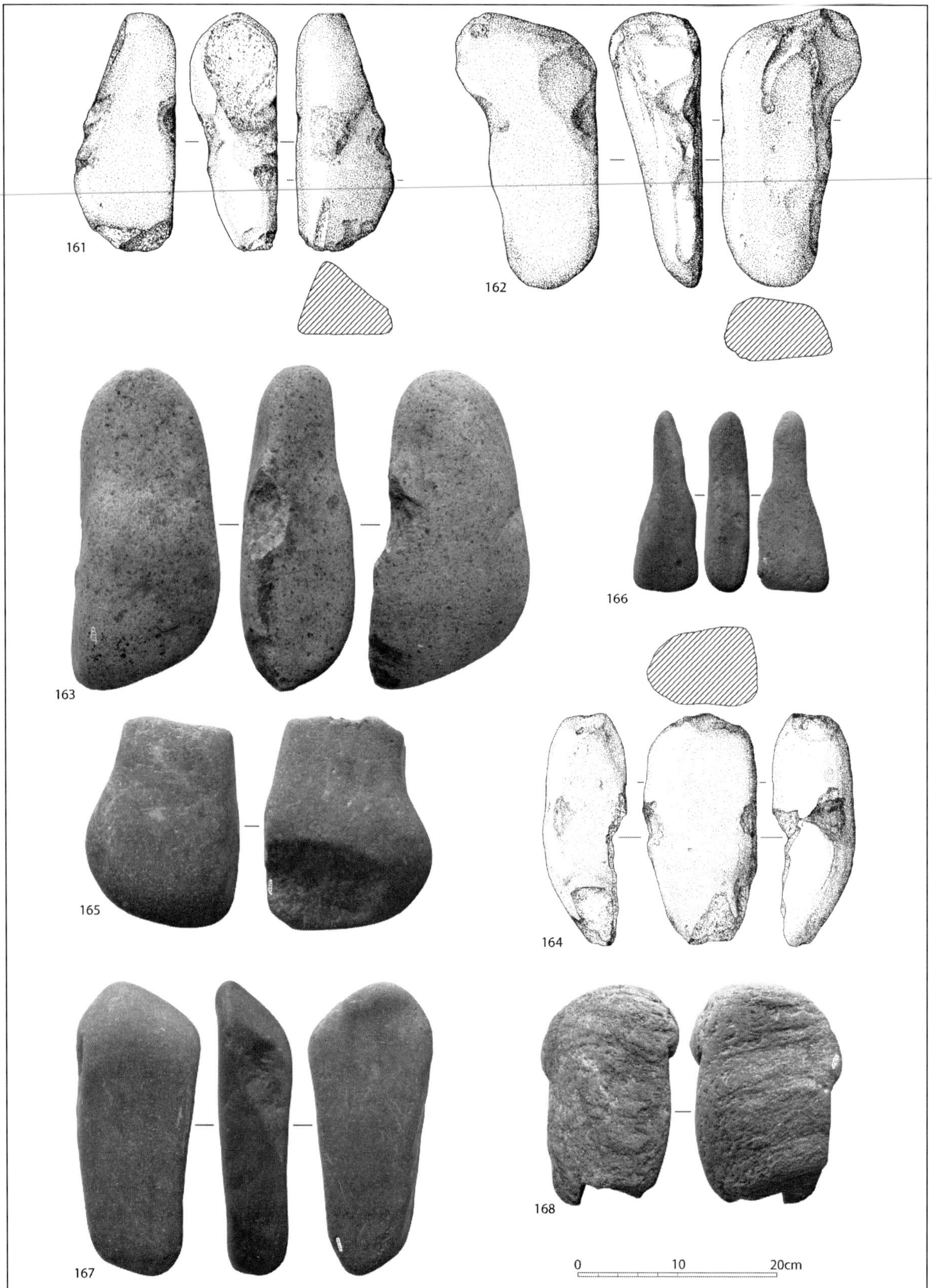

Plate 34. Trench III. – 161–168 grooved hammers. Scale 1:5.

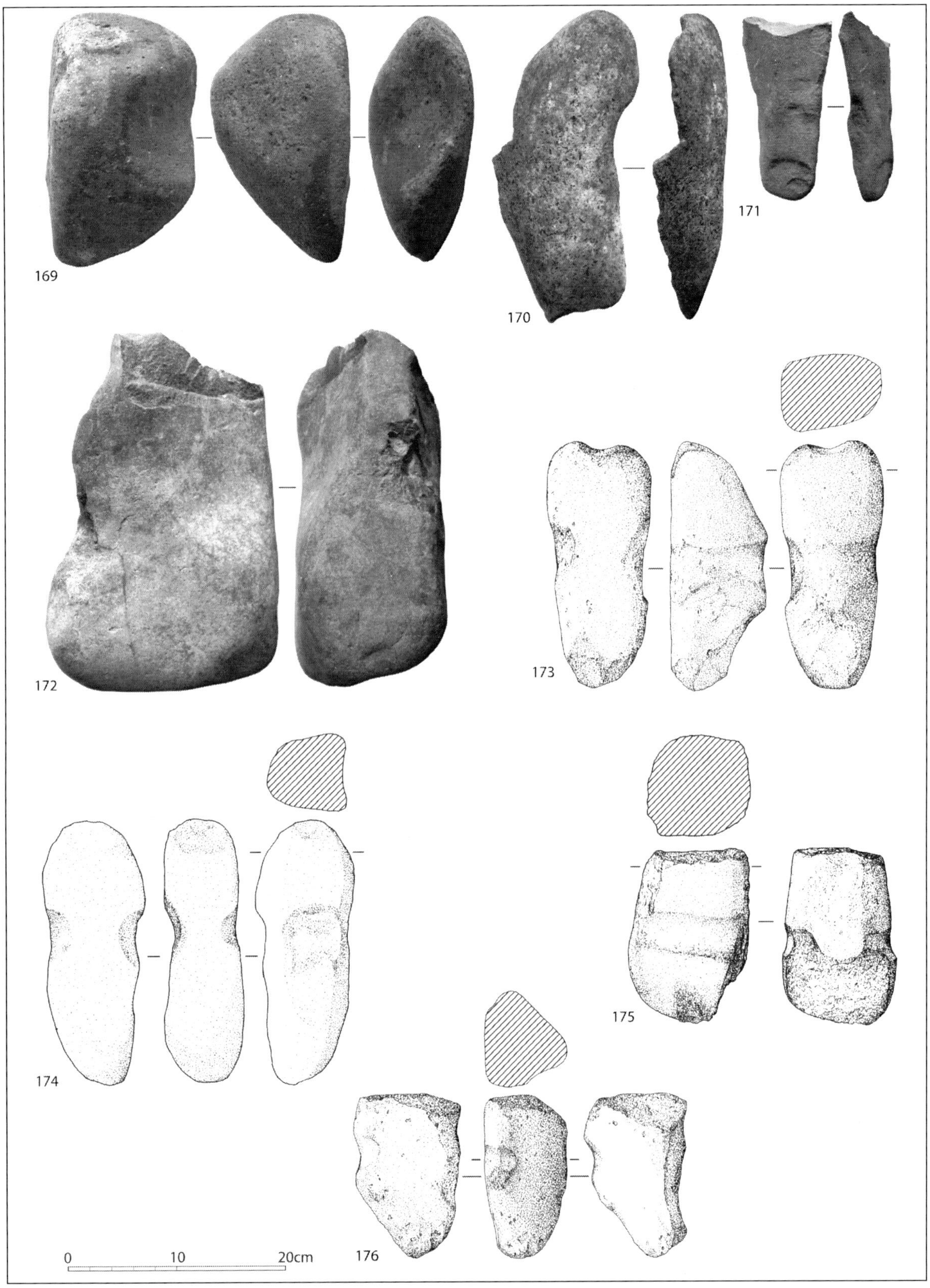

Plate 35. Trench III: 169–171; Trench XXII: 172; without context: 173–176. – 169–176 grooved hammers. Scale 1:5.

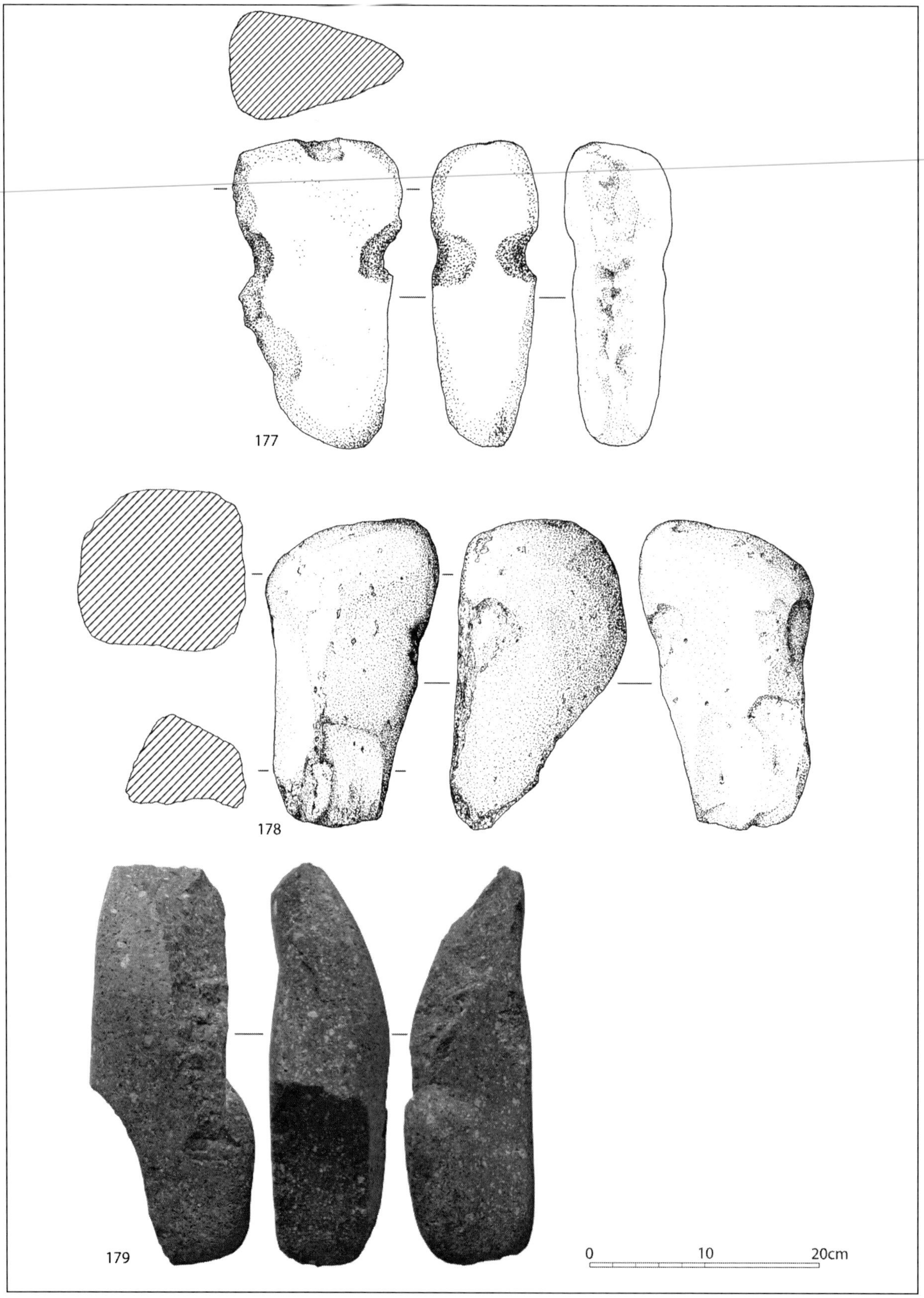

Plate 36. Without context. – 177–179 grooved hammers. Scale 1:5.

Plate 37. Without context. – 180–185 grooved hammers. Scale 1:5.

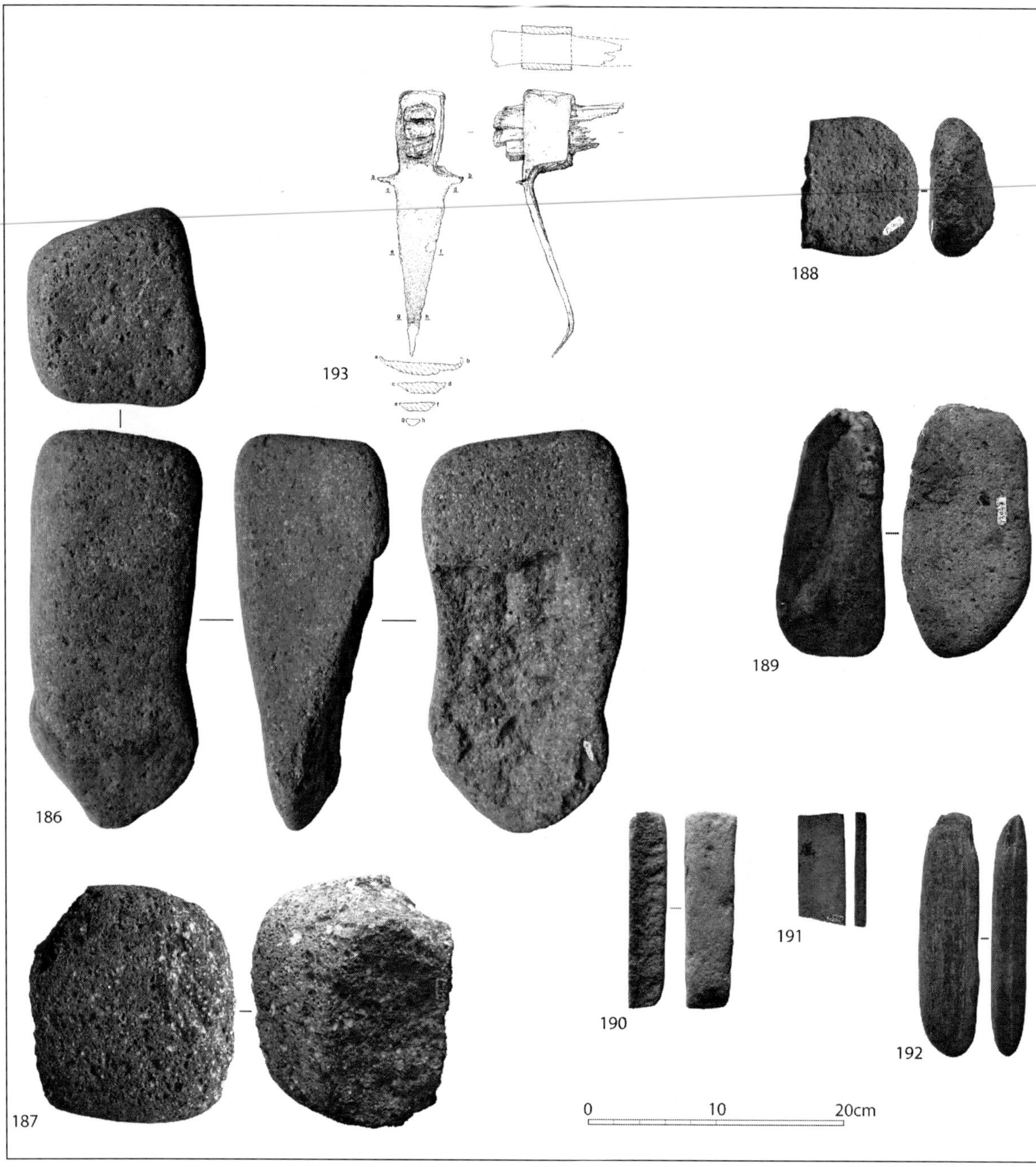

Plate 38. Trench I: 190; Trench XV: 187. 191. 192; without context: 186. 188. 189. – 186 grooved hammer. – 187–189 hammers. – 190–192 whetstones (?). 193 iron pick. Scale 1:5.

List of Authors

Steven J. Allen, York Archaeological Trust, Conservation Laboratories, 47 Aldwark, York, Y01 7BX, UK. Email: sallen@yorkat.co.uk

Eszter Bánffy, Institute of Archaeology, Research Centre for the Humanities, Hungarian Academy of Sciences, H-1014 Budapest, Országház u. 30. Email: banffy.eszter@btk.mta.hu

Richard Brunning, Somerset Historic Environment Record, Somerset Heritage Centre, Brunel Way, Norton Fitzwarren, Taunton, TA2 6BT, UK. Email: RBrunning@somerset.gov.uk

Krzysztof Bukowski, Faculty of Geology, Geophysics and Environment Protection, AGH University of Science and Technology, Kraków, Poland. Email: buk@agh.edu.pl

Dan Buzea, Muzeul Național al Carpaților Răsăriteni, str. Gábor Áron 16, 520008 Sfîntu Gheorghe, Romania. Email: buzealuci@yahoo.com

Andrea Chiricescu, Muzeul Național al Carpațilo Răsăriteni, str. Gábor Áron 16, 520008 Sfîntu Gheorghe, Romania. Email: andrea.chiricescu@gmail.com

Rowena Gale, Bachefield House, Kimbolton, Leominster, HR6 OEP, UK. Email: rowena@jimgale.eclipse.co.uk

Anthony Harding, Department of Archaeology, University of Exeter, Laver Building, North Park Road, Exeter, EX4 4QE, UK. Email: a.f.harding@exeter.ac.uk

Valerii Kavruk, Muzeul Național al Carpațilo Răsăriteni, str. Gábor Áron 16, 520008 Sfîntu Gheorghe, Romania. Email: valer.kavruk@mncr.ro

Adela Kovács, Muzeul Național al Carpațilo Răsăriteni, str. Gábor Áron 16, 520008 Sfîntu Gheorghe, Romania. Email: adelina_ab@yahoo.com

George G. Marinescu, Complexul Muzeal Bistrița-Năsăud, Gen. Grigore Bălan street 19, 420016 Bistrița, Romania. Email: georgemarinescu_bn@yahoo.com

Nyssa Mildwaters, Royal Armouries Museum, Armouries Drive, Leeds, LS10 1LT, UK. Email: nyssa.mildwaters@armouries.org.uk

Ionel Popa, Forest Research and Management Institute, Forest Research Station for Norway Spruce Silviculture, Calea Bucovinei 73 bis, Câmpulung Moldovenesc, Romania. Email: popaicas@gmail.com; popa.ionel@icassv.ro

Anca Semeniuc, Forest Research and Management Institute, Forest Research Station for Norway Spruce Silviculture, Calea Bucovinei 73 bis, Câmpulung Moldovenesc, Romania. Email: ank.yonela@yahoo.com

Dan Ştefan, 'Vasile Parvan' Institute of Archaeology, 11 Henry Coanda Str., 010667, Bucharest, and SC Digital Domain srl, 5 Garibaldi Str, Ap. 23, 020221, Bucharest, Romania. Email: danstefan00@gmail.com

Gelu Ţiculeanu, Laboratorul Zonal de Restaurare, Complexul Muzeal Bucovina, Suceava, Romania. Email: ticuleanug@yahoo.com

Marion Uckelmann, Department of Archaeology, University of Exeter, Laver Building, North Park Road, Exeter, EX4 4QE, UK. Email: m.uckelmann@exeter.ac.uk

Tomasz Ważny, University of Arizona, Laboratory of Tree-Ring Research, 1215 E. Lowell Street, Box 210045, Tucson, AZ 85721, USA. Email: twazny@email.arizona.edu

ARCHAEOLINGUA

Edited by
ERZSÉBET JEREM and WOLFGANG MEID

Main Series

1. **Cultural and Landscape Changes in South-East Hungary. I: Reports on the Gyomaendrőd Project.** Edited by Sándor Bökönyi. 1992. 384 pp. € 36.-. ISBN 963 7391 60 6.
2. Stefan Schumacher: **Die rätischen Inschriften. Geschichte und heutiger Stand der Forschung.** 1992. 2. vermehrte Auflage 2004. 375 pp. € 62.-. ISBN 963 8046 53 8.
3. **Onomasticon Provinciarum Europae Latinarum.** Herausgegeben von Barnabás Lőrincz und Ferenc Redő. Vol. I. 1994. XIV, 364 pp. € 50.-. ISBN 963 8046 01 5 Ö. ISBN 963 8046 02 3 K.
5. **Cultural and Landscape Changes in South-East Hungary. II.** Edited by Sándor Bökönyi. 1996. 453 pp. € 36.-. ISBN 963 8046 04 X.
7. **Die Osthallstattkultur. Akten des Internationalen Symposiums, Sopron, 10. – 14. Mai 1994.** Herausgegeben von Erzsébet Jerem und Andreas Lippert. 1996. 588 pp. € 88.-. ISBN 963 8046 10 4.
8. **Man and the Animal World. Studies in Archaeozoology, Archaeology, Anthropology and Palaeolinguistics in memoriam Sándor Bökönyi.** Edited by Peter Anreiter, László Bartosiewicz, Erzsébet Jerem and Wolfgang Meid. 1998. 720 pp. € 92.-. ISBN 963 8046 15 5.
9. **Archaeology of the Bronze and Iron Age – Environmental Archaeology, Experimental Archaeology, Archaeological Parks. Proceedings of the International Archaeological Conference, Százhalombatta, 3–7 October, 1996.** Edited by Erzsébet Jerem and Ildikó Poroszlai. 1999. 488 pp. € 68.-. ISBN 963 8046 25 2.
10. **Studia Celtica et Indogermanica.** Festschrift für Wolfgang Meid. Herausgegeben von Peter Anreiter und Erzsébet Jerem. 1999. 572 pp. € 78.-. ISBN 963 8046 28 7.
11. **From the Mesolithic to the Neolithic. Proceedings of the International Archaeological Conference held in the Damjanich Museum of Szolnok, September 22–27, 1996.** Edited by Róbert Kertész and János Makkay. 2001. 461 pp. € 72.-. ISBN 963 8046 35 X.
12. Garrett Olmsted: **Celtic Art in Transition during the First Century BC. An Examination of the Creations of Mint Masters and Metal Smiths, and an Analysis of Stylistic Development during the Phase between La Tène and Provincial Roman.** 2001. 340 pp., with 142 plates. € 72.-. ISBN 963 8046 37 6.
13. **The Archaeology of Cult and Religion.** Edited by Peter F. Biehl and François Bertemes with Harald Meller. 2001. 288 pp. € 68.-. ISBN 963 8046 38 4.
14. Gerhard Tomedi: **Das hallstattzeitliche Gräberfeld von Frög. Die Altgrabungen von 1883 bis 1892**. 2002. 706 S., mit 118 Karten, € 76.- ISBN 963 8046 42 2.
15. **Morgenrot der Kulturen. Frühe Etappen der Menschheitsgeschichte in Mittel- und Südosteuropa. Festschrift für Nándor Kalicz zum 75. Geburtstag.** Herausgegeben von Erzsébet Jerem und Pál Raczky. 2003. 570 pp. € 78.-. ISBN 963 8046 46 5.
16. **The Geohistory of Bátorliget Marshland.** Edited by Pál Sümegi and Sándor Gulyás. 2004. 360 pp. € 66.-. ISBN 963 8046 47 3.
17. **Nord-Süd, Ost-West. Kontakte während der Eisenzeit in Europa. Akten der Internationalen Tagungen der AG Eisenzeit in Hamburg und Sopron 2002.** Herausgegeben von Erzsébet Jerem, Martin Schönfelder und Günther Wieland. 2010. 350 pp. € 84.-. ISBN 978 963 8046 57 4.

18. Raimund Karl: **Altkeltische Sozialstrukturen.** 2006. 609 pp. € 78.-. ISBN 963 8046 69 4.

19. Martin Hannes Graf: **Schaf und Ziege im frühgeschichtlichen Mitteleuropa. Sprach- und kulturgeschichtliche Studien.** 2006. 320 pp. € 60.-. ISBN 963 8046 70 8.

20. **Anthropology of the Indo-European World and Material Culture. Proceedings of the 5th International Colloquium of Anthropology of the Indo-European World and Comparative Mythology.** Edited by Marco V. García Quintela, Francisco J. González García and Felipe Criado Boado. 2006. 368 pp. € 64.-. ISBN 963 8046 72 4.

21. Ferenc Gyulai: **Archaeobotany in Hungary. Seed, Fruit, Food and Beverage Remains in the Carpathian Basin from the Neolithic to the Late Middle Ages.** 2010. 480 pp. with CD-ROM. € 84.-. ISBN 978 963 8046 93 2.

22. **Vingt ans après Georges Dumézil (1898–1986) – Veinte años después de Georges Dumézil (1898–1986). Mythologie comparée indo-européenne et idéologie trifonctionelle: bilans, perspectives et nouveaux domaines – Mitología comparada indoeuropea e ideología trifuncional: balance, perspectivas y nuevos campos.** Edited by François Delpech et Marco V. García Quintela. 2009. 352 pp. € 64.-. ISBN 978 963 9911 05 5.

23. **The Archaeology of People and Territoriality.** Edited by George Nash and Dragos Gheorghiu. 2009. 340 pp. € 68.-. ISBN 978 963 9911 06 2.

24. **The Master of Animals in Old World Iconography.** Edited by Derek B. Counts and Bettina Arnold. 2010. 256 pp. € 56.-. ISBN 978 963 9911 14 7.

25. **Archaeological, Cultural and Linguistic Heritage. Festschrift for Erzsébet Jerem in Honour of her 70th Birthday.** Edited by Peter Anreiter, Eszter Bánffy, László Bartosiewicz, Wolfgang Meid and Carola Metzner-Nebelsick. 2012. 640 pp. € 88.-. ISBN 978 963 9911 28 4.

26. **The Levantine Question/ El problema "Levantino" (bilingual).** Edited by José Julio García Arranz, Hipólito Collado Giraldo and George Nash. 2012. 432 pp. € 74.-. ISBN 978 963 9911 31 4.

27. **The Medieval Royal Palace at Visegrád.** Edited by Gergely Buzás and József Laszlovszky. 2013. 398 pp. € 72.-. ISBN 978-963-9911-39-0.

Series Minor

6. Marija Gimbutas: **Das Ende Alteuropas. Der Einfall von Steppennomaden aus Südrußland und die Indogermanisierung Mitteleuropas.** 1994. 2. Aufl. 2000. 135 pp. € 32.-. ISBN 963 8046 09 0.

7. Eszter Bánffy: **Cult Objects of the Neolithic Lengyel Culture. Connections and Interpretation.** 1997. 131 pp. € 26.-. ISBN 963 8046 16 3.

8. Wolfgang Meid: **Die keltischen Sprachen und Literaturen. Ein Überblick.** 1997. 2. Aufl. 2005. 94 pp. € 20.-. ISBN 963 8046 65 1.

9. Peter Anreiter: **Breonen, Genaunen und Fokunaten. Vorrömisches Namengut in den Tiroler Alpen.** 1997. 173 pp. € 30.-. ISBN 963 8046 18 X.

10. Nándor Kalicz: **Figürliche Kunst und bemalte Keramik aus dem Neolithikum Westungarns.** 1998. 156 pp. € 30.-. ISBN 963 8046 19 8.

11. **Transhumant Pastoralism in Southern Europe. Recent Perspectives from Archaeology, History and Ethnology**. Edited by Haskel J. Greenfield and László Bartosiewicz. 1999. 245 pp. € 36.-. ISBN 963 8046 11 2.

12. Francisco Marco Simón: **Die Religion im keltischen Hispanien**. 1998. 168 pp. € 32.-. ISBN 963 8046 24 4.

13. Peter Raulwing: **Horses, Chariots and Indo-Europeans. Problems of Chariotry Research from the Viewpoint of Indo-European Linguistics.** 2000. 210 pp. € 36.-. ISBN 963 8046 26 0.

14. John Chapman: **Tension at Funerals – Micro-Tradition Analysis in Later Hungarian Prehistory.** 2000. 184 pp. € 32.-. ISBN 963 8046 29 5.

15. Eszter Bánffy: **A Unique Prehistoric Figurine of the Near East**. 2001. 106 pp. € 24.-. ISBN 963 8046 36 8.

16. Peter Anreiter: **Die vorrömischen Namen Pannoniens.** 2001. 316 pp. € 36.-. ISBN 963 8046 39 2.

17. Paul Gaechter: **Die Gedächtniskultur in Irland.** 2003. 116 pp. € 20.-. ISBN 963 8046 45 7.

18. **The Geoarchaeology of River Valleys.** Edited by Halina Dobrzańska, Erzsébet Jerem and Tomasz Kalicki. 2004. 214 pp. € 32.-. ISBN 963 8046 48 1.

19. Karin Stüber: **Schmied und Frau. Studien zur gallischen Epigraphik und Onomastik**. 2004. 125 pp. € 24.-. ISBN 963 8046 55 4.

20. Wolfgang Meid: **Keltische Personennamen in Pannonien**. 2005. 350 pp. € 40.- ISBN 963 8046 56 2.

21. **The Archaeology of Cult and Death. Proceedings of the Session "The Archaeology of Cult and Death" Organized for the 9th Annual Meeting of the European Association of Archaeologists, 11th September 2003, St. Petersburg, Russia.** Edited by Mercourios Georgiadis and Chrysanthi Gallou. 2006. 194 pp. € 32.-. ISBN 963 8046 67 8.

22. **Landscape Ideologies.** Edited by Thomas Meier. 2006. 260 pp. € 34.-. ISBN 963 8046 71 6.

23. **The Archaeology of Fire. Understanding Fire as Material Culture.** Edited by Dragos Gheorghiu and George Nash. 2007. 261 pp. € 34.-. ISBN 978-963-8046-79-6.

24. Erika Gál: **Fowling in Lowlands. Neolithic and Chalcolithic Bird Exploitation in South-East Romania and the Great Hungarian Plain**. 2007. 149 pp. € 28.-. ISBN 978-963-8046-85-7.

25. Anthony Harding: **Warriors and Weapons in Bronze Age Europe**. 2007. 224 pp. € 36.-. ISBN 978-963-8046-86-4.

26. Patrice Lajoye: **Des Dieux gaulois. Petits essais de mythologie.** 2008. 240 pp. € 36.-. ISBN 978-963-8046-92-5.

27. István Zalai-Gaál: **Zur Herkunft des Schädekults im Neolithikum des Karpatenbeckens**. 2009. 138 pp. € 28.-. ISBN 978-963-9911-08-6.

28. Sofia Pescarin: **Reconstructing Ancient Landscape**. 2009. 264 pp. € 38.-. ISBN 978-963-9911-09-3.

29. **Situating Gender in European Archaeologies.** Edited by Liv Helga Dommasnes, Tove Hjørungdal, Sandra Montón-Subías, Margarita Sánchez Romero and Nancy L. Wicker. 2010. 309 pp. € 36.-. ISBN 978-963-9911-15-4.

30. **A History of Central European Archaeology.** Edited by Alexander Gramsch and Ulrike Sommer. 2011. 219 pp. € 24.-. ISBN 978 963 9911 23 9.

31. Mária Bondár: **Prehistoric Wagon Models in the Carpathian Basin.** 2012. 142 pp. € 28.-. ISBN 978 963 9911 34 5.

Eastern Alpine Iron Age Studies

1. **Die Kelten in den Alpen und an der Donau. Akten des Internationalen Symposions St. Pölten, 14.–18. Oktober 1992.** Herausgegeben von Erzsébet Jerem, Alexandra Krenn-Leeb, Johannes-Wolfgang Neugebauer und Otto H. Urban. 1996. Second, revised ed. 2004. 462 pp. € 62.-. ISBN 963 8046 21 X.

Praehistoria

1. **Praehistoria. International prehistory journal of the University of Miskolc.** Edited by Árpád Ringer, Zsolt Mester and Erzsébet Jerem. **Volume 1, 2000.** 188 pp. € 40.-. **Volume 2, 2001.** 201 pp. € 20.-. **Volume 3, 2002.** 338 pp. € 24.-. **Volume 4–5, 2003–2004.** 247 pp. € 40.-. **Volume 6, 2005.** 130 pp. € 30.-. **Volume 7-8, 2006-2007.** 271 pp. € 24.-. **Volume 9–10, 2008–2009.** 346 pp. € 40.-. HU ISSN 1586 7811.

Studia Aegyptiaca

1. **The Mortuary Monument of Djehutymes (TT 32)** Vols. I–II. Edited by László Kákosy, Tamás A. Bács, Zoltán Bartos, Zoltán I. Fábián and Ernő Gaál. 2004. 384 pp. € 124.-. Vol. I.: ISBN 963 8046 51 1; Vol. II.: ISBN 963 8046 52 X.
2. Gábor Schreiber: **The Mortuary Monument of Djehutymes II.** 2008. 223 pp. € 88.-. ISBN 978 963 9911 02 4.
3. **Bi'r Minayh. Report on the Survey 1998–2004.** Edited by Ulrich Luft. 2010. 340 pp. € 116.-. ISBN 978 963 9911 11 6.

Other Titles

1. **On the Road to reconstructing the Past. Computer Applications and Quantitative Methods in Archaeology. Proceedings of the 36th International Conference.** Budapest 2008. (2011) Edited by E. Jerem, F. Redő and V. Szeverényi. 340 pp. € 36.-. ISBN 978-963-9911-30-7.
2. **Itinerarium Hungaricum II. Prehistoric Monuments and Collections in Hungary.** 2010. 264 pp. € 28.-. ISBN 978 963 8046 91 8.
3. Dóra Mérai: **The True and Exact Dresses and Fashion. Archaeological Clothing Remains and their Social Contexts in Sixteenth- and Seventeenth-century Hungary.** 2010. 97 pp. € 52.-. ISBN 978 1 4073 0555 4.
4. Nándor Kalicz: **Méhtelek. The First Excavated Site of the Méhtelek Group of the Early Neolithic Körös Culture in the Carpathian Basin.** 2011. 128 pp. € 30.-. ISBN 978 1 4073 0904 0.
5. EAC Occasional Paper No. 5. **Remote Sensing for Archaeological Heritage Management.** Edited by David C. Cowley. 2011. 312 pp. € 40.-. ISBN 978 963 9911 20 8.
6. EAC Occasional Paper No. 6. **Large Scale Excavations in Europe. Fieldwork Strategies and Scientific Outcome.** 2012. Edited by Jörg Bofinger and Dirk Krausse. 208 pp. € 35.-. ISBN 978-963-9911-29-1.
7. EAC Occasional Paper No. 7. **Heritage Reinvents Europe.** 2013. Edited by Dirk Callebaut, Jan Mařík and Jana Maříková-Kubková. 224 pp. € 36.-. ISBN 978-963-9911-41-3.
8. EAC Occasional Paper No. 8. **Who cares? Perspectives on Public Awareness, Participation and Protection in Archaeological Heritage Management.** 2013. Edited by Agneta Lagerlöf. 136 pp. € 28.-. ISBN 978-963-9911-42-0.

Please address orders to:

ARCHAEOLINGUA

H-1250 Budapest, Pf. 41.
Fax: (+361) 3758939
e-mail: kovacsr@archaeolingua.hu
http://www.archaeolingua.hu www.hungarianarchaeology.hu